ELSEVIER'S
DICTIONARY OF
AUTOMOTIVE
ENGINEERING

ELSEVIER'S DICTIONARY OF AUTOMOTIVE ENGINEERING

in
English, German, French, Dutch and Polish

compiled by

A. SCHELLINGS
Amsterdam, The Netherlands

1998
ELSEVIER
Amsterdam – Lausanne – New York – Oxford – Shannon – Singapore – Tokyo

ELSEVIER SCIENCE B.V.
Sara Burgerhartstraat 25
P.O. Box 211, 1000 AE Amsterdam, The Netherlands

Library of Congress Cataloging in Publication Data.
A catalog record from the Library of Congress has been applied for.

ISBN: 0-444-82159-7

PREFACE

This dictionary of automotive engineering comprises technical terms relating to vehicles, materials, assemblies and components parts. Terminology of tools used in garages, with their own specific jargon, is also included.

It is the first technical dictionary that covers the three international languages English, French and German, as well as Dutch and Polish. The Basic Table lists the entries in alphabetical order of the English terms. Alphabetical indexes in the other languages, referring to the Basic Table, are also included so each language can be used as a source and a target language. The work is intended for translators and others working with technical automotive documentation.

I am very grateful to all specialists for the useful information and documentation they supplied, in particular Citroën Nederland, and to my husband J.G.M. Schellings for his assistance in technical matters. Special thanks are due to Drs. J. Tur, sworn translator and managing director of a translation office in Poland, and Drs. Irena and Andrzej Kaczmarek, for their technical assistance and help with the Polish translation.

A. Schellings

Basic Table

A

1 abrasion indicator
d Abtriebindikator *m*
f témoin *m* d'usure
n slijtage-indicator
p wskaźnik ścierania

2 abrasion of lining
d Belagverschleiss *m*
f usure *f* de garniture
n slijtage van een voering; voeringslijtage
p ścieranie okładzin

3 abrasion pattern
d Verschleissbild *n*
f image *f* d'usure
n slijtagepatroon
p profil zużycia

4 abrasive compound
d Schleifpaste *f*
f pâte *f* abrasive
n schuurpasta
p pasta ścierna

5 abrasive stem
d Reibstange *f*
f pierre *f* abrasive
n slijtstift
p pręt ścierny

6 abrasive stem file
d Karborundfeile *f*
f lime *f* carborundum
n carborundumvijl
p pilnik ścierny

* abrasive wheel dresser → 2543

7 ABS brake
d Bremse *f* mit einer Antiblockieranlage
f frein *m* autostable
n rem met antiblokkeerremsysteem
p hamulec z urządzeniem przeciwblokującym

8 ABS failure indicator
d ABS-Ausfallanzeigeleuchte *f*
f témoin *m* de défaillance du système
   d'antiblocage
n controlelicht voor defect aan ABS
p lampka kontrolna systemu ABS

9 absolute temperature
d absolute Temperatur *f*
f température *f* absolue
n absolute temperatuur
p temperatura absolutna

10 absorb
d absorberen; dämpfen
f absorber; amortir
n absorberen; opnemen
p absorbować; tłumić

11 absorption coefficient
d Absorptionskoeffizient *m*
f coefficient *m* d'absorption; absorptivité *f*
n absorptiecoëfficiënt
p współczynnik absorbcji; współczynnik
   pochłaniania; absorbcyjność

12 absterge
d reinigen
f nettoyer
n reinigen
p czyścić

13 abutting surface
d Auflägeflache *f*
f plan *m* tangent
n raakvlak; aanligvlak
p powierzchnia stykowa; powierzchnia
   przylegania

14 accelerated motion
d beschleunigte Bewegung *f*
f mouvement *m* accéléré
n versnelde beweging
p ruch przyśpieszony

15 accelerating power
d Beschleunigungskraft *f*
f force *f* d'accélération
n versnellingskracht
p siła pzyśpieszenia

16 acceleration resistance
d Beschleunigungswiderstand *m*
f résistance *f* due à l'accélération
n acceleratieweerstand
p opór przyśpieszenia

17 acceleration skid control
d Antischleifkontrolle *f*
f antidérapeur *m* automatique
n antidoorslipregelsysteem
p kontrola poślizgu

18 accelerator pedal
d Fussgashebel *m*
f pédale *f* d'accélérateur
n gaspedaal
p pedał przyśpieszenia

19 accelerator pedal hinge
d Gaspedalgelenk *n*
f articulation *f* de pédale
n gaspedaalscharnier
p przegub pedału

20 accelerator pedal mounting bracket
d Gaspedalgrundplatte *f*
f support *m* de pédale
n gaspedaalbevestigingsbeugel
p wspornik pedału przyśpieszenia

21 accelerator pedal return spring; pedal return spring
d Gaspedalfeder *f*; Rückzugfeder *f*
f ressort *m* de rappel de pédale
n gaspedaalveer
p sprężyna pedału przyśpieszenia

22 accelerator pump
d Beschleunigungspumpe *f*
f pompe *f* de reprise
n versnellingspomp
p pompa przyśpieszenia

23 accelerator pump discharge nozzle
d Zerstaüber *m* der Beschleunigungspumpe
f injecteur *m* de pompe d'accélération
n acceleratiesproeier; versnellingspompsproeier
p rozpylacz pompy przyśpieszenia

24 accelerator pump lever
d Pumpenhebel *m*
f levier *m* de pompe d'accélération
n acceleratiepomphefboom; versnellingspomphefboom
p dźwignia pompy przyśpieszającej

25 accelerator pump piston
d Kolben *m* der Beschleunigungspumpe
f piston *m* de pompe d'accélération
n versnellingspompzuiger
p tłoczek pompy przyśpieszającej

26 accelerator spring
d Feder *f* für Gasbetätigung
f ressort *m* accélérateur
n gasklepveer; acceleratieveer
p sprężyna przyśpieszacza

27 accelerometer
d Beschleunigungsmesser *m*
f accéléromètre *m*
n acceleratiemeter; versnellingsmeter
p przyśpieszeniomierz; akcelerometr

28 access by road
d Zufahrt *f*
f accès *m*
n oprit
p dojazd

29 access hole
d Zulauföffnung *f*
f orifice *m* d'accès
n toegangsopening
p otwór doprowadzający

30 accessories
d Zubehör *n*
f accessoires *mpl*
n accessoires; toebehoren
p akcesoria

31 accessory drive
d Hilfsaggregatenantrieb *m*
f commande *f* des organes auxiliaires
n aandrijving van aangekoppelde aggregaten
p napęd osprzętu

32 accident prevention
d Unfallverhütung *f*
f prévention *f* des accidents
n ongevallenpreventie
p zapobieganie wypadkom

33 accommodation
d Anpassung *f*
f adaptation *f*
n aanpassing
p dostosowanie

34 accumulation of heat
d Wärmespeicherung *f*
f accumulation *f* de chaleur
n accumulatie van warmte
p akumulowanie ciepła

35 accuracy of form
d Formgenauigkeit *f*
f exactitude *f* de forme
n vormnauwkeurigheid
p dokładność kształtu

36 acicular
d spitz
f pointu

n spits; puntig
p spiczasty

37 acid
d Säure *f*
f acide *m*
n zuur
p kwas

* acid tester → 42

38 acid free
d sauerfrei
f neutre d'acide
n zuurvrij
p bezkwasowy

39 acid funnel
d Säuretrichter *m*
f entonnoir *m* à acide
n trechter voor zuur
p lejek do kwasu

40 acidity
d Säuregrad *m*
f acidité *f*
n zuurgraad
p kwasowość

41 acid number
d Säurezahl *f*
f indice *m* d'acide
n zuurgetal
p liczba kwasowa

42 acidometer; acid tester
     (apparatus to measure the contents of acidity
     in liquid)
d Säuremesser *m*
f pèse-acide *m*
n zuurmeter
p kwasomierz

43 acid proof
d säurebeständig
f à l'épreuve de l'acide
n zuurvast
p kwasoodporny; odporny na działanie kwasów

44 acoustic pressure
d Geräuschdruck *m*
f pression *f* acoustique
n geluidsdruk
p ciśnienie akustyczne; ciśnienie dźwięku

45 acoustic wave
d Schallwelle *f*
f onde *f* sonore

n geluidsgolf
p fala dźwiękowa; fala akustyczna

46 acrylic resin lacquer
d Akrylharzlack *m*
f vernis *m* à résines acryliques
n acrylharslak
p żywica akrylowa

47 active height; working height
d wirksame Länge *f*
f hauteur *f* comprimée
n werklengte
p długość czynna

48 active power
d Wirkleistung *f*
f puissance *f* active
n werkzaam vermogen
p moc czynna

49 actuating lever
d Betätigungshebel *m*
f levier-poussoir *m*
n aandrijfhefboom
p dźwignia uruchamiająca

50 actuating link
d Betätigungsstange *f*
f tige *f* de commande
n bedieningsstang
p cięgło włączające

51 actuating pin
d Mitnehmerbolzen *m*
f doigt *m* d'entraînement
n meeneempen
p kołek zabierakowy

52 actuating rod
d Federungsstössel *m*
f tige *f* poussoir
n drukstang
p drążek naprężajacy

53 adapter; adaptor
d Passstück *n*
f raccord *m*
n verbindingsstuk
p złączka; łącznik; adapter

54 adapter cable
d Adapterkabel *n*
f câble *m* d'adaptation
n adapterkabel
p kabel przystawki

55 adaptive ride control
d mikroprozessorgesteuerte Radaufhängung *f*

f suspension *f* à roues électronique
n elektronisch geregelde wielophanging
p elektronicznie sterowane zawieszenie kół

* adaptor → 53

56 addendum modification
d Profilverschiebung *f* der Zahnräder
f déport *m* de pignon
n profielverschuiving van tandwiel
p przesunięcie zarysu zębów

57 additional brake
d Hilfsbremsanlage *f*
f frein *m* auxiliaire
n hulprem
p hamulec pomocniczy

* additional brake light → 4268

58 additional heating
d Zusatzheizung *f*
f chauffage *m* supplémentaire
n aanvullende verwarming
p ogrzewanie dodatkowe

59 adhesion
d Adhäsion *f*; Haftung *f*
f adhésion *f*
n adhesie
p adhezja; przyleganie

60 adhesive foil
d Klebefolie *f*
f feuille *f* autocollante
n zelfklevende folie
p folia lepka

61 adhesive masking tape
d Abdeckband *n*
f bande *f* collante
n afplakband
p taśma osłaniająca

62 adhesive sealant
d Dichtflüssigkeit *f*
f joint *m* fluide
n vloeibare pakking
p masa uszczelniająca płynna

63 adjustable
d verstellbar
f réglable
n verstelbaar
p nastawny

64 adjustable axle stands
d nachstellbare Achsstützen *fpl*

f chandelles *fpl* réglables
n verstelbare assteunen
p nastawne podpory osi

65 adjustable friction damper
d verstellbarer Reibungsstossdämpfer *m*
f amortisseur *m* à friction réglable
n zelfregelbare frictiedemper
p amortyzator cierny nastawny

66 adjustable headrest
d verstellbare Kopfstütze *f*
f appui-tête *m* réglable
n verstelbare hoofdsteun
p zagłówek nastawny

67 adjustable level
d nachstellbare Wasserwage *f*
f niveau *m* ajustable
n bijstelbare waterpas
p poziomnica nastawcza

68 adjustable oil filter wrench
d Spannschlüssel *m* für Ölfiltergehäuse
f clé *f* de filtre à huile réglable
n verstelbare oliefiltersleutel; afstelbare oliefiltertang
p klucz nastawny do filtra oleju

69 adjustable pivot
d einstellbarer Zapfen *m*
f pivot *m* réglable
n afstelnok
p czop nastawny

70 adjustable spanner; shifting spanner
d verstellbarer Schlüssel *m*
f clé *f* à ouverture réglable
n verstelbare steeksleutel; Engelse sleutel; verstelbare sleutel
p klucz nastawny

71 adjustable stop
d einstellbarer Anschlag *m*
f vis *f* de ralenti; vis *f* de richesse
n verstelbare stuitnok
p kołek oporowy nastawny

72 adjustable wheelbase semitrailer
d ausziehbarer Auflieger *m*
f semi-remorque *f* extensible
n oplegger met verstelbare wielbasis
p naczepa teleskopowa wydłużalna

73 adjustable wire stripping pliers
d einstellbare Abisolierzange *f*
f pince *f* à dénuder à vis
n instelbare afstriptang
p kleszcze nastawne do ściągania izolacji

74 adjusting
d Einregulierung *f*
f réglage *m*
n afstelling
p regulacja; nastawianie; regulowanie

75 adjusting block
d Einstellblock *m*
f bloc *m* de réglage
n stelblok
p blok regulacyjny

76 adjusting bow
d Verstellbügel *m*
f étrier *m* ajustable
n verstelbeugel
p jarzmo nastawne

77 adjusting button
d Einstellknopf *m*
f bouton *m* de réglage
n afstelknop
p gałka do nastawiania

78 adjusting file
d Justierfeile *f*
f lime *f* d'usinage
n justeervijl
p pilnik regulacyjny

79 adjusting instruction
d Einstellvorschrift *f*
f instruction *f* de réglage
n afstelinstructie
p instrukcja regulacji

80 adjusting lever
d Nachstellhebel *m*
f levier *m* de réglage
n regelhefboom
p dźwignia nastawcza

81 adjusting nut
d Einstellmutter *f*
f écrou *m* de réglage
n bijstelmoer
p nakrętka regulująca; nakrętka nastawcza

82 adjusting pin
d Nachstellbolzen *m*
f goupille *f* de position
n stelpen; stelstift
p kołek ustalający; trzpień oporowy

83 adjusting procedure
d Einstellvorgang *m*
f procédure *f* de réglage
n afstelprocedure
p procedura regulacji

84 adjusting screw
d Einstellschraube *f*
f vis *f* de réglage
n afstelschroef
p śruba regulacyjna

85 adjusting shim
d Nachstellplatte *f*
f plaquette *f* de réglage
n afstelshim
p podkładka regulacyjna; podkładka nastawcza

86 adjusting sleeve; control sleeve
d Einstellhülse *f*
f manchon *m* de réglage
n afstelhuls
p tuleja regulacyjna

87 adjusting socket
d Regulierhülse *f*
f douille *f* de réglage
n regelhuls
p nasadka regulacyjna; tulejka regulacyjna

88 adjusting spring
d Einstellfeder *f*
f ressort *m* d'ajustage
n regelveer
p sprężyna regulacyjna

*   admissible clearance → 3719

*   admissible velocity → 179

89 advanced ignition
d Frühzündung *f*
f avance *f* d'allumage
n voorontsteking
p zapłon przedwczesny

90 advance plate; driving plate
d Antriebsflansch *m*
f disque *m* mené
n aandrijfflens
p tarcza sprzęgająca

91 aerial
d Antenne *f*
f antenne *f*
n antenne
p antena

92 aerial conductor wire
d Antennenkabel *n*
f câble *m* d'antenne
n antennekabel
p linka antenowa

93 **aerodynamic body**
d stromlinienförmiger Aufbau *m*
f carrosserie *f* aérodynamique
n stroomlijncarrosserie
p nadwozie o kształcie opływowym

94 **aerodynamic resistance**
d aerodynamischer Widerstand *m*
f résistance *f* aérodynamique
n luchtweerstand
p opór aerodynamiczny

95 **aerosol**
d Sprühdose *f* mit Treibgas
f aérosol *m*
n spuitbus met drijfgas
p aerozol

96 **afterburning**
d Nachbrennen *n*
f combustion *f* retardée
n naverbranding
p dopalanie

97 **afterburning space**
d Nachverbrennungsraum *m*
f espace *m* de postcombustion
n naverbrandingsruimte
p przestrzeń dopalania

98 **aging**
d Alterung *f*
f vieillissement *m*
n veroudering
p starzenie

99 **aiming of headlamp**
d Scheinwerfereinstellung *f*
f réglage *m* des projecteurs
n koplichtenafstelling
p ustawianie reflektorów

100 **air actuated friction clutch**
d pneumatische Reibungskupplung *f*
f embrayage *m* à friction à commande
   pneumatique
n pneumatisch bediende frictiekoppeling
p pneumatyczne sprzęgło cierne

101 **air baffle**
d Luftleitblech *n*; Luftleitplatte *f*
f déflecteur *m* d'air; tôle *f* directrice d'air
n luchtgeleidingsplaat
p deflektor powietrza

102 **air bag control light**
d Kontrolleuchte *f* von Airbagsystem

f témoin *m* du système air bag
n controlelicht van airbagsysteem
p lampa kontrolna systemu Airbag

103 **air bag saving sensor**
d Schutzfühler *m* von Airbagsystem
f sensor *m* de sécurité du système air bag
n beveiligingssensor van airbagsysteem
p czujnik bezpieczeństwa systemu Airbag

104 **air bleeder hole**
d Entlüftungsbohrung *f*
f orifice *m* de désaération
n ontluchtingsgat; ontluchtingsopening
p otwór odpowietrznika

105 **air box; wind box**
d Luftgehäuse *n*; Windkasten *m*
f boîtier *m* d'air
n windkast
p skrzynia dmuchowa; skrzynia wiatrowa

* **air brake → 154**

106 **air brake force limiter**
d Druckluftbremskraftbegrenzer *m*
f limiteur *m* d'intensité des freins pneumatiques
n luchtdrukremkrachtbegrenzer
p ogranicznik siły hamowania
   aerodynamicznego

107 **air brake governor**
d Druckluftbremsenregler *m*
f régulateur *m* de frein pneumatique
n luchtdrukremregelaar
p regulator hamulca pneumatycznego

108 **air bypass; supplementary air**
d Nebenluft *f*
f air *m* additionnel
n extra lucht
p powietrze dodatkowe

109 **air chamber**
d Stauluftraum *m*
f chambre *f* à air
n luchtkamer; luchtkast
p zasobnik powietrza

110 **air chipper**
d Druckluftmeissel *m*
f burin *m* pneumatique
n persluchtbeitel
p przecinak pneumatyczny

111 **air choke; air strangler**
d Lufteinlassklappe *f*; Luftklappe *f*

f papillon *m* d'air; volet *m* d'air
n luchtgeleider
p zasuwa wlotu powietrza

112 air chuck
d Druckluftspanndorn *m*
f mandrin *m* de serrage pneumatique
n pneumatische spanklauw
p uchwyt pneumatyczny

113 air circulation
d Luftumsatz *m*
f circuit *m* d'air
n luchtcirculatie
p obieg powietrza

* **air cleaner** → **133**

114 air collector
d Luftverteiler *m*
f répartiteur *m* d'air
n luchtkamer; luchtverzamelaar
p rozdzielacz powietrza

115 air compressor
d Luftverdichter *m*
f compresseur *m* d'air
n luchtcompressor
p sprężarka powietrza

116 airconditioner
d Klimaanlage *f*
f installation *f* d'air conditionné
n airconditioning
p klimatyzator; urządzenie klimatyzacyjne

117 air conditioning condenser
d Kondensator *m* für Klimatisierung
f condensateur *m* de climatisation
n condensator van airconditioning
p kondensator klimatyzacji

118 air conditioning switch
d Schalter *m* für Klimatisierung
f commutateur *m* de climatisation
n schakelaar van airconditioning
p włącznik klimatyzacji

119 air control valve
d Luftklappe *f*
f volet *m* d'air
n luchtregelklep
p zawór regulacyjny powietrza

120 air cooled brake
d luftgekühlte Bremse *f*
f frein *m* à refroidissement par air
n luchtgekoelde rem
p hamulec chłodzony powietrzem

121 air cooled cylinder
d luftgekühlter Zylinder *m*
f cylindre *m* refroidi par air
n luchtgekoelde cilinder
p cylinder chłodzony powietrzem

122 air cooled cylinder head
d luftgekühlter Zylinderkopf *m*
f culasse *f* refroidie par air
n luchtgekoelde cilinderkop
p głowica cylindra chłodzona powietrzem

123 air cooling
d Luftkühlung *f*
f refroidissement *m* par air
n luchtkoeling
p chłodzenie powietrzem

124 aircraft refueller
d Flughafentankwagen *m*
f citerne *f* d'aérodrome
n vliegtuigtankauto
p cysterna lotniskowa

125 air crimper
d pneumatische Würgezange *f*
f pince *f* à soyer pneumatique
n pneumatische kantzettang
p pneumatyczne szczypce do obciskania

126 air current
d Luftströmung *f*
f courant *m* d'air
n luchtstroom
p prąd powietrza

127 air cushion
d Luftpolster *n*
f sac *m* gonflable
n airbag; luchtzak
p poduszka ochronna

128 air deflector
d Luftleiteinsatz *m*
f déflecteur *m* d'air
n luchtgeleidingsplaat
p przesłona powietrza

129 air demand
d Luftbedarf *m*
f demande *f* d'air
n luchtverbruik
p zapotrzebowanie powietrza

130 air dome
d Windkessel *m*; Windkammer *f*
f dôme *m* d'air
n luchtkoepel
p komora powietrza

131 air drill
d Pressluftbohrer *m*
f perforatrice *f* à air comprimé
n luchtdrukboor
p świder pneumatyczny

132 air duct
d Luftkanal *m*
f conduit *m* d'air
n luchtkanaal; luchtafvoerkanaal
p kanał powietrzny

133 air filter; air cleaner
d Luftfilter *m*
f filtre *m* à air
n luchtfilter
p filtr powietrza

134 air filter cartridge
d Einsatz *m* des Luftfilters
f élément *m* de filtre à air
n luchtfilterelement
p wkład filtru powietrza

135 airflow meter
d Luftmengenmesser *m*
f débitmètre *m* d'air à volet
n luchtdebietmeter; luchthoeveelheidmeter
p dmuchomierz

136 airflow rate
d Ansaugluftverbrauch *m*
f débit *m* d'air d'admission
n inlaatluchtverbruik
p zużycie powietrza zasysanego

137 air fuel mixture
d Kraftstoffluftgemisch *n*
f mélange *m* d'air et d'essence
n brandstof-luchtmengsel
p mieszanka paliwowo-powietrzna

138 air gas
d Luftgas *n*
f gaz *m* à l'air
n luchtgas
p gaz powietrzny

139 air guide sheet
d Luftleitblech *n*; Luftleitplatte *f*
f déflecteur *m* d'air
n luchtgeleidingsplaat
p prowadnica powietrza; osłona kierująca powietrze

140 air hammer
d Drucklufthammer *m*
f marteau *m* pneumatique

n luchtdrukhamer
p młotek pneumatyczny

141 air heating
d Luftheizung *f*
f chauffage *m* à air
n luchtverwarming
p ogrzewanie powietrzem

142 air hose
d Luftschlauch *m*
f tuyau *m* à air
n luchtslang
p wąż powietrza

143 air inlet; air intake
d Lufteinlass *m*
f entrée *f* d'air; admission *f* d'air
n luchtinlaatopening
p otwór wlotu powietrza

144 air inlet cover
d Lufteinlassklappe *f*; Lufteinlassdeckel *m*
f couvercle *m* de prise d'air; trappe *f* de prise d'air
n luchtklep; deksel van luchtinlaat
p pokrywa wlotu powietrza; pokrywa filtru powietrza

* **air intake → 143**

* **air jet → 147**

145 air leak
d Falschluft *f*
f fuite *f* d'air
n luchtlekkage
p nadmiar powietrza

146 air line connector and gauge
d Handreifendruckmesser *m*
f contrôleur *m* de pression de pneu
n bandenspanningsmeter
p ciśnieniomierz do ogumienia; aparat do sprawdzania ciśnienia w ogumieniu

147 air nozzle; air jet
d Luftdüse *f*
f buse *f* d'air
n mondstuk voor luchttoevoer; luchtuitstroomopening
p nawietrznik

148 air operated servo unit
d Druckluftzylinder *m*
f cylindre *m* pneumatique
n luchtdrukcilinder
p siłownik pneumatyczny

149 **air operated valve grinder**
  d Druckluftventileinschleifgerät *n*
  f rodoir *m* de soupapes pneumatique
  n pneumatische klepslijper
  p pneumatyczna szlifierka do zaworów

150 **air outlet**
  d Luftauslass *m*
  f sortie *f* d'air
  n luchtuitstroomopening
  p otwór wylotu powietrza

151 **air passage**
  d Luftweg *m*
  f passage *m* d'air
  n luchtdoorlaat
  p pasaż powietrza

152 **air plenum chamber**
  d Luftkammer *f* von Katalysator
  f chambre *f* à catalyseur
  n luchtkamer van katalysator
  p komora powietrza katalizatora

153 **air pollution control standards**
  d Luftreinhaltungsnormen *fpl*
  f normes *fpl* antipollution
  n luchtverontreinigingsnormen
  p normy dopuszczalnego skażenia atmosfery

154 **air pressure brake; air brake**
  d Druckluftbremse *f*; Luftdruckbremse *f*
  f frein *m* pneumatique
  n luchtdrukrem
  p hamulec pneumatyczny; hamulec
    nadciśnieniowy

155 **air pressure gauge**
  d Luftdruckprüfer *m*
  f manomètre *m* à air
  n luchtdrukmeter; manometer
  p ciśnieniomierz

156 **air pressure reducer**
  d Luftdruckminderer *m*
  f réducteur *m* de pression d'air
  n luchtdrukbegrenzer
  p reduktor ciśnienia

157 **air pump**
  d Luftpumpe *f*
  f pompe *f* à air
  n luchtpomp
  p pompa próżniowa

158 **air refreshing tube**
  d Belüftungsrohr *n*
  f tube *m* de refroidissement d'air

  n luchtverversingspijp
  p przewód napowietrzający

159 **air resistance**
  d Luftwiderstand *m*
  f résistance *f* de l'air
  n luchtweerstand
  p opór powietrza

160 **air resistance coefficient**
  d Luftwiderstandsbeiwert *m*
  f coefficient *m* de pénétration dans l'air
  n luchtweerstandcoëfficiënt
  p współczynnik oporu powietrza

161 **air spring**
  d Luftfeder *f*
  f coussin *m* pneumatique
  n luchtveer
  p sprężyna pneumatyczna

162 **air spring piston**
  d Luftfederkolben *m*
  f piston *m* façonné
  n luchtveerzuiger
  p nurnik

163 **air spring rubber diaphragm**
  d Gummirollbalg *m*
  f diaphragme *m* en caoutchouc
  n rubberen rolbalg
  p przepona gumowa

  * **air spring suspension** → **3826**

164 **air storage reservoir; air tank**
  d Luftbehälter *m*
  f réservoir *m* d'air comprimé
  n drukluchtreservoir; luchtvoorraadtank
  p zbiornik sprężonego powietrza

  * **air strangler** → **111**

165 **air suction bellows**
  d Luftansaugbalg *m*
  f soufflet *m* de ventilation
  n luchtaanzuigbalg
  p mieszek zasysania powietrza

166 **air supply**
  d Luftzufuhr *f*; Luftzuführung *f*
  f alimentation *f* d'air
  n luchttoevoer
  p dopływ powietrza

* **air suspension** → 3826

167 air swirl
d Luftkreisel *m*; Luftwirbel *m*
f turbulence *f* d'air
n luchtturbulentie
p wir powietrzny

* **air tank** → 164

168 air temperature sensor
d Lufttemperaturfühler *m*
f capteur *m* de température d'air
n luchttemperatuursensor
p czujnik temperatury powietrza

169 airtight
d luftdicht
f étanche à l'air
n luchtdicht
p hermetyczny; powietrznoszczelny

170 airtight lines
d luftdichte Leitungen *fpl*
f canalisations *fpl* étanches
n luchtdichte leidingen
p przewody hermetyczne

171 airtight seal
d luftdichter Anschluss *m*
f fermeture *f* étanche
n luchtdichte afsluiting
p zamknięcie hermetyczne

* **air valve** → 3827

172 air vent duct
d Entlüftungskanal *m*
f trou *m* d'air
n ontluchtingskanaal
p kanał odpowietrzający

173 alarm chain
d Alarmkette *f*
f chaîne *f* d'alarme
n alarmketting
p łańcuch alarmowy

174 alarm signal; warning signal
d Alarmsignal *n*
f signal *m* de détresse
n alarmsignaal
p sygnał alarmowy

175 alidade
(the component part of the measure
instruments which provide the angular
measurement)
d Alhidade *f*

f alidade *f*
n alhidade
p alidada

176 alimentation
d Stromversorgung *f*
f alimentation *f* électrique
n elektrische voeding
p zasilanie elektryczne

177 allen wrench
d Sechskantstiftschlüssel *m*
f clé *f* mâle
n inbussleutel
p klucz do wkrętów z sześciokątnym
gniazdkiem

178 alligator clip
d Krokodilklemme *f*
f pince *f* crocodile
n krokodillenklem
p krokodylek

179 allowable speed; admissible velocity
d zulässige Geschwindigkeit *f*
f vitesse *f* admissible
n toelaatbare rijsnelheid
p prędkość dopuszczalna

180 all steel body
d Ganzstahlaufbau *m*
f caisse *f* tout acier
n geheel stalen carrosserie
p stalowe nadwozie

181 all steel piston
d Stahlgusskolben *m*
f piston *m* tout acier
n gietstalen zuiger
p tłok ze staliwa

182 all terrain vehicle
d Allradgeländewagen *m*
f véhicule *m* tout terrain
n terreinwagen
p samochód terenowy

183 all wheel drive
d Allradantrieb *m*
f commande *f* de toutes les roues
n vierwielaandrijving
p napęd na wszystkie koła

184 alternating current motor
d Wechselstrommotor *m*
f électromoteur *m* à courant alternatif
n wisselstroommotor
p silnik prądu przemiennego

185 **alternating current rectifier**
d Wechselstromgleichrichter *m*
f redresseur *m* de courant alternatif
n wisselstroomgelijkrichter
p prostownik prądu zmiennego

186 **alternating load**
d Wechselbelastung *f*
f effort *m* alternatif
n wisselbelasting
p obciążenie przemienne

187 **alternator**
d Wechselstromgenerator *m*
f alternateur *m*
n wisselstroomdynamo
p alternator; prądnica prądu zmiennego

188 **alternator output**
d Drehstromgeneratorleistung *f*
f puissance *f* d'alternateur
n prestatie van de alternator
p moc alternatora

189 **aluminium chromium cylinder**
d Aluminiumchromzylinder *m*
f cylindre *m* en aluminium chromé
n aluminium-verchroomde cilinder
p cylinder aluminiowo-chromowy

190 **aluminium compensating sleeve**
d Leichtmetallausgleichhülse *f*
f entretoise *f* de compensation en aluminium
n aluminium compensatiehuls
p aluminiowa tuleja wyrównawcza

191 **aluminium steel**
d Aluminiumstahl *m*
f acier *m* à l'aluminium
n aluminium staal
p stal aluminiowa

192 **aluminium wheel**
d Leichtmetallrad *n*
f roue *f* en alliage léger
n lichtmetalen wiel
p koło z metalu lekkiego

193 **ambient temperature switch**
d Umgebungstemperaturschalter *m*
f thermostat *m* d'ambiance
n omgevingstemperatuurschakelaar
p wyłącznik cieplny

194 **ambulance**
d Krankenwagen *m*
f voiture *f* de malades
n ambulance; ziekenauto
p karetka pogotowia

195 **amperage**
d Stromstärke *f*
f ampèrage *m*
n stroomsterkte
p natężenie prądu

196 **ampere hour**
d Amperestunde *f*
f ampère-heure *m*
n ampere-uur
p amperogodzina

197 **amplification**
d Abstützung *f*; Unterstützung *f*
f appui *m* d'ajustement
n ondersteuning; versterking
p podparcie; podpora

198 **amplifier**
d Verstärker *m* von Soundpaket
f amplificateur *m*
n versterker van geluidsinstallatie
p wzmacniacz instalacji dźwiękowej

199 **amplitude modulation**
d Amplitudenmodulation *f*
f modulation *f* d'amplitude
n amplitudemodulatie
p modulacja amplitudowa

200 **analogue indication**
d Analoganzeige *f*
f affichage *m* analogique
n analoge aanwijzing
p wskazanie analogowe

* **anchorage** → 3121

201 **anchorage piece**
d Verankerungsstück *n*
f ancrage *m*
n ankerstuk
p element do zakotwiczenia

202 **anchor arm**
d Ankerarm *m*
f bras *m* d'ancrage
n ankerarm
p ramię twornika

203 **anchor bolt**
d Ankerschraube *f*
f boulon *m* d'ancrage
n ankerbout
p śruba kotwowa

204 **anchor link**
d Ankerstange *f*; Ankerhülse *f*
f tige *f* d'ancrage

n ankerstang; ankerstaaf
p pręt twornika

205 anchor pin
d Ankerbolzen *m*
f axe *m* de mâchoires
n ankerbout
p sworzeń oporowy

206 anchor plate
d Ankerplatte *f*
f plaque *f* d'ancrage
n ankerplaat
p płyta kotwowa

207 anchor ring
d Einhängering *m*
f anneau *m* d'accrochage
n ankerring
p pierścień zaczepowy

* **angle bar → 211**

208 angle brace
d Winkelband *n*
f contrefiche *m*
n hoekverband
p zawiasa kątowa

209 angled spoon
d Schwanenhalstafel *m*
f palette *f* coudée
n zwanenhalstafel
p wygięta łyżka blacharska

210 angle grinder
d Winkelschleifer *m*
f meuleuse *f* d'angle
n haakse slijpmachine
p szlifierka ręczna z końcówką kątowa

211 angle iron; angle bar
d Winkeleisen *n*
f fer *m* cornière
n hoekstaal
p kątownik stalowy

212 angle meter
d Winkelmesser *m*
f mesureur *m* d'angles
n hoekmeter
p kątomierz

213 angle of advance
d Voreilungswinkel *m*
f angle *m* d'avance
n vervroegingshoek
p kąt przodovania

214 angle of departure
d hinterer Überhangwinkel *m*
f angle *m* de fuite
n afloophoek
p kąt zejścia

215 angle of incidence
d vorderer Überhangwinkel *m*
f angle *m* d'approche
n invalshoek
p kąt natarcia

216 angle of inclination
d Neigungswinkel *m*
f angle *m* d'inclination
n hellingshoek
p kąt nachylenia

217 angle of phase displacement
d Phasenversatzwinkel *m*
f angle *m* de déphasage
n faseverschuivingshoek
p kąt przesunięcia fazowego

218 angle of rotation
d Kurbelwinkel *m*
f angle *m* de rotation
n rotatiehoek; draaiingshoek
p kąt obrotu

219 angular acceleration
d Winkelbeschleunigung *f*
f accélération *f* angulaire
n hoekversnelling
p przyśpieszenie kątowe

220 angular adjustment
d Winkeleinstellung *f*
f réglage *m* angulaire
n hoekverstelling
p nastawianie kątowe

221 angular contact ball bearing
d Schrägkugellager *n*
f roulement *m* radial de butée à billes
n hoekcontactlager
p łożysko kulkowe skośne

222 angular cutter
d Winkelfräser *m*
f fraise *f* angulaire
n hoekfrees
p frez kątowy

223 angular drilling machine
d Winkelbohrmaschine *f*
f perceuse *f* angulaire
n haakse boormachine
p wiertarka kątowa

224 **angular error**
  d Winkelabweichung *f*
  f déclinaison *f*
  n hoekafwijking
  p przesunięcie kątowe

225 **angular reamer**
  d Winkelreibahle *f*
  f alésoir *m* angulaire
  n haakse handruimer
  p rozwiertak kątowy

226 **angular speed; angular velocity**
  d Winkelgeschwindigkeit *f*
  f vitesse *f* angulaire
  n hoeksnelheid
  p prędkość kątowa

227 **angular support frame**
  d Koppeltragachse *f*
  f traverse *f* d'angle
  n koppeldraagas
  p kątowa oś nośna

  * **angular velocity** → 226

228 **annular ball bearing**
  d Rillenkugellager *n*
  f roulement *m* radial à billes
  n groefkogellager
  p łożysko kulkowe zwykłe

229 **annular combustor**
  d Ringbrennkammer *f*
  f chambre *f* de combustion annulaire
  n ringvormige verbrandingskamer
  p pierścieniowa komora spalania

230 **annular rubber spring**
  d Gummiringfeder *f*
  f ressort *m* annulaire en caoutchouc
  n rubberen ringveer
  p pierścieniowy resor gumowy

231 **annular valve**
  d Ringventil *n*
  f soupape *f* annulaire
  n ringventiel
  p zawór pierścieniowy

232 **annulus; internal gear wheel**
  d Zahnrad *n* mit Innenverzahnung
  f roue *f* dentée intérieure
  n wiel met inwendige vertanding
  p koło o uzębieniu wewnętrznym

233 **anti-icing additive**
  d Vereisungsschutzadditiv *n*
  f additif *m* antigivre
  n anti-ijsdope
  p środek zapobiegający oblodzeniu

234 **anticorrosion lacquer**
  d Korrosionsschutzlack *m*
  f vernis *m* anticorrosif
  n anticorrosielak
  p lakier przeciwkorozyjny

235 **anticorrosion oil**
  d Korrosionsschutzöl *n*
  f huile *f* anticorrosive
  n antiroestolie
  p olej antykorozyjny

236 **anticorrosion warranty**
  d Rostgarantie *f*
  f garantie *f* de la corrosion
  n roestgarantie
  p gwarancja antykorozyjna

237 **antidazzle light; low beam**
  d Abblendlicht *n*
  f éclairage *m* code
  n dimlicht
  p światła mijania

  * **antidazzle screen** → 4878

238 **antidazzle switch; dimmer switch**
  d Abblendlichtschalter *m*
  f interrupteur *m* code
  n dimlichtschakelaar
  p przełącznik świateł mijania

239 **antidetonant**
  d Antiklopfmittel *n*
  f antidétonant *m*
  n antiklopmiddel
  p antydetonator; środek przeciwstukowy

240 **antidrumming compound**
  d Antidröhnmasse *f*; Antidröhnpaste *f*
  f enduit *m* antibruit
  n antidreunpasta
  p pasta głusząca; masa dźwiękoszczelna

241 **antifreeze**
  d Frostschutzmittel *n*
  f antigel *m*
  n antivries
  p ciecz niskokrzepnąca; płyn przeciw zamarzaniu

242 **antifreeze hydrometer**
  (tool to measure the amount of antifreeze in engine coolant)
  d Frostschutzspindel *m*

 f pèse-antigel *m*
 n antivriesweger
 p miernik cieczy niezamarzającej

243 antifreeze pump
 d Frostschutzpumpe *f*
 f pompe *f* antigel
 n antivriespomp
 p pompa mrozoodporna

244 antifriction lining
 d Laufschicht *f*
 f couche *f* antifriction
 n antifrictievoering; antifrictielaag
 p okładzina przeciwcierna

245 antiknock fuel
 d klopffester Kraftstoff *m*
 f supercarburant *m*
 n superbrandstof
 p paliwo przeciwstukowe

246 antiknock value
 d Klopffestigkeitswert *m*
 f valeur *f* antidétonante
 n klopvastheid van benzine
 p odporność detonacyjna

247 antilock braking system
 d Antiblockiersystem *n*
 f système *m* d'antipatinage; système *m* de
   freinage antibloquant
 n antiblokkeerremsysteem
 p system przeciwblokady hamulców

248 antioxidant
 d Antioxidans *n*; Antioxidationsmittel *n*
 f agent *m* antioxydant
 n antioxidatiedope
 p antyutleniacz

249 antirattler
 d Geräuschdämpfer *m*
 f antigrincement *m*
 n ruisonderdrukker
 p tłumik szumu

250 antirattle spring
 d Dämpfungsfeder *f*
 f ressort *m* d'amortisseur
 n demperveer
 p sprężyna tłumik hałasu

251 antiskid chain
 d Gleitschutzkette *f*
 f chaîne *f* à neige; chaîne *f* antidérapante
 n sneeuwketting
 p łańcuch przeciwślizgowy

252 antitheft device; antitheft protection
 d Diebstahlsicherung *f*
 f antivol *m*
 n antidiefstalinrichting
 p urządzenie zabezpieczajace przed
   kradzieżą

253 antitheft nut
 d diebstahlgesicherte Radmutter *f*
 f antivol *m* pour roue
 n antidiefstalmoer
 p śruba zabezpieczająca koło przed kradzieżą

 * **antitheft protection** → 252

254 anvil block
 d Ambossklotz *m*; Ambossuntersatz *m*
 f billot *m* d'enclume
 n aanbeeldsstok; aanbeeldsblok
 p podstawa kowadła

255 API norm
 d API-Norm *f*
 f normes *fpl* API
 n API-normen
 p norma APN

256 application
 d Anwendung *f*
 f emploi *m*; usage *m*
 n toepassing
 p zastosowanie

257 apron
 d Windschutzhaube *f*; Windschutz *m*
 f tablier *m*
 n voorplaat
 p osłona przed wiatrem

258 armature
 d Anker *m*
 f induit *m*
 n anker
 p twornik

259 armature core
 d Ankerkern *m*
 f noyau *m* d'induit
 n ankerkern
 p rdzeń twornika

260 armature groove
 d Ankernut *f*
 f encoche *f* d'induit
 n ankermoer
 p żłobek twornika

261 armature shaft
(armature shaft is part of a dynamo; electrical
equipment)
d Ankerwelle *f*
f arbre *m* d'induit
n ankeras
p wał twornika

262 armature stop
d Ankerbegrenzer *m*
f limiteur *m* de course d'armature
n ankeraanslag
p ogranicznik ruchu zwory

263 armature winding
(electrical equipment)
d Ankerwicklung *f*
f bobinage *m* d'induit
n ankerwikkeling
p uzwojenie twornika

264 armoured vehicle
d Panzerwagen *m*
f voiture *f* blindée
n pantserwagen
p samochód pancerny

265 arm rest washer
d Armstützeunterlage *f*
f plaquette *f* d'accoudoir
n bevestiging van armsteun
p podkładka podłokietnika

266 arrangement of cylinders
d Zylinderanordnung *f*
f groupement *m* des cylindres
n rangschikking van de cilinders
p usytuowanie cylindrów

267 articulated bit drilling machine
d Gelenkspindelbohrmaschine *f*
f perceuse *f* à broches articulées
n boormachine met gelede spillen
p wiertarka z wałkami przegubowymi

268 articulated bus
d Gelenkbus *m*
f autobus *m* articulé
n gelede bus
p autobus członowy

269 articulated jack; scissors jack
d Scherenheber *m*
f cric *m* à parallélogramme
n schaarkrik
p dźwignik nożycowy

270 articulated joint
d gelenkige Verbindung *f*

f joint *m* articulé
n beweegbare verbinding
p połączenie przegubowe

271 artificial leather
d Kunstleder *n*
f similicuir *m*
n kunstleder
p skóra sztuczna

272 artillery wheel
d Artillerierad *n*
f roue *f* type artillerie
n gesloten wielschijf
p koło artyleryjskie

273 asbestos washer
d Asbestunterlegscheibe *f*
f rondelle *f* en amiante
n asbestsluitring
p podkładka azbestowa

274 asphalt
d Asphalt *m*
f asphalte *m*
n asfalt
p asfalt

275 aspirator
d Ansauger *m*
f aspirateur *m*
n aanzuiger
p zasysacz

276 assemblage
d Montage *f*
f montage *m*
n montage
p montaż

277 assembling clearance
d Montagespiel *n*
f jeu *m* de montage
n montagespeling
p luz montażowy

278 assembling line
d Montageband *n*; Montagelinie *f*
f courroie *f* de montage
n montagelijn
p linia montażowa

279 assembling plate
d Verbindungsstück *n*
f renfort *m* d'assemblage
n verbindingsplaat
p element łączący**

280 **assembly; unit; set**
 d Baugruppe *f*
 f groupe *m*; ensemble *m*
 n samenstel; set
 p zespół

281 **assembly instruction**
 d Einbauvorschrift *f*
 f instruction *f* de montage
 n montagevoorschrift
 p instrukcja montażowa

282 **asymmetric**
 d asymmetrisch
 f asymétrique
 n asymmetrisch
 p asymetryczny

283 **atmospheric pressure**
 d Luftdruck *m*
 f pression *f* atmosphérique
 n atmosferische druk
 p ciśnienie atmosferyczne

284 **atmospheric valve**
 d selbsttätiges Saugventil *n*
 f soupape *f* atmosphérique
 n luchtklep
 p zawór atmosferyczny

285 **atomization**
 d Zerstäubung *f*
 f atomisation *f*
 n verstuiving
 p atomizacja; rozpylanie

286 **attaching nipple**
 d Anschlussnippel *m*
 f douille *f* de raccord
 n aansluitnippel
 p złączka łącząca

287 **attachment screw**
 d Befestigungsschraube *f*
 f vis *f* de fixation
 n bevestigingsschroef
 p śruba mocująca

288 **attended car park**
 d bewachter Parkplatz *m*
 f parking *m* gardé
 n bewaakte parkeerplaats
 p parking strzeżony

289 **audible wear indicator**
 d akustische Verschleissanzeige *f*
 f indicateur *m* d'usure acoustique
 n akoestische slijtage-indicator
 p wskaźnik starcia akustycznego

290 **autogenous welding**
 d autogene Schweissung *f*
 f soudure *f* autogène
 n autogeen lassen
 p spawanie gazowe

291 **autoinduction**
 d Selbstinduktion *f*
 f self-induction *f*
 n zelfinductie
 p samoindukcja

292 **auto lift**
 d Hebelbühne *f*; Hebelbrücke *f*
 f pont *m* élévateur
 n hefbrug
 p podnośnik mechaniczny

293 **automatic**
 d automatisch
 f automatique
 n automatisch
 p automatyczny; samoczynny

294 **automatic clutch**
 d automatische Kupplung *f*
 f embrayage *m* automatique
 n automatische koppeling
 p sprzęgło automatyczne

295 **automatic control box**
 d Steuerblock *m*
 f boîte *f* de commande
 n regelgroep
 p skrzynka rozrządcza

296 **automatic control device**
 d Selbstnachstellungsanlage *f*
 f dispositif *m* d'ajustement automatique
 n automatische verstelinrichting
 p urządzenie regulacji automatycznej

297 **automatic gear change**
 d automatische Gangschaltung *f*
 f changement *m* de vitesse automatique
 n automatische gangwissel
 p automatyczna zmiana biegów

298 **automatic load responsive regulating valve**
 d automatisch lastabhängiges Bremskraftregelventil *n*
 f répartiteur *m* automatique de la force de freinage en fonction du chargement
 n automatische lastafhankelijke remkrachtregelaar
 p zawór automatyczny regulujący obciążenie

299 **automatic timing advance device**
 d automatischer Spritzversteller *m*

f variateur *m* d'avance
n automatische inspuitversteller
p regulator wyprzedzenia wtryskiwania

300 automatic timing control
  d Selbstverstellung *f* der Zündung
  f commande *f* automatique d'avance
  n automatische ontstekingsafstelling
  p samoczynna regulacja momentu zapłonu

301 automatic transmission
  d Getriebeautomat *m*
  f transmission *f* automatique
  n automatische transmissie
  p przekładnia automatyczna

302 automatic transmission electronically piloted control
  d elektronisch vorgesteuerte Automatikgetriebeschaltung *f*
  f pilotage *m* électronique de transmission automatique
  n elektronisch bestuurde schakeling van automatische transmissie
  p elektroniczne sterowanie przekładni hydromechanicznej

303 automatic transmission hydraulic control
  d hydraulische Automatikgetriebesteuerung *f*
  f commande *f* hydraulique de transmission automatique
  n hydraulisch bestuurde schakeling van automatische transmissie
  p hydrauliczne sterowanie przekładni hydromechanicznej

304 automatic variable transmission
  d automatische stufenlose Kraftübertragung *f*
  f transmission *f* automatique à variation continue
  n automatische traploze transmissie
  p samoczynna przekładnia bezstopniowa

305 automatic volume control
  d Lautstärkenregelung *f*
  f réglage *m* du volume sonaire automatique
  n automatische geluidssterkteregeling
  p automatyczne regulowanie głośności

306 autoscope
  (tool to pinpoint vibrations or leaks)
  d Autoskop *m*
  f autoscope *m*
  n autoscope
  p autoskop

307 autothermic piston
  d Stahlstreifenkolben *m*
  f piston *m* autothermique

  n autothermische zuiger
  p tłok ze stalowymi wstawkami; tłok ze stalowymi pierścieniami

308 auxiliary
  d zusätzlich; behelfsmässig
  f auxiliaire
  n aanvullend
  p dodatkowy

309 auxiliary brake
  d Zusatzbremse *f*
  f frein *m* auxiliaire
  n hulprem
  p hamulec pomocniczy

310 auxiliary frame
  d Fahrschemel *m*
  f cadre *m* auxiliaire
  n subframe; hulpchassis
  p podwozie pomocnicze

  * **auxiliary gearbox** → 4058

311 auxiliary jet
  d Hilfsdüse *f*
  f gicleur *m* auxiliaire
  n hulpsproeier
  p dysza pomocnicza

312 auxiliary shaft
  d Nebenwelle *f*; Vorgelegewelle *f*
  f arbre *m* auxiliaire
  n hulpas
  p wał pomocniczy; wałek pośredniczący

313 auxiliary tank
  d Reservetank *m*
  f réservoir *m* auxiliaire
  n hulptank
  p zbiornik dodatkowy

314 average speed; average velocity
  d Durchschnittsgeschwindigkeit *f*
  f vitesse *f* moyenne
  n gemiddelde snelheid
  p prędkość średnia

  * **average velocity** → 314

315 axial bearing
  d Axiallager *n*
  f roulement *m* de butée
  n aslager
  p łożysko wzdłużne

316 axial piston motor
  d Axialkolbenmotor *m*

f moteur *m* hydraulique à piston axial
n axiale hydraulische zuigermotor
p silnik wielotłokowy

317 axial piston pump
d Axialkolbenpumpe *f*
f pompe *f* à piston axial
n axiale zuigerpomp
p pompa wielotłoczkowa osiowa

318 axial play
d Längsspiel *n*
f jeu *m* axial
n axiale speling
p luz półosiowy

319 axial thrust
d achsrechter Schub *m*
f poussée *f* axiale
n langsdruk
p nacisk osiowy

320 axle
d Achse *f*
f axe *m*
n as
p oś

321 axle breakage
d Achsenbruch *m*
f rupture *f* d'arbre; rupture *f* d'essieu
n asbreuk
p złamanie osi

322 axle collision
d Heckstoss *m*; Heckaufprall *m*
f choc *m* arrière
n asbotsing
p uderzenie osi

323 axle end
d Achszapfen *m*; Wellenzapfen *m*
f bout *m* d'arbre
n asstomp
p czop osi; czop wału

324 axle housing
d Achsgehäuse *n*
f corps *m* de pont
n ashuis
p obudowa mostu

325 axle load
d Achsdruck *m*
f charge *f* par essieu
n asdruk; asbelasting
p obciążenie osi

326 axle nut
d Achsmutter *f*
f écrou *m* d'axe
n asmoer
p nakrętka osi

327 axle shaft
d Achswelle *f*
f arbre *m* de roue
n wielaandrijfas
p półoś pędna

328 axle shaft bearing
d Achswellenlager *n*
f roulement *m* d'arbre de roue
n aandrijfaslager
p łożysko półosi

329 axle shaft flange
d Achswellenflansch *m*
f flasque *m* d'arbre de roue
n aandrijfasflens
p kołnierz półosi pędnej

330 axle shaft tube
d Achswellengehäuse *n*
f trompette *f* de pont
n aandrijfashuis
p rura pochwy półosi

331 axle shaft tube
d Achswellengehäuserohr *n*
f tube *m* de carter d'arbre de roue
n buis van het achterashuis
p rura pochwy półosi

332 axle suspension
d Achsenaufhängung *f*
f suspension *f* d'essieu
n asophanging
p zawieszenie osi

B

333 babbit metal
(bearing metal)
d Lagermetall *n*
f métal *m* blanc
n witmetaal
p babbit (stop łożyskowy)

334 back-framed erasing saw
d Abgleichsäge *f* mit Lehne
f scie *f* à araser à dos
n wiszaag met achterframe
p piła wykończeniowa z oparciem

335 back current switch
d Rückstromschalter *m*
f interrupteur *m* de courant de détour
n tegenstroomschakelaar
p wyłącznik prądu wstecznego

336 backfire
d Rückzündung *f*
f retour *m* d'allumage; retour *m* de flamme
n terugslag
p zapłon ponowny

337 backflash
d Zahnspiel *n*
f jeu *m* entre dents
n speling tussen tanden
p luz międzyzębny

338 back flow filter
d Rücklauffilter *m*
f filtre *m* de retour
n terugloopfilter; terugstroomfilter
p filtr przepływu wstecznego

339 backing pad
d Einlegerung *m*
f plateau *m* souple
n steunschijf
p pierścień wkładowy; pierścień wstawiany

340 backing up lamp
d Rückfahrtleuchte *f*
f phare *m* de recul
n achteruitrijlampen
p reflektor do jazdy wstecz

341 back pressure
d Rückdruck *m*
f contre-pression *f*
n tegendruk
p ciśnienie wsteczne

342 back pressure valve; check valve
d Rückschlagventil *n*
f soupape *f* de retenue
n terugslagklep
p zawór zwrotny

343 back shield
d Hinterschirm *m*
f panneau *m* arrière
n achterscherm
p osłona tylna

344 back spoiler
d Heckspoiler *m*
f becquet *m* arrière
n achterspoiler
p spoiler tylni

345 back squab
d Rückenpolster *n*
f dossier *m* rembourré
n lendesteun; zijsteun in de rug
p poduszka oparcia

346 back up lock ratchet
d Rückfahrtsperrhebel *m*
f manivelle *f* de verrouillage de la marche arrière
n spernok voor achteruitrijden
p dźwignia ryglująca jazdy wstecz

347 back up system
d Reserveanlage *f*
f système *m* de réserve
n reservesysteem
p system rezerwowy

348 back up washer
d Stützring *m*
f rondelle *f* d'appui
n steunring
p pierścień oporowy

349 backward
d gegenläufig
f étiré
n tegenlopend
p przeciwbieżny

* back window → 4039

350 baffled piston
d Nasenkolben *m*
f piston *m* à déflecteur
n profielkopzuiger
p tłok z garbem

351 baffle plate
d Ablenkscheibe *f*
f disque *m* déflecteur
n scheidingsplaat
p pierścień zaporowy

352 bag air spring
d Luftkissen *n*
f coussin *m* d'air
n luchtkussen
p resor poduszkowy

353 baggage
d Gepäck *n*
f bagage *m*
n bagage
p bagaż

354 baked enamel
d Einbrennlack *m*
f vernis *m* séchant au four; peinture *f* cuite au four
n moffellak
p lakier piecowy

355 bakelite
d Bakelit *n*
f bakélite *f*
n bakeliet
p bakelit

356 balance
d Gleichgewicht *n*
f équilibre *m*
n evenwicht
p równowaga

357 balance flap
d Kompensationsklappe *f*
f volet *m* d'équilibrage; volet *m* d'amortissement
n compensatieklep
p przysłona amortyzująca

358 balance hole
d Ausgleichbohrung *f*
f perçage *m* de compensation
n compensatieboring
p otwór wyrównawczy

359 balance spring
d Reglerschraubenfeder *f*
f ressort *m* équilibrant
n veer van regelschroef
p sprężyna rozpierająca

360 balance weight
d Gegengewicht *n*
f contrepoids *m*
n contragewicht
p przeciwciężar

361 balancing machine
d Auswuchtmaschine *f*
f machine *f* d'équilibrage
n balanceerapparaat
p wyważarka

362 balancing transformer
d Ausgleichtransformator *m*
f transformateur-compensateur *m*
n symmetrietrafo
p transformator wyrównawczy

363 bald tyre
d verschlissener Reifen *m*
f pneu *m* lisse
n versleten band
p opona "łysa"

364 ball
d Kugel *f*
f bille *f*
n bal; kogel
p kulka

365 ballast resistor
d Vorwiderstand *m*; Belastungswiderstand *m*
f résistance *f* de choc; résistance *f* ballast
n voorschakelweerstand
p rezystor obciążeniowy

366 ball bearing
d Kugellager *n*
f roulement *m* à billes
n kogellager
p łożysko kulkowe

367 ball bearing grease
d Kugellagerfett *n*
f graisse *f* pour roulements à billes
n lagervet
p smar do łożyska kulkowego

368 ball bearing housing
d Kugellagergehäuse *n*
f cage *f* de roulement à billes
n kogellagerhuis
p osłona łożyska kulkowego

369 ball bearing puller
d Kugellagerabzieher *m*
f extracteur *m* de roulement à billes
n kogellagertrekker
p ściągacz łożyska kulkowego

23   **ball thrust**

370 ball bearing seat
d Kugellagersitz *m*
f palier *m* de roulement à billes
n kogellagerzitting
p gniazdo łożyska kulkowego

371 ball cage
d Kugelkäfig *m*
f cage *f* à billes
n kogellagerkooi
p koszyczek łożyska

372 ball circulating nut
d Kugelumlaufmutter *f*
f écrou *m* à billes
n kogelomloopmoer
p nakrętka toczna

373 ball circulating screw
d Kugelumlaufspindel *f*
f vis *f* à billes
n kogelomloopspil
p śruba pociągowa toczna

374 ball cup stop
d Lagersteinanschlag *m*
f arrêt *m* de cuvette
n aanslag
p zderzak kamienia

375 ball end rod
d Kugelgelenkstange *f*
f barre *f* à rotule
n stang met kogeleind
p trzpień przegubu kulkowego

376 ball gear change
d Kugelschaltung *f*
f commande *f* par levier à rotule
n kogeloverschakelorgaan
p kulkowy układ przełączenia

377 ball gudgeon; ball pivot
d Kugelzapfen *m*
f pivot *m* rotule
n kogeltap
p czop kulisty

378 ball headed screw
d Einstellgelenk *n*
f rotule *f* de positionnement
n instelstuk
p przegub nastawczy

379 ball jointed arm
d Kugelgelenkarm *m*
f bras *m* d'articulation à rotule
n arm van kogelgewricht
p ramię przegubu kulkowego

380 balloon tyre
d Ballonreifen *m*
f pneu *m* ballon
n ballonband
p opona balonowa

381 ball peening
d Kugelstrahlen *n*
f billage *m*
n kogelstralen
p śrutowanie

382 ball pin
d Kugelbolzen *m*
f boulon *m* à tête sphérique
n kogelnootbout
p sworzeń kulisty

383 ball pin spring
d Feder *f* für Mittelbolzen
f ressort *m* de rotule
n kogelnootboutveer
p sprężyna sworznia kulistego

* **ball pivot** → 377

384 ball race
d Laufring *f* des Kugellagers
f bague *f* du roulement à billes
n kooi van kogellager
p bieżnia łożyska kulkowego

385 ball race groove
d Kugellaufrille *f*
f gorge *f* de roulement à billes
n kogelbaan van kogellager
p bieżnia łożyska tocznego

386 ball recirculating steering
d Kugelumlauflenkung *f*
f direction *f* à recirculation de billes
n kogelkringloopbesturing
p kulkowa przekładnia kierownicza

387 ball seat
d Kugelzapfenlager *n*
f siège *m* de rotule
n kogeltaplager
p gniazdo sworznia

388 ball seat nut
d Kugelbundmutter *f*
f écrou *m* à face d'appui sphérique
n moer voor kogelzitting
p kulkowa nakrętka wieńcowa

389 ball thrust
d Kugeldruck *m*
f bille *f* de butée

   n kogeldruk
   p nacisk kulki

390 ball thrust bearing
   d Druckkugellager *n*
   f roulement *m* de butée à billes
   n drukkogellager
   p oporowe łożysko kulkowe

391 ball valve
   d Kugelventil *n*
   f clapet *m* à bille
   n kogelklep
   p zawór kulkowy

392 band brake
   d Bandbremse *f*
   f frein *m* à bande
   n bandrem
   p hamulec taśmowy

393 banjo axle
   d Banjoachse *f*
   f pont *m* banjo
   n banjoas
   p most pędny o jednolitej budowie; most banjo

394 banjo connection
   d Ringstück *n*
   f raccord *m* banjo
   n banjoverbinding
   p połączenie banjo

395 banjo type axle housing
   d Banjoachsgehäuse *n*
   f corps *m* de pont banjo
   n banjoashuis
   p obudowa mostu banjo

396 bank
   d Kai *m*
   f quai *n*
   n kade
   p nabrzeże

397 bar; rod
   d Stange *f*
   f barre *f*
   n stang; staaf
   p pręt

398 barleycorn cross chisel
   d Gerstenkornkreuzmeissel *m*
   f bédane *f* à grain d'orge
   n gerstekorrelmotiefbeitel
   p wycinak z motywem ziarn jęczmienia

399 bar locking seat
   d Stabverankerung *f*

   f ancrage *m* de barre de torsion
   n borgvlak van stang
   p obsada drążka

400 barred road
   d Strassensperre *f*
   f route *f* barrée
   n wegversperring
   p barykada

401 barrel and valve assembly
   d Einspritzelement *n*
   f élément *m* d'injection
   n inspuitelement
   p element wtrysku

402 barrel shaped roller
   d Tonnenrolle *f*
   f racleau *m* tonneau
   n tonvormige rol
   p wałeczek baryłkowy

403 barrette file
   d Barrettfeile *f*
   f barrette *f*
   n afgeplatte driekante vijl
   p pilnik trójkątny dolnotnący

404 barrier layer
   d Isolierschicht *f*
   f couche *f* isolante
   n isolatielaag
   p warstwa izolacyjna; powłoka izolacyjna

405 base plate
   d Grundplatte *f*
   f tablier *m*
   n voetplaat
   p podstawa

406 basic engine
   d Grundmotor *m*
   f moteur *m* de base
   n standaardmotor
   p silnik bez osprzętu

407 basic research
   d Grundlagenforschung *f*
   f recherche *f* fondamentale
   n fundamenteel onderzoek
   p badania podstawowe

408 bastard file
   d Vorfeile *f*
   f bâtarde *m*
   n bastaardvijl
   p zdzierak

409 bath lubrication
d   Badschmierung *f*
f   graissage *m* par bain
n   spatsmering
p   smarowanie kąpielowe; smarowanie zanurzeniowe

410 battery box
d   Batteriekasten *m*; Batteriebehälter *m*
f   coffre *m* à batterie; bac *m* d'accumulateur
n   accubak
p   skrzynka akumulatora

411 battery capacity
d   Ladefähigkeit *f* der Batterie
f   capacité *f* de batterie
n   accucapaciteit
p   ładowność akumulatora

412 battery carrier
d   Batterieträger *m*
f   support *m* de batterie
n   accudrager
p   dźwigar akumulatorowy

413 battery cell tester
d   Batteriezellenprüfgerät *n*
f   appareil *m* d'essai d'élément de batterie d'accumulateurs
n   accuceltester
p   woltomierz widełkowy

414 battery charger
d   Batterieladevorrichtung *f*
f   chargeur *m* pour batteries
n   acculaadapparaat
p   urządzenie do ładowania akumulatora

415 battery charge warning light
d   Warnanzeige *f* für Batterieladung
f   témoin *m* de charge batterie
n   waarschuwingslicht van acculading
p   światło ostrzegawcze ładowania akumulatora

416 battery charging speed
d   Geschwindigkeit *f* der Batterieladung
f   vitesse *f* de charge de la batterie
n   laadstroomsnelheid
p   prędkość ładowania akumulatora

417 battery clamp
d   Batterieklemme *f*
f   pince *f* pour batterie
n   accuklembeugel
p   zacisk akumulatora

418 battery condition
d   Ladezustand *m*
f   état *m* de charge
n   laadtoestand
p   stan załadowania

419 battery element
d   Batteriezelle *f*
f   élément *m* de la batterie
n   accucel
p   ogniwo akumulatora

420 battery filler
d   Zellenfüller *m*
f   remplisseur *m* de batterie
n   accuvulbal
p   napełniacz elementu baterii akumulatorowej

421 battery ignition
d   Batteriezündung *f*
f   allumage *m* par batterie
n   batterij-ontsteking
p   zapłon akumulatorowy

422 battery jump cable; starter cable; jumper cable
d   Starthilfekabel *n*
f   câble *m* de démarrage
n   startkabel
p   kabel ułatwiający rozruch; kabel rozruchowy

423 battery lifter
d   Batteriehebegerät *n*
f   dispositif *m* de levage et de transport de batterie
n   accubatterijlichter
p   urządzenie do zdejmowania i transportu baterii akumulatorów

424 battery main switch
d   Batteriehauptschalter *m*
f   robinet *m* de batterie
n   hoofdschakelaar
p   wyłącznik główny baterii

425 battery pliers
d   Batteriezange *f*
f   pince *f* batterie
n   accutang
p   szczypce do akumulatora

426 battery positive terminal
d   Batteriepluspol *m*
f   borne *f* positive de batterie
n   accupluspool
p   dodatni czop biegunowy akumulatora

427 battery support
d   Batteriestütze *f*
f   support *m* de batterie
n   accusteun
p   podpora akumulatora

428 **battery terminal post clamp**
 d Batteriepolklemme *f*
 f borne *f* à vis d'accumulateur
 n accupoolklem
 p zacisk akumulatora

429 **battery voltage sensor**
 d Batteriespannungsfühler *m*
 f capteur *m* de voltage d'accumulateur
 n accuspanningsensor
 p czujnik napięcia akumulatora

430 **baulk ring**
 d Synchronscheibenring *m*
 f bague *f* de synchronisation
 n synchromeshring
 p pierścień synchronizatora

431 **bayonet socket**
 d Bajonettverschluss *m*
 f douille *f* à baïonnette
 n bajonetfitting
 p zaczep bagnetowy

432 **bead chafing**
 d Wulstscheuerung *f*
 f frottement *m* de tringle
 n afschuring van hiel van autoband
 p ścieranie się stopki opony

433 **bead core**
 d Wulstkern *m*
 f tringle *f*
 n hielkern; hieldraad
 p drutówka

434 **beading cutter**
 d Viertelstabfräser *m*
 f fraise *f* 1/4 sphérique
 n kwartrondfrees
 p frez profilowy półokrągły wklęsły

435 **bead insulation**
 d Drahtkernisolierung *f*
 f isolement *m* de tringle
 n draadkernisolatie
 p izolacja drutówki

436 **bead plies**
 d Wulstlagen *fpl*
 f bandelette *f* d'accrochage
 n koordlagen
 p wczep

437 **bead saw**
 d kleine Rückensäge *f*
 f petite scie *f* à dosseret
 n kleine rugzaag
 p mała piła grzbietnica

438 **bead tyre**
 d Reifenwulst *m*
 f bandelette *f*
 n hiel van autoband
 p stopka opony

439 **bead wire**
 d Drahtkern *m*; Wulstdraht *m*
 f tringle *f*
 n hieldraad
 p drutówka

440 **bead wires guard**
 d Drahtkernüberzug *m*
 f protection *f* de tringle
 n hielkernbescherming
 p osłona drutówki

441 **beam**
 d Balken *m*
 f corps *m*; traverse *f*
 n balk
 p belka

442 **beam of light**
 d Lichtstrahl *m*; Strahlbündel *n*
 f faisceau *m* lumineux
 n lichtbundel; lichtstraal
 p wiązka światła

443 **beam trammel**
 d Stangenzirkel *m*
 f compas *m* à verge
 n stokpasser
 p cyrkiel drążkowy

444 **bearing**
 d Lager *n*
 f palier *m*
 n lager
 p łożysko

445 **bearing axle; carrying axle; supporting axle**
 d Tragachse *f*
 f essieu *m* porteur; axe *m* porteur
 n draagas
 p oś nośna

446 **bearing base**
 d Lagerbock *m*
 f poteau *m* d'appui
 n lagerstoel
 p kozioł łożyskowy

447 **bearing bushing**
 d ungeteilte Lagerbuchse *f*
 f bague *f* de palier
 n lagerbus
 p tuleja łożyskowa

448 **bearing cap**
d Lagerdeckel *m*
f chapeau *m* de palier
n lagerdeksel
p pokrywa łożyska

449 **bearing clearance**
d Lagerspiel *n*
f jeu *m* de palier
n lagerspeling
p luz łożyskowy

450 **bearing cone**
d Lagerkegel *m*
f cône *m* de roulement à rouleaux coniques
n lagerconus
p pierścień wewnętrzny łożyska stożkowego

451 **bearing cover gasket**
d Lagerdeckeldichtung *f*
f joint *m* de chapeau de palier
n lagerdekselpakking
p uszczelka pokrywy łożyska

452 **bearing extractor; bearing puller**
d Lagerabziehvorrichtung *f*
f extracteur *m* de roulements
n lagertrekker
p ściągacz do łożysk

453 **bearing face**
d Bezugsebene *f*
f plan *m* de base
n grondvlak; draagvlak
p płaszczyzna odniesienia; płaszczyzna
   wyjściowa

454 **bearing failure**
d Lagerauslaufen *n*
f coulage *f* d'une bielle
n uitlopen van een lager
p uszkodzenie łożyska

455 **bearing half ring**
d Druckhalbring *m*
f semi-bague *f* de butée
n drukhalfring
p półpierścień oporowy

456 **bearing half shell locating lip**
d Verdrehklaue *f* der Lagerschale
f nez *m* de fixation de demi-coussinet
n neus van lagerschaalnok
p zaczep ustalający półpanewkę

457 **bearing housing**
d Gehäuse *n* für Kugellager
f boîtier *m* de roulement

n lagerhuis
p osłona łożyska

458 **bearing metal**
d Lagermetall *n*
f métal *m* pour coussinets
n lagermetaal
p stop łożyskowy

459 **bearing needle**
d Lagernadel *f*
f aiguille *f* de rouleau
n lagernaald
p igła łożyska

460 **bearing pin**
d Ankerbolzen *m*
f axe *m* d'ancrage
n ankerbout
p sworzeń łożyskowy

461 **bearing placement**
d Lagerung *f*
f roulement *m*
n lagerplaatsing
p ułożyskowanie

462 **bearing preload**
d Lagervorspannung *f*
f précharge *m* de roulement
n lagervoorspanning
p wstępne obciążenie łożyska

* **bearing puller → 452**

463 **bearing separator**
d Abzieh- und Trennvorrichtung *f*
f extracteur-décolleur *m* à vis
n lagerdemontageapparaat
p aparat do demontażu łożysk

464 **bearing shell**
d Lagerbuchse *f*
f coussinet *m*
n lagerschaal; lagervoering
p panew

465 **bearing support**
d Lagerstütze *f*
f support *m* de palier
n lagersteun
p wspornik łożyska

466 **bearing surface**
d Lageroberfläche *f*
f surface *f* de portée
n draagvlak; loopvlak
p bieżnia łożyska

467 **beat** *v*
d hämmern
f battre
n hameren
p klepać

* **beetle** → 3263

468 **bell crank**
d Winkelhebel *m*
f renvoi *m* de sonnette
n kniehefboom
p dźwignia kątowa

469 **Belleville spring**
d Tellerfeder *f*
f ressort *m* de Belleville
n schotelveer
p sprężyna talerzowa

470 **bellows air spring**
d Faltenbalgluftfeder *f*
f soufflet *m* pneumatique
n balg van luchtveersysteem
p bębnowy resor pneumatyczny

471 **bellow stretcher**
d konische Staubkappe *f*
f expandeur *m* de soufflets
n stofhoesconus
p stożek rozpierający

472 **belt**
d Riemen *m*
f courroie *f*
n riem
p pas

473 **belt drive**
d Riemenantrieb *m*
f transmission *f* par courroie
n riemaandrijving
p napęd pasowy

474 **belt drive with a tensioner**
d Spannrollentrieb *m*
f transmission *f* par courroie en tension
n riemoverbrenging met spanrol
p mechanizm napędowy z rolką naciągową

475 **belt pulley**
d Riemenscheibe *f*
f poulie *f* pour courroie
n riempoelie; riemschijf
p koło pasowe

476 **belt punch; hollow punch**
d Locheisen *n*
f emporte-pièce *m* rond
n holpijp
p przebijak ręczny

477 **belt retract system**
d Gurtaufrollvorrichtung *f*
f rétracteur *m*
n oprolmechanisme van gordel
p pas z mechanizmem wciągająco-napinającym

478 **belt slip**
d Riemenschlupf *m*
f glissement *m* de la courroie
n slip van riem
p poślizg pasa

479 **belt striker**
d Riemengabel *f*
f fourchette *f* de courroie
n riemvork
p widełki pasowe

480 **belt torque converter**
d Riemenscheibengetriebe *n*
f variateur *m* à courroies
n poelieoverbrenging
p pasowy przetwornik momentu

481 **belt wear**
d Riemenabnützung *f*; Riemenverschleiss *m*
f usure *f* de courroie
n riemslijtage
p zużycie pasa

482 **bench clamp**
d Werktischklemme *f*
f valet *m* d'établi
n werkbankklem
p imadło stolarskie

483 **bench drilling machine**
d Tischbohrmaschine *f*
f perceuse *f* d'établi
n tafelboormachine
p wiertarka stołowa

484 **bench grinder**
d Tischschleifmaschine *f*
f touret *m* d'établi
n werkbankslijpmachine
p szlifierka stołowa

485 **bench hook**
d Bankeisen *n*
f valet *m* d'établi; crochet *m*
n klemhaak; bankhaak
p hak do mocowania

486 bench seat
d Sitzbank *f*
f banquette *f*
n zitbank
p siedzisko

487 bending
d Biegung *f*
f flexion *f*
n buiging
p zginanie; uginanie

488 bending machine
d Biegemaschine *f*
f plieuse *f*
n buigmachine
p giętarka

489 bending moment
d Biegungsmoment *n*
f moment *m* fléchissant; moment *m* de flexion
n buigmoment
p moment zginający; moment gnący

490 bending strength
d Biegungsfestigkeit *f*
f résistance *f* à la flexion
n buigkracht
p wytrzymałość na zginanie

491 bending stress
d Biegungsbeanspruchung *f*
f effort *m* de flexion
n buigspanning; buigbelasting
p naprężenie zginania

492 bendix drive
d Bendixantrieb *m*
f lanceur *m* de démarreur Bendix
n bendix-aandrijving van startmotor
p bezwładnościowe urządzenie sprzęgające

493 bendix screw
d Bendixschraube *f*
f vis *f* Bendix
n bendix-schroef
p śruba typu Bendix

494 bend test
d Biegeprobe *f*
f essai *m* de flexion
n buigtest
p próba zginania

495 bent chisel
d gekrümmter Beitel *m*
f ciseau *m* courbé
n gebogen beitel
p dłuto zakrzywione

496 bent needle nose pliers
d Radiozangen *fpl* mit Seitenschneider und gebogenen Backen
f pinces *fpl* radio coupantes coudées
n gebogen puntbektang
p szczypce ze zgiętym nosem; szczypce z zakrzywionym nosem

497 bent rim; curved rim
d verbogene Felge *f*
f jante *f* incurvée
n kromme velg
p zgięta obręcz

498 bent shaft
d verdrehte Welle *f*
f arbre *m* tordu
n verbogen as
p wał przekręcony

499 bent sheet iron
d gebogenes Eisenblech *n*
f tôle *f* pliée
n gebogen plaatijzer
p zgięta blacha stalowa

500 bent spoon
d Löffeleisen *n* zurückgebogen
f cuillère *f* recourbée
n teruggebogen uitdeuklepel
p łyżka wygięta

501 bent valve
d Schlangenventil *n*
f valve *f* contrecoudée
n kromme klep
p zawór powietrzny wygięty

502 benzene; benzole
d Benzol *n*
f benzol *m*
n benzeen; benzol
p benzen; benzol

* benzole → 502

503 bevel angle
d Öffnungswinkel *m*
f angle *m* du chanfrein
n openingshoek
p kąt ukosu

504 bevel edge chisel
d Beitel *m* mit abgeschrägten Kanten
f ciseau *m* à bords biseautés
n beitel met afgeschuinde kanten
p dziobak

505 bevel edges
  d Schrägkanten *fpl*
  f chanfreins *mpl*
  n schuine zijkanten
  p skosy

506 bevel gear drive
  d Kegelradantrieb *m*
  f transmission *f* par pignons coniques
  n kegelwielenaandrijving; aandrijving door middel van conische tandwielen
  p przekładnia kątowa; przekładnia stożkowa

507 bevel gear of differential
  d Kegelraddifferential *n*
  f différentiel *m* à roues d'angle
  n differentieel met conische tandwielen
  p stożkowy mechanizm różnicowy

508 bevel gear wheel
  d Kegelzahnrad *n*
  f roue *f* conique
  n conisch tandwiel
  p koło stożkowe zębate

509 bevel hub
  d Kegelnabe *f*
  f moyeu *m* conique
  n conische naaf
  p stożek piasty

510 bevel point flat file
  d Flachfeile *f* mit Spitzenverjüngung
  f lime *f* plate galbée
  n platte vijl met afgeschuinde punt
  p pilnik płaski z ukośnym punktem

511 bevel seated valve
  d Kegelsitzventil *n*
  f soupape *f* à portée conique
  n klep met afgeschuinde zetel
  p zawór z gniazdem stożkowym

  * **bevel type final drive** → 4639

512 bias belted tyre
  d Gürtelreifen *m*
  f pneumatique *m* à carcasse diagonale ceinturée
  n gordelband met diagonaalkarkas
  p opasana opona diagonalna

513 bias cutting
  d Schrägschnitt *m*
  f coupe *f* en biais
  n schuine snede
  p przekrój skośny

514 bibendum rim
  d Bibendum-Felge *f*
  f jante *f* Michelin
  n bibendumrand
  p obręcz koła

515 bicoloured
  d zweifarbig
  f bicolore
  n tweekleurig
  p dwukolorowy

516 bimetallic piston
  d Zweimetallkolben *m*
  f piston *m* bimétal
  n bimetalen zuiger
  p tłok bimetalowy

517 bimetallic shell
  d Verbundschale *f*
  f coussinet *m* bimétal
  n lagerschaal van bimetaal
  p panew z bimetalu

518 bimetallic spring
  d Bimetallfeder *f*
  f ressort *m* bimétallique
  n bimetalen veer
  p sprężyna bimetalowa

519 bimetallic valve
  d Bimetallventil *n*
  f soupape *f* bimétallique
  n bimetalen klep
  p zawór bimetalowy

520 binary fuel
  d Kraftstoffzweiergemisch *n*
  f carburant *m* binaire
  n binaire brandstof
  p paliwo dwuskładnikowe

521 bind *v*
  d klemmen
  f gripper
  n vastklemmen
  p przymocować

522 binding agent
  d Bindemittel *n*; Bindestoff *m*
  f agent *m* de liaison; liant *m*
  n bindmiddel
  p spoiwo; środek wiążący; śruba zaciskająca

523 binding screw
  d Klemmschraube *f*
  f vis *f* de serrage

n klemschroef
p śruba zaciskowa; wkręt zaciskowy; śruba
zaciskająca

524 bituminous
d bituminös
f bitumineux
n bitumineus
p bitumiczny

* **blade bit → 2178**

525 blade connector
d Flachstecker *m*; Messerstecker *m*
f fiche *f* à couteau
n messtekker
p łącznik łopatkowy

526 blade row; set of blades
d Schaufelkranz *m*
f distributeur *m* d'aubes
n schoepenkrans
p wieniec łopatek

527 blade shock absorber
d Flügerstossdämpfer *m*
f amortisseur *m* à volet
n vleugelschokdemper
p amortyzator skrzydełkowy

528 bleed *v*
d entlüften
f purger
n ontluchten
p odpowietrzać

529 bleeding pipe
d Entlüftungsleitungbeule *f*
f reniflard *m*
n ontluchtingspijp
p przewód odpowietrzający

530 bleed screw
d Entlüftungsschraube *f*
f vis *f* de purge d'air
n ontluchtingsnippel
p śruba odpowietrzająca; wkręt do
odpowietrzania

531 blind end cylinder
d geschlossener Zylinder *m*
f cylindre *m* borgne
n gesloten cilinder
p cylinder zamknięty

532 blind hole
d Sackloch *n*
f trou *m* de borgne

n blind gat
p otwór ślepy; otwór nieprzelotowy

* **blind nut → 615**

533 blind rivet
d Blindniet *m*
f rivet *m* faux
n blindklinknagel
p nit jednostronnie zamykany

534 blind tenon and mortise joint
d blinde Verzapfung *f*
f assemblage *m* aveugle à mortaise et tenon
n blinde pen-en-gatverbinding
p ślepe złącze na czop

535 blinker
d Blinker *m*
f clignoteur *m*
n richtingaanwijzer
p kierunkowskaz migowy

536 blister
d Sandbeule *f*
f bulle *f*; soufflure *f*
n luchtbel
p pęcherz; bąbel

537 block
d Block *m*
f bloc *m*
n blok
p blok

538 block engine
d Blockmotor *m*
f moteur *m* monobloc
n monoblokmotor
p silnik jednokadłubowy

539 blocking diode
d Sperrdiode *f*
f diode *f* de verrouillage
n sperdiode
p dioda blokująca

540 blocking relay
d Sperrelais *n*
f relais *m* de blocage
n sperrelais
p przekaźnik blokujący

541 block radiator
d Teilblockkühler *m*
f radiateur *m* à éléments séparés
n blokradiateur
p chłodnica sekcyjna

542 block spring
 d Gummifeder *f*
 f ressort *m* en caoutchouc
 n bladveer
 p sprężyna gumowa

543 blower
 d Gebläse *n*
 f soufflante *f*
 n ventilator
 p dmuchawa

544 blower casing
 d Gebläsegehäuse *n*
 f boîtier *m* de soufflante
 n ventilatorhuis
 p obudowa dmuchawy

545 blower impeller
 d Gebläselaufrad *n*
 f rotor *m* de soufflante
 n ventilatorloopwiel
 p wirnik dmuchawy

546 blower stator
 d Gebläseleitrad *n*
 f directrice *f* de soufflante
 n ventilatorschoepenwiel
 p kierownica dmuchawy

547 blow gun
 d Abblasepistole *f*
 f pistolet *m* à air
 n luchtpistool
 p pistolet pneumatyczny

* **blowing iron** → 550

548 blow lamp
 d Lötlampe *f*
 f lampe *f* à souder
 n soldeerlamp
 p lampa lutownicza

549 blow off valve
 d Ausblasventil *n*
 f soupape *f* de purge
 n afblaasklep
 p zawór wydmuchowy

550 blowpipe; blowing iron
 d Glasbläserpfeile *f*
 f canne *f* de verrier
 n blaaspijp
 p piszczel

551 blueprint
 d Blaupause *f*
 f bleu *m* ozalid
 n blueprint
 p światłokopia

552 blunt chisel
 d stumpfer Beitel *m*
 f ciseau *m* obtus
 n botte beitel
 p tępe dłuto

553 body
 d Karosserie *f*
 f carrosserie *f*
 n carrosserie
 p nadwozie; karoseria

554 body and fender tool set
 d Ausbeulsortiment *n*
 f coffret *m* de tôlier
 n gereedschapsassortiment voor plaatwerk
 p zestaw narzędzi blacharskich

555 body bracket; body support
 d Aufbaustütze *f*
 f support *m* de carrosserie
 n carrosseriesteun
 p wspornik nadwozia

556 body builder; tinman
 d Karosseriehersteller *m*
 f carrossier *m*
 n koetswerkhersteller; plaatwerker
 p blacharz

557 body cleaning compound
 d Karosseriepflegemittel *n*
 f produit *m* d'entretien de la carrosserie
 n carrosserieonderhoudsmiddel
 p środek do konserwacji karoserii

558 body colour
 d Deckfarbe *f*
 f peinture *f* couvrante
 n dekverf
 p farba kryjąca

559 body decoration
 d Karosseriedekoration *f*
 f habillage *m* de carrosserie
 n carrosseriedecoratie
 p dekoracja karoserii

560 body design
 d Aufbauform *f*
 f forme *f* de la carrosserie
 n carrosserievorm
 p forma nadwozia

33

561 body elements
d Karosserieteile *npl*
f éléments *mpl* de carrosserie
n carrosseriedelen
p elementy karoserii; elementy nadwozia

562 body hinge
d Kastengelenk *n*
f gond *m* de basculement
n carrosseriescharnier
p przegub nadwozia

563 body hydraulics
d Komforthydraulik *f*
f hydraulique *f* de carrosserie
n hydraulisch systeem van carrosserie
p układ hydrauliczny karoserii

564 body insulation
d Isolation *f* des Aufbaus
f isolement *m* de carrosserie
n carrosserie-isolatie
p izolacja nadwozia

565 body lacquer
d Decklack *m*
f vernis *m* couvrant; vernis m final
n deklak; aflaklaag
p lakier kryjący; lakier nawierzchniowy

566 body number
d Karosserienummer *f*
f numéro *m* de carrosserie
n carrosserienummer
p numer nadwozia; numer karoserii

567 body rim
d Umrahmung *f*
f encadrement *m*
n omlijsting
p obramowanie

568 body spring
d Tragfeder *f*
f ressort *m* de carrosserie
n draagveer
p sprężyna nośna

569 body straightening tool
d Karosserierichtwerkzeug *n*
f outils *mpl* pour réparation de carrosserie
n carrosseriegereedschap
p urządzenie do prostowania karoserii

* **body support → 555**

570 body weight
d Gewicht *n* der Karosserie
f poids *m* de carrosserie
n carrosseriegewicht
p ciężar nadwozia

571 bodywork lock grip pliers
d Karosseriegripzangen *fpl*
f pinces *fpl* étaux
n griptangen
p szczypce do nadwozia

572 boilermaker's hammer
d Kesselschmiedehammer *m*
f marteau *m* de chaudronnier
n ketelmakershamer
p młotek kotlarski

573 bolt
d Durchsteckschraube *f*
f boulon *m*
n bout; schroef
p śruba; sworzeń

574 bolt chisel
d Lochbeitel *m*
f ciseau *m* à grignoter
n hoekbeitel
p dłuto gniazdowe

575 bolt cutter
d Bolzenschneider *m*
f coupe-boulons *m*
n boutenschaar
p szczypce przegubowe do prętów

576 bolt joint
d Durchsteckschraubenverbindung *f*
f assemblage *m* boulonné
n boutverbinding
p połączenie śrubowe

577 bolt spring
d Feder *f* für Zapfen
f ressort *m* de doigt
n boutveer
p sprężyna czopu

578 bonnet
d Motorhaube *f*
f capot-moteur *m*
n motorkap
p maska silnika

579 bonnet catch
d Motorhaubenverschluss *m*
f fermeture *f* de capot
n motorkapsluiting
p zamek maski

580 bonnet hinge
 d Motorhaubenscharnier *n*
 f charnière *f* de capot
 n motorkapscharnier
 p zawiasa maski silnika

581 bonnet lock
 d Motorhaubenschloss *n*
 f serrure *f* de capot
 n motorkapslot
 p zamek maski silnika

582 boost control unit
 d Ladedruckregler *m*
 f correcteur *m* à pression de charge
 n vuldrukregelaar
 p regulator ciśnienia ładowania

583 boost pressure gauge
 d Ladedruckmesser *m*
 f manomètre *m* de suralimentation
 n laaddrukmeter
 p manometr ciśnienia ładowania

584 boot floor
 d Kofferboden *m*
 f fond *m* de malle arrière
 n vloer van bagageruimte
 p podłoga bagażnika

585 boot lamp
 d Kofferraumleuchte *f*
 f éclaireur *m* de coffre
 n kofferbakverlichting; bagageruimteverlichting
 p oświetlenie bagażnika

586 boot lid; trunk lid
 d Kofferraumdeckel *m*; Kofferdeckel *m*
 f couvercle *m* de coffre à bagages
 n kofferdeksel
 p pokrywa bagażnika; drzwi bagażnika

587 boot lid hinge
 d Deckelscharnier *n*
 f charnière *f* de couvercle
 n kofferdekselscharnier
 p zawiasa pokrywy bagażnika

588 boot lid support
 d Deckelstütze *f*
 f bras *m* d'appui de couvercle
 n kofferdekselsteun
 p wspornik pokrywy bagażnika

589 boot lining
 d Kofferraumbelag *m*
 f garniture *f* de coffre
 n bagageruimtebekleding; kofferbekleding
 p wykładzina bagażnika

590 boot lock
 d Kofferraumschloss *m*
 f verrou *m* de coffre
 n bagageruimteslot
 p zamek bagażnika

591 bording socket
 d Bördeleisen *n*
 f bordoir *m*
 n pijlerbus
 p zaginadło; zawijak

592 bore
 d Höhlung *f*
 f alésage *m*
 n inwendige diameter
 p średnica wewnętrzna otworu

593 boring tool
 d Bohrwerkzeug *n*
 f outil *m* à percer
 n boorgereedschap
 p narzędzie wiertnicze

594 boss
 d Nabe *f*
 f bossage *m*
 n aangegoten verdikking
 p zgrubienie; nadlew

595 bottle holder
 d Flaschenhalter *m*
 f porteur *m* de flacon
 n flessenhouder
 p uchwyt na butelki

596 bottom arm
 d Unterarm *m*
 f levier *m* inférieur
 n onderarm
 p ramię dolne

597 bottom cover
 d unterer Deckel *m*
 f couvercle *m* inférieur
 n onderdeksel
 p pokrywa dolna

* bottom gear → 3155

598 bottom plate
 d Bodenblech *n*
 f plaque *f* inférieure
 n bodemplaat
 p blacha podłogowa**

599 boundary layer
d Grenzschicht *f*
f couche *f* limitée
n grenslaag
p warstwa graniczna; warstwa przyścienna

600 boundary lubrication
d Grenzschmierung *f*
f limite *f* de graissage
n grenssmering
p smarowanie graniczne

601 bourdon tube
d Bourdonrohr *n*
f tube *m* de bourdon
n Bourdon-buis
p rurka Bourdona; manometr rurkowy Bourdona

602 bow
d Spriegel *m*
f arceau *m*
n boog; daktoog
p łuk; pałąk

603 bowden brake
d Bowdenzugbremse *f*
f frein *m* bowden
n door bowdenkabel bediende rem
p hamulec cięgła Bowdena

604 bowden cable
d Bowdenkabel *m*
f câble *m* bowden
n bowdenkabel
p linka Bowdena

605 bowden control
d Bowdenzug *m*; Bowdenkabel *n*
f commande *f* bowden
n bowdenkabel
p cięgło elastyczne opancerzone

606 bow drill
d Drillbohrer *m*
f drille *m*
n drilboor; spilboor
p furkadło

607 bowl cover
d Filterdeckel *m*
f couvercle *m* de pompe à combustible
n luchthelm
p pokrywa osadnika

608 bowl retaining clip
d Bügelhalter *m*
f étrier *m* de filtre à combustible
n klembeugel
p jarzmo osadnika

609 bow saw
d Bogensäge *f*
f scie *f* courbe
n beugelzaag
p piła łukowa; piła kabłąkowa

610 bow separator
d Spriegeltrennstück *n*
f entretoise *f* d'arceau
n beugelseparator
p element rozdzielczy pałąka

611 bow supporter
d Spriegelhalter *m*
f support *m* d'arceau
n boogsteun
p uchwyt pałąka

612 box body
d Kastenaufbau *m*
f caisse *f* forme boîte
n opbouw van vrachtauto
p nadwozie furgonowe

613 box girder
d Kastenträger *m*; Kastenform *f*
f poutre *f* en caisson
n kokerbalk
p podłużnica skrzynkowa

614 boxing tenon
d Verbindungszapfen *m*
f téton *m*
n verbindingstap; verbindingspen
p czop łączący

615 box nut; blind nut
d Hutmutter *f*
f écrou *m* de borgne
n dopmoer
p nakrętka kołpakowa; nakrętka kapturkowa

616 box section side member
d Kastenträger *m*
f longeron *m* en caisson
n kokerlangsdrager
p podłużnica skrzynkowa

617 box sill
d Schwelle *f*; Schweller *m*
f bas *m* de caisse
n drempelkokerbalk
p próg

618 box spanner; box wrench
d Steckschlüssel *m*
f clé *f* à douille
n dopsleutel
p klucz nasadowy rurowy

* box wrench → 618

* brace → 1753

619 brace rod
  d Strebe *f*
  f tige *f* de renfort
  n versterkingsstang
  p podpora ukośna

620 bracket
  d Halter *m*; Konsole *f*; Bügel *m*; Klammer *f*;
    Stütze *f*
  f support *m*; console *f*; bride *f*; collet *m*
  n console; beugel; oplegsteun
  p wspornik; konsola

621 bracket bearing
  d Konsollager *n*
  f palier *m* de support
  n steunlager
  p wspornik wzmacniający podparcie łożyska;
    podparcie wspornikowe łożyska

622 bracket plate
  d Halteblech *n*
  f plaquette *f* de support
  n steunplaat
  p tarcza nośna

623 brake
  d Bremse *f*
  f frein *m*
  n rem
  p hamulec

624 brake action
  d Bremswirkung *f*
  f action *f* du frein
  n remwerking
  p działanie hamulca

625 brake adjuster
  d Bremsnachsteller *m*; Bremseinstellvorrichtung
    f
  f dispositif *m* de réglage de frein
  n remafsteller
  p nastawnik luzu szczęk hamulca

626 brake adjuster wrench
  d Bremseinstellschlüssel *m*
  f clé *f* de réglage de freins
  n remstelsleutel
  p klucz nastawczy do hamulców

627 brake air compressor
  d Bremskompressor *m*
  f compresseur *m* d'air
  n remcompressor
  p sprężarka hamulcowa

628 brake band
  d Bremsband *n*
  f bande *f* de frein
  n remband
  p taśma hamulca

629 brake band anchorage
  d Bremsbandverankerung *f*
  f ancrage *m* de bande de frein
  n verankering van de remband
  p zawieszenie taśmy hamulca

630 brake band anchor arm
  d Bremsbandankerarm *m*
  f bras *m* d'ancrage de bande de frein
  n rembandarm
  p zaczep taśmy hamulca

631 brake band tensioner
  d Bremsbandspannvorrichtung *f*
  f tendeur *m* de bande de frein
  n rembandspanner; spaninrichting van de
    remband
  p napinacz taśmy hamulca

632 brake beam
  d Bremsausgleich *m*
  f palonnier *m* de timonerie de frein
  n remcompensatie
  p kompensacja hamulca

633 brake bleeding unit
  d Druckentlüfter *m*; Entlüftergerät *n*
  f appareil *m* purgeur sous pression
  n remontluchtingsapparaat
  p urządzenie odpowietrzające

634 brake bolt
  d Bremsbolzen *m*
  f boulon *m* de frein
  n rembout
  p śruba hamulca

635 brake booster
  d Bremskraftverstärker *m*
  f frein *m* assisté
  n rembekrachtiger; remservo
  p urządzenie wspomagające hamulca

636 brake buffer
  d Bremsanschlag *m*
  f butée *f* de frein
  n stootrubber
  p ogranicznik ruchu hamulca

637 brake cable
  d Bremszugseil *n*

f câble *m* de frein
n remkabel
p linka hamulca

638 brake calculation
d Bremsberechnung *f*
f calcul *m* des freins
n remkrachtberekening
p obliczanie siły hamowania

639 brake cam
d Bremsnocke *f*
f came *f* de frein
n remnok
p rozpieracz krzywkowy szczęk hamulca

640 brake cam lever
d Bremsnockenhebel *m*
f levier *m* de came de frein
n hefboom die remnokas bedient
p dźwignia rozpieracza hamulcowego

641 brake camshaft
d Bremsnockenwelle *f*
f arbre *m* à cames de freinage
n remnokas
p wałek rozpieracza hamulca

642 brake circuit
d Bremskreis *m*
f circuit *m* de freins
n remkring
p układ hamulcowy

643 brake compensating device
d Bremsausgleichvorrichtung *f*
f dispositif *m* d'équilibrage des freins
n remcompensatie-inrichting
p urządzenie kompensacyjne hamulca

644 brake control
d Bremsbetätigung *f*
f commande *f* de frein
n bediening van reminrichting
p sterowanie hamulców

**645 brake control light; brake system control
light**
d Bremsleuchtenausfallanzeigeleuchte *f*
f témoin *m* de défaillance du freinage
n controlelicht voor defect aan remsysteem
p lampka kontrolna systemu hamulcowego

646 brake coupling
d Bremskupplung *f*
f accouplement *m* de frein
n aanhangwagenremkoppeling
p sprzęg hamulcowy

647 brake cross shaft
d Bremswelle *f*
f arbre *m* transversal de frein
n bedieningsas van remsysteem
p wał hamulcowy poprzeczny

648 brake cylinder
d Bremszylinder *m*
f cylindre *m* de frein
n remcilinder
p cylinder hamulcowy

649 brake cylinder body
d Bremszylinderkörper *m*
f corps *m* de cylindre de roue
n wielremcilinderhuis
p korpus cylindra hamulcowego

650 brake deceleration
d Bremsverzögerung *f*
f décélération *f* de freinage
n remvertraging
p opóźnienie przy hamowaniu

651 brake disc
d Bremsscheibe *f*
f disque *m* de freinage
n remschijf
p tarcza hamulcowa

652 brake drum
d Bremstrommel *f*
f tambour *m* de frein
n remtrommel
p bęben hamulcowy

653 brake drum flange
d Bremstrommelwulst *m*
f bord *m* de tambour
n remtrommelkraag
p obrzeże bębna hamulcowgo

654 brake fading
d Bremsfading *n*
f affaiblissement *m* des freins
n remfading
p zmniejszona skuteczność hamulców

655 brake flange
d Bremsschild *m*
f plateau *m* de frein
n remankerplaat
p tarcza hamulcowa

656 brake fluid
d Bremsflüssigkeit *f*
f liquide *m* pour freins
n remvloeistof
p płyn hamulcowy

657 brake fluid change
d Bremsflüssigkeitswechsel *m*
f changement *m* de liquide de freins
n remvloeistofverversing
p wymiana płynu hamulcowego

658 brake fluid lever
d Bremsflüssigkeitsniveau *n*
f hauteur *f* du liquide de frein
n remvloeistofniveau
p poziom płynu hamulcowego

659 brake fluid tank
d Bremsflüssigkeitsbehälter *m*
f réservoir *m* à liquide de frein
n remvloeistofreservoir;
   remvloeistofvoorraadtank
p zbiornik płynu hamulcowego

660 brake force balance
d Bremskraftregelung *f*
f distribution *f* du freinage
n remkrachtverdeling
p ograniczenie ciśnienia hamowania

661 brake grease
d Bremszylinderpaste *f*
f graisse *f* de frein
n remvet
p smar do cylinderków hamulcowych

662 brake horsepower
d Bremsleistung *f*
f puissance *f* de freinage
n rempaardenkracht
p moc hamowania

663 brake hub
d Bremsnabe *f*
f moyeu *m* de frein
n remnaaf
p piasta bębna hamulcowego

664 brake latch
d Bremshebelsperrklinke *f*
f cliquet *m* d'arrêt du levier de frein
n remplaat
p zapadka dźwigni hamulcowej

665 brake lever handle
d Bremshebelgriff *m*
f poignée *f* de frein
n remhefboomhandgreep
p uchwyt dźwigni hamulcowej

666 brake light
d Bremslicht *n*
f feu *m* de freinage; feu *m* de stop
n remlicht
p światło hamowania; światło stop

667 brake lights test
d Bremslichtenprüfung *f*
f test *m* feux stop
n remlichtentest
p próba świateł hamowania

668 brake lining
d Bremsbelag *m*
f garniture *f*
n remvoering
p wykładzina hamulcowa; okładzina hamulcowa

669 brake lining area
d Bremsbackenbelagoberfläche *f*
f surface *f* de garniture de frein
n remvoeringoppervlak
p powierzchnia okładzin ciernych szczęk
   hamulca

670 brake lining rivet
d Bremsbelagniet *m*
f rivet *m* de garniture de frein
n holnagel
p nit okładziny ciernej

* brake lining wear → 700

671 brake linkage
d Bremsgestänge *n*
f timonerie *f* de frein
n stangenstelsel voor rembediening
p zespół dźwigni i drążków hamulca

672 brake mechanism
d Bremsanlage *f*
f mécanisme *m* de freinage
n remmechanisme
p mechanizm hamulcowy

673 brake operating
d Bremsbetätigung *f*
f commande *f* du frein
n rembediening
p uruchomienie hamulca

674 brake pad; friction pad
d Bremsklotz *m*
f plaquette *f* de frein; patin *m* de frein
n remblok
p klocek hamulcowy

675 brake pad remover
d Schlagzieher *m* für Bremsklotze
f extracteur *m* de plaquettes de freins
n slagtrekker voor remblokken
p usuwacz płytki ciernej hamulca

676 brake pad wear
d Bremsklotzabnützung *f*

f usure *f* de patin de frein
n remblokslijtage
p zużycie klocków hamulcowych

* **brake path** → 713

677 **brake pedal**
d Bremspedal *n*
f pédale *m* de frein
n rempedaal
p pedał hamulca

678 **brake pedal free movement**
d Leerweg *m* des Bremspedals
f course *f* à vide de pédale de frein
n vrije slag in rempedaal
p skok hamulca

679 **brake pedal lever**
d Bremspedalhebel *m*
f levier *m* de pédale de frein
n rempedaalhefboom
p dźwignia pedału hamulca

680 **brake pedal pad**
d Bremspedalgummi *n*
f caoutchouc *m* de pédale de frein
n rempedaalrubber
p guma pedału hamulca

681 **brake pipe**
d Bremsleitung *f*
f tuyau *m* de frein
n remleiding
p przewód hamulcowy

682 **brake piston**
d Bremskolben *m*
f piston *m* de frein
n remzuiger
p tłoczek hamulca

683 **brake piston bumper**
d Bremskolbenpuffer *m*
f pare-chocs *m* de piston de frein
n remzuigerstop
p zderzak tłoczka

684 **brake piston cup**
d Bremsmanchette *f*
f coupelle *f* de piston de frein
n remcup
p uszczelka tłoczka hamulcowego

685 **brake piston tong**
d Zange *f* zum Zurückschieben und Drehen der Bremssattelkolben
f pince *f* pour repouser et tourner les pistons

d'étrier
n remzuigertang
p szczypce do tłoków hamulcowych

686 **brake point**
d Bremspunkt *m*
f point *m* de freinage
n rempunt
p punkt hamowania

687 **brake power distributor**
d Bremskraftverteiler *m*
f compensateur *m* de force de freinage
n remkrachtverdeler
p rozdzielacz siły hamowania

688 **brake power limiter**
d Bremskraftbegrenzer *m*
f limiteur *m* de force de freinage
n remkrachtbegrenzer
p ogranicznik siły hamowania

689 **brake pressure**
d Bremsdruck *m*
f pression *f* de freinage
n remdruk
p ciśnienie hamowania

690 **brake pressure switch**
d Bremswarnlichtschalter *m*
f manocontact *m* de freins
n remlichtschakelaar
p wyłącznik świateł hamowania

691 **brake pulley**
d Bremskabelleitrolle *f*
f poulie *f* de frein
n rempoelie
p koło hamulcowe

692 **brake ratio**
d Übersetzung *f* der Bremsanlage
f rapport *m* de couple de freinage
n remverhouding
p przełożenie hamulca

693 **brake reliner**
d Bremsbelagauflegegerät *n*
f appareil *m* de regarnissage des mâchoires de frein
n remvoering
p urządzenie do wymiany okładziny hamulca

694 **brake rigging**
d Bremsbelag *m*
f frottement *m* des freins
n remvoering
p okładzina hamulcowa

695 brake servo
 d Bremskraftverstärkung *f*
 f amplification *f* de frein
 n rembekrachtiging
 p wspomaganie hamowania

696 brake shaft
 d Bremswelle *f*
 f axe *m* de frein
 n geremde as; remas
 p wał hamulcowy

697 brake shoe
 d Bremsschuh *m*
 f mâchoire *f* de frein
 n remschoen
 p szczęka hamulca

698 brake shoe adjusting device
 d Bremsnachsteller *m*
 f compensateur *m* de frein; compensateur *m* du jeu des mâchoires
 n remafsteller
 p wyrównywacz luzu szczęk hamulca

699 brake shoe lever
 d Bremsklotzhebel *m*
 f levier *m* de mâchoire de frein
 n hefboom met remblok
 p dźwignia klocka hamulcowego

700 brake shoe lining wear; brake lining wear
 d Abnutzung *f* der Bremsbeläge; Bremsbelagabnützung *f*
 f usure *f* des garnitures de freins
 n remvoeringslijtage
 p zużycie okładzin hamulcowych

701 brake shoe link pin
 d Lenkerbolzen *m*
 f axe *m* d'articulation
 n gewrichtbout
 p sworzeń przegubu

702 brake shoe return spring
 d Backenrückzugfeder *f*
 f ressort *m* de rappel
 n remschoenterugtrekveer
 p sprężyna odwodząca szczęk hamulca

703 brake shoe thrust
 d Bremsbackendruck *m*
 f pression *f* des mâchoires de frein
 n remschoendruk
 p nacisk szczęk hamulcowych

704 brake spider
 d Bremsspinne *f*
 f étoile *f* de frein

 n remkrachtverdeler
 p jarzmo hamulca

705 brake spring
 d Bremsfeder *f*
 f ressort *m* de frein
 n remveer
 p sprężyna hamulca

706 brake spring pliers
 d Bremsfederzangen *fpl*
 f pinces *fpl* pour ressorts de segments de freins
 n remveertang
 p szczypce do sprężyny hamulca

707 brake support
 d Bremsstütze *f*
 f support *m* de frein
 n remzadeldrager
 p wspornik mechanizmu hamulcowego

708 brake supporting collar
 d Bremsträgerflansch *m*
 f flasque *m* de frein
 n flens van de remankerplaat
 p kołnierz mechanizmu hamulcowego

* brake system control light → 645

709 brake test
 d Bremsprobe *f*
 f essai *m* de frein
 n remmentest
 p próba hamulców

710 brake testing bench
 d Bremsenprüfstand *m*
 f banc *m* d'essai de freins
 n remmentestbank
 p hamownia; stanowisko badawcze hamulców

711 brake toggle
 d Bremsdaumen *m*
 f came *f* de freinage
 n remnok
 p rozpieracz szczęk hamulca bębnowego

712 brake valve
 d Bremsventil *n*
 f soupape *f* de frein
 n remklep
 p zawór hamowania

713 braking distance; brake path
 d Bremsweg *m*
 f distance *f* de freinage
 n remafstand; remweg
 p droga hamowania

714 **braking force**
 d Bremskraft *f*
 f effort *m* de freinage
 n remkracht
 p siła hamowania

715 **braking force limiter**
 d Anhängerbremskraftregler *m*
 f régulateur *m* de freinage
 n remkrachtregelaar
 p regulator skuteczności hamowania

716 **braking resistance**
 d Bremswiderstand *m*
 f résistance *f* de freinage
 n remweerstand
 p opór hamowania

717 **braking surface**
 d Bremsfläche *f*
 f surface *f* de freinage
 n remoppervlakte
 p powierzchnia hamowania

718 **braking system**
 d Bremsanlage *f*
 f système *m* de freinage
 n remsysteem
 p układ hamulcowy

719 **braking time**
 d Bremszeit *f*
 f temps *m* de freinage
 n remtijd
 p czas hamowania

720 **braking torque**
 d Bremsmoment *n*
 f couple *m* de freinage
 n remmoment
 p moment hamujący

721 **branch road**
 d Nebenstrasse *f*
 f route *f* secondaire
 n zijstraat
 p droga boczna

722 **branch terminal**
 d Abzweigklemme *f*
 f borne *f* de dérivation
 n aansluitklem van aftakleiding
 p złączka odgałęźna

723 **branded lubricant**
 d Markenschmierstoff *m*
 f lubrifiant *m* de marque
 n merksmeermiddel
 p smar firmowy

724 **branded oil**
 d Markenöl *n*
 f huile *f* de marque
 n merkolie
 p olej firmowy

725 **brand name**
 d Firmenschriftzug *m*
 f empreinte *f* de marque
 n merknaam
 p nazwa firmowa

726 **brass sheet**
 d Messingblech *n*
 f tôle *f* de laiton
 n messingplaat
 p blacha mosiężna

727 **brazing hearth**
 d Hartlötofen *m*
 f foyer *m* à soudure forte
 n hardsoldeeroven
 p piec do lutowania twardego

728 **breakaway brake**
 d Abreissbremse *f*
 f frein *m* de sécurité pour remorque
 n losbreekreminrichting van aanhangwagen
 p samoczynny hamulec przyczepy

729 **breakdown; defect**
 d Störung *f*; Panne *f*
 f panne *f*
 n defect; storing
 p defekt; usterka

730 **breakdown lamp**
 d Pannenlampe *f*
 f lampe *f* de dépannage
 n pechlamp
 p lampa awaryjna

731 **breakdown lorry; breakdown truck**
 d Abschleppwagen *m*
 f dépanneuse *f*
 n service-wagen
 p platforma transportowa do uszkodzonych
    pojazdów; laweta

732 **breakdown repair set**
 d Pannenbeseitigungskoffer *m*
 f coffret *m* de dépannage
 n pechkoffer
 p awaryjny zestaw naprawczy

 *   **breakdown truck** → 731

733 **breakdown voltage**
 d Durchbruchspannung *f*

f tension *f* de claquage
n doorslagspanning
p napięcie przebicia

734 breaker cam
d Unterbrechernocken *m*
f came *f* de rupteur
n onderbrekernok
p krzywka przerywacza

735 breaker cam angle
d Schliesswinkel *m* des Unterbrechers
f angle *m* de came
n contacthoek
p kąt zwarcia styków przerywacza

736 breaker drive shaft
d Unterbrecherantriebswelle *f*
f axe *m* de rupteur
n onderbrekeras
p wał przerywacza

737 breaker edge looseness
d Kissenkantenlösung *f*
f séparation *f* gomme de bordage de toile de division
n loszitten van onderbrekerrand
p zluzowanie skrzyni poduszkowej

738 breaker gap angle
d Abreisswinkel *m* des Unterbrechers
f angle *m* d'ouverture
n openingshoek
p kąt rozwarcia styków przerywacza

739 breakerless transistorized ignition
d kontaktlos gesteuerte Transistorzündung *f*
f allumage *m* transistorisé sans contacts
n contactpuntloze transistorontsteking
p zapłon tranzystorowy bezstykowy

740 breaker moving contact
d beweglicher Unterbrecherkontakt *m*
f contact *m* mobile de rupteur
n bewegend contactpunt
p ruchomy styk przerywacza

741 breaker plate
d Unterbrecherplatte *f*
f plaque *f* de rupteur
n grondplaat van de onderbreker
p ruchoma tarcza przerywacza

742 breaker point
d Unterbrecherkontakt *m*
f contact *m* de rupteur
n contactpunt
p styk przerywacza

743 breaker point gap
d Abstand *m* zwischen den Unterbrecherkontakten
f écartement *m* des contacts de rupteur
n contactpuntafstand
p odstęp pomiędzy stykami przerywacza

744 breaker resistor
d Vorschaltwiderstand *m*
f résistance *f* ballast
n voorschakelweerstand
p opornik wstępny; opornik szeregowy

745 breaker spring
d Unterbrecherfeder *f*
f ressort *m* de rupteur
n onderbrekerveer
p sprężyna przerywacza

746 breaker strip
d Leinwandschicht *f* der Reifenlauffläche
f protecteur *m*
n loopvlakversterking van autoband
p pasmo bieżnika opony

747 break even point
d Rentabilitätsgrenze *f*
f point *m* de transition
n omslagpunt
p granica rentowności

748 breaking strength; bursting strength
d Bruchfestigkeit *f*; Berstfestigkeit *f*
f charge *f* de rupture
n breuksterkte; breeksterkte
p wytrzymałość na rozerwanie

749 break in prevention
d Einbruchverhinderung *f*
f précautions *fpl* antivol
n inbraakpreventie
p zabezpieczenie przed włamaniem

750 break line
d Leitlinie *f*
f ligne *f* interrompue
n onderbroken lijn
p linia przerywana

751 breakneck speed
d waghalsige Fahrt *f*
f roulement *m* à tombeau ouvert
n dolle vaart
p nadmierna szybkość

752 breather
d Entlüfter *m*
f reniflard *m*
n ontluchter
p odpowietrznik

753 breather cap
  d Entlüfterkappe *f*
  f capuchon *m* de reniflard
  n ontluchterkap
  p klapa odpowietrzająca

754 brick road
  d Klinkerstrasse *f*
  f route *f* à revêtement de briques
  n klinkerstraat
  p ulica klinkierowa

755 bridge
  d Brücke *f*
  f pont *m*
  n brug
  p most

756 bridge connection; bridge network
  d Brückenschaltung *f*
  f connexion *f* de pont
  n brugschakeling
  p układ mostkowy

  * **bridge network** → 756

757 brightness of colours
  d Farbenglanz *m*
  f éclat *m* des couleurs
  n kleurglans
  p połysk farby

758 Brinell Hardness Number
  d Brinellhärte *f*
  f dureté *f* Brinell
  n Brinell-hardheid
  p twardość Brinella

759 brittleness
  d Brüchigkeit *f*
  f fragilité *f*
  n breekbaarheid
  p łamliwość

760 broach; reamer
  d Reibahle *f*; Räumbahle *f*
  f alésoir *m*
  n ronde vijl; ruimer
  p przeciągacz; rozwiertak

761 broad beam headlamp
  d Breitstrahlscheinwerfer *m*
  f projecteur *m* à faisceau large
  n breedstraler
  p reflektor szerokopromieniowy

762 bronze
  d Bronze *f*
  f bronze *m*

  n brons
  p brąz

763 bronze bush
  d Bronzebüchse *f*
  f bague *f* en bronze
  n bronzen huls
  p tuleja z brązu

764 bronze bushing
  d Bronzelagerschale *f*
  f coussinet *m* en bronze
  n bronslagerschaal
  p brązowa panewka łożyska

765 brush
  d Bürste *f*
  f balai *m*
  n borstel
  p szczotka

766 brush holder
  d Bürstenhalter *m*
  f porte-balai *m*
  n borstelhouder
  p szczotkotrzymacz

767 brush holder plate
  d Bürstenhalterplatte *f*
  f plateau *f* porte-balai
  n borstelhouderplaat
  p tarcza szczotkotrzymaczy

768 brush sparking
  d Kontaktfunk *f*
  f étincelle *f* au balai
  n vonk van de borstels
  p iskra zetknięcia

769 brush spring
  d Kohlenbürstenfeder *f*
  f ressort *m* de balai
  n borstelveer
  p sprężyna szczotki

770 bucket seat
  d Muldensitz *m*; Schalensitz *m*
  f siège *m* contouré; siège *m* moulé sur mesure
  n kuipstoel
  p siedzenie wywrotne

771 buckled casing
  d gestauchte Karkasse *f*
  f carcasse *f* plisée
  n gebogen kast
  p karkas z zamkiem

772 buffer
  d Buffer *m*

f tampon *m* de choc
n buffer
p zderzak amortyzacyjny

773 buffing compound
d Polierpaste *f*
f pâte *f* à polir
n polijstpasta
p pasta polerska

774 built on area
d bebautes Gebiet *n*
f terrain *m* bâti
n bebouwde kom
p obszar zabudowany

775 bulb adapter
d Glühlampenfassung *f*
f douille *f* d'ampoule
n lampfitting
p oprawka żarówki

776 bulb control
d Glühlampenkontrolle *f*
f témoin *m* de contrôle d'ampoule
n gloeilampencontrole
p kontrola żarówki

777 bulb holder
d Fassung *f* für Glühbirne
f douille *f* de lampe
n lamphouder
p oprawka żarówki

778 bulb replacement
d Glühlampenwechsel *m*
f remplacement *m* de l'ampoule
n vervangen van gloeilampen
p wymieniać żarówkę

779 bulb shield
d Glühlampenabdeckschirm *m*
f écran *m* de lampe
n afscherming van de lamp
p przesłona żarówki

780 bulb socket
d Glühlampensockel *m*
f socle *m* d'ampoule
n lamphuls
p cokół żarówki

781 bulb socket tag
d Sockelhalter *m*
f baïonnette *f* de douille
n bajonetstrip
p zaczep cokołu żarówki

782 bulker
d Behälterwagen *m*

f camion *m* trémie
n vrachtauto voor bulkgoederen
p samochód zbiornikowy

783 bumper bracket; bumper support; bumper stay; bumper holder
d Träger *f* für Stossstangen
f support *m* de pare-chocs
n bumpersteun
p wspornik zderzaka

784 bumper clamp
d Stossstangenhalter *m*
f attache *f* de pare-chocs
n bumperklem
p uchwyt zderzaka

* **bumper holder** → 783

785 bumper horn
d Stossstangenhorn *n*
f butoir *m* de pare-chocs
n bumperhoorn
p nakładka zderzaka

786 bumper jack
d Stossstangenheber *m*
f cric *m* de pare-chocs
n krik aan bumper
p wciągarka zderzakowa

787 bumper plate
d Aufprallplatte *f*
f plaque *f* pare-chocs
n bumperplaat
p miękka nakładka na zderzaki

* **bumper stay** → 783

* **bumper support** → 783

788 bumping hammer
d Spann- und Schlichthammer *m*
f marteau *m* postillon
n postillonhamer
p młotek do klepania blachy

789 bump pressure space
d Druckölarbeitsraum *m*
f chambre *f* de travail de compression
n schok-drukruimte
p przestrzeń robocza obciążania

790 bump valve; compression valve
d Druckstufendrosselventil *n*
f clapet *m* de choc; clapet *m* de compression
n druksmoorklep
p zawór obciążania

791 **burning point**
  d Zündtemperatur *f*
  f température *f* d'allumage
  n ontstekingstemperatuur
  p temperatura samozapłonu

792 **burnt contacts**
  d angebrande Kontakte *mpl*
  f contacts *mpl* grillés
  n aangebrande contacten
  p nadpalone styki

793 **burring reamer**
  d Vorreibahle *f*
  f alésoir *m* d'ébarbage
  n voorruimer
  p rozwiertak zdzierak

  * **bursting strength** → 748

794 **bush hammer**
  d Charierhammer *m*
  f boucharde *m*
  n hamer om moffen te plaatsen
  p młotek do groszkowania

795 **bushing chain**
  d Hülsenkette *f*
  f chaîne *f* tubulaire
  n busketting
  p łańcuch panwiowy

796 **bushing stud chain**
  d Stahlbolzenkette *f*
  f chaîne *f* mouflée
  n damketting
  p łańcuch rolkowy; łańcuch tulejowy

797 **bush seat**
  d Sitz *m* der Lagerbuchse
  f siège *m* de bague
  n lagerbuszitting
  p gniazdo tulei

798 **butt collar**
  d Anlaufbuchse *f*
  f collier *m* d'appui
  n drukkraag; drukring
  p tulejka prowadzona

799 **butt corner joint**
  d Eckstumpfstoss *m*
  f assemblage *m* d'angle en about
  n stompe hoeknaad
  p połączenie na styk

  * **butterfly nut** → 5471

800 **butterfly valve**
  d Drosselklappe *f*
  f papillon *m* de gaz
  n vlinderklep; vleugelklep; gasklep
  p przepustnica

801 **butting ring**
  d Anlaufscheibe *f*
  f rondelle *f* de butée
  n drukring
  p podkładka regulacyjna

802 **button**
  d Knopf *m*
  f bouton *m*
  n knop; drukknop
  p guzik; przycisk

803 **button switch**
  d Druckknopfschalter *m*
  f interrupteur *m* à poussoir
  n drukknopschakelaar
  p łącznik przyciskowy

804 **butt strap**
  d Stossblech *n*; Deckleiste *f*
  f sangle *f* limitation de débattement
  n afdeklat; aftimmerlat
  p nakładka

805 **butt weld**
  d Stumpfnaht *f*
  f soudure *f* bout à bout
  n stompe naad
  p spoina doczołowa

806 **butt welding**
  d Stumpfschweissung *f*
  f souder *m* bout à bout
  n stomplassen
  p spawanie doczołowe

807 **buzzer**
  d Summer *m*
  f vibreur *m*; bruiteur *m*
  n zoemer
  p brzęczyk

808 **bypass filter**
  d Nebenstromfilter *m*
  f filtre *m* en dérivation
  n nevenstroomfilter
  p filtr bocznikowy

C

809 cabinet file
d Kabinettfeile *f*
f lime *f* d'ébéniste
n kabinetvijl
p pilnik szeroki półokrągły

810 cabinet rasp
d Kabinettraspel *f*
f râpe *f* d'ébéniste
n kabinetrasp
p tarnik szeroki półokrągły

811 cable
d Kabel *m*
f câble *m*
n kabel
p kabel; linka

812 cable box
d Kabelanschlusskasten *m*
f boîte *f* de junction
n bedradingsaansluitdoos
p skrzynka kablowa

813 cable bridge
d Kabelbrücke *f*
f pont *m* de câble
n kabelbrug
p most wiszący kablowy

814 cable clamp; cable clip
d Kabelklemme *f*
f serre-câble *m*
n kabelklem
p zacisk kabla; zacisk liny

* **cable clip → 814**

815 cable collar
d Kabelschelle *f*
f collier *m* pour câble
n kabelklem
p uchwyt kablowy

816 cable conduit
d Kabelhülle *f*
f gaîne *f* de câble
n kabelmantel
p płaszcz kabla; powłoka kabla

817 cable cutters
d Kabelschneider *m* für elektrische Kabel
f pinces *fpl* coupe-câbles pour câbles électriques
n kabelmes
p przecinak kabli

818 cable grip
d Kabelkralle *f*
f cosse *f* à crochet
n kabelklem
p zacisk linowy

819 cable grommet
d Kabeltülle *f*; Kabeldurchgang *m*
f passe-câble *m*
n kabeltule; doorvoertule
p przelotka cięgna giętkiego

820 cable guide
d Seilhülle *f*; Kabelhülle *f*
f gaîne *f* de câble
n kabelgeleider
p prowadnica linki

821 cable holder
d Kabelhalter *m*
f coller *m* de fixation
n kabelhouder
p uchwyt przewodu

822 cable insulating tube
d Isolierschlauch *m*
f tube *m* isolant
n kabeldoos; kabelgoot
p rurka izolacyjna przewodów

823 cable lead through
d Kabelführung *f*
f passe-câble *m*
n kabelleiding
p prowadnica kabla

824 cable nipple
d Kabelnippel *m*
f nipple *f* de raccord
n kabelnippel
p złączka kablowa

825 cable outlet
d Kabelaustritt *m*
f sortie *f* de câble
n kabeldoorvoer
p wyjście kabla

826 cable plug
d Kabelstecker *m*
f connecteur *m* avec câbles
n kabelstekker
p wkładka z przewodami

827 **cable protective tube**
  d Kabelisolierschlauch *m*
  f tube *m* protecteur des câbles
  n kabelisolatiemantel
  p rurka ochronna przewodów

828 **cable reeling unit**
  d Kabeltrommel *m*
  f prolongateur-enrouleur *m*
  n kabelhaspel
  p bęben do nawijania; bęben kablowy

829 **cable sheath**
  d Schutzschlauch *m*
  f gaîne *f* de câble
  n kabelmantel
  p pancerz cięgna

830 **cable shoe**
  d Kabelschuh *m*
  f borne *f* de câble
  n kabelschoen
  p zacisk przewodu

831 **cable sleeve**
  d Kabelmuffe *f*
  f douille *f* de raccord
  n kabelmof
  p mufa kablowa

832 **cable socket**
  d Polklemme *f*
  f casse *f* de câble
  n kabelsok
  p końcówka przewodu; zacisk przewodu

833 **cable stop**
  d Seilsperre *f*
  f bride *f* de câble
  n kabeleindplaat
  p docisk kabla

834 **cable terminal**
  d Kabelklemme *f*
  f cosse *f* pour câble
  n kabelaansluiting
  p gniazdo przewodu

835 **cage**
  d Anlage *f*
  f accouplement *m*
  n kooiconstructie
  p złożenie

836 **cage pilot**
  d Kugelkäfigführungskappe *f*
  f coupelle *f* guide
  n kogelkooigeleider
  p prowadnica koszyczka łożyska

837 **calculating pressure**
  d Berechnungsdruck *m*
  f pression *f* de calcul
  n theoretische druk
  p ciśnienie obliczone

838 **calibrated spring**
  d kalibrierte Feder *f*
  f ressort *m* taré
  n gekalibreerde veer
  p sprężyna kalibrowana

839 **calibrating tool**
  d Kalibrierinstrument *n*
  f jaugeur *m*
  n ijkgereedschap; kalibreergereedschap
  p przyrząd pomiarowy

840 **calibration**
  d Kalibrierung *f*
  f calibrage *m*
  n kalibrering
  p kalibrowanie

841 **caliper compass; outer compass**
  d Greifzirkel *m*; Aussenzirkel *m*
  f compas *m* extérieur
  n krompasser; buitenpasser
  p macki zewnętrzne

842 **caliper gauge; snap gauge**
  d Schiebelehre *f*; Messschieber *m*
  f pied *m* à coulisse
  n schuifmaat
  p sprawdzian szczękowy

843 **calorific power**
  d Heizwert *m*
  f puissance *f* calorifique
  n calorische waarde
  p moc cieplna

844 **calorific value**
  d Heizwert *m*; Wärmewert *m*
  f valeur *f* thermique
  n calorische waarde
  p wartość termiczna

845 **calory**
  d Kalorie *f*
  f calorie *f*
  n calorie
  p kaloria

846 **cam**
  d Nocken *m*
  f came *f*
  n kam; nok
  p krzywka

847 camber angle
d Sturzwinkel *m*
f angle *m* de carrossage
n wielvluchthoek
p kąt pochylenia koła

848 camber of spring
d Blattfedersprengung *f*
f flèche *f* d'un ressort à lames
n gebogen stand van bladveer
p strzałka ugięcia resoru

849 cam disc
d Nockenscheibe *f*
f disque *m* à cames
n nokkenschijf
p krzywka tarczowa

850 cam drum
d Nockentrommel *f*
f tambour *m* à cames
n nokkentrommel
p bęben krzywkowy

851 camelback
d Rohlaufstreifen *m*
f bande *f* de rechapage
n gecoverde band
p taśma bieżnikowa

852 cam lever shaft
d Nockenhebelwelle *f*
f arbre *m* de levier à came
n as van de nokarm
p wałek krzywki

853 cam lift
d Nockenhub *m*
f levée *f* de came
n lichthoogte van nok
p wznios krzywki

854 cam profile
d Nockenform *f*
f profil *m* de came; rampe *f* de came
n nokprofiel
p profil krzywki

855 cam ring
d Nockenring *m*
f anneau *m* à cames
n nokkenring
p pierścień krzywkowy

856 cam roller
d Rollenstössel *m*
f gallet *m*
n nokrol
p popychacz krążkowy; popychacz rolkowy

857 camshaft
d Nockenwelle *f*
f arbre *m* à cames
n nokkenas
p wał rozrządu

858 camshaft arrangement
d Lage *f* der Nockenwelle
f position *f* d'arbre à cames
n ligging van de nokkenas
p usytuowanie wału rozrządu

859 camshaft bearing bush
d Lagerbuchse *f* der Nockenwelle
f bague *f* de palier d'arbre à cames
n nokkenaslager
p tuleja łożyskowa wału rozrządu

860 camshaft bearing bush seat
d Sitz *m* der Nockenwellenlagerbuchse
f logement *m* de palier d'arbre à cames
n zitting van nokkenaslager
p gniazdo tulei łożyskowej

861 camshaft drive
d Nockenwellenantrieb *m*
f commande *f* de l'arbre à cames
n nokkenasaandrijving
p napęd wału rozrządu

862 camshaft drive gear cover
d Nockenwellenantriebsgehäuse *n*
f couvercle *m* de distribution
n distributiedeksel
p obudowa napędu wału rozrządu

863 camshaft eccentric
d Nockenwellenexzenter *m*
f excentrique *m* de l'arbre à cames
n excentrische nokkenas
p wałek rozrządu; wałek krzywkowy

864 camshaft gear wheel
d Nockenwellenzahnrad *n*
f roue *f* de distribution
n nokkenastandwiel
p koło zębate wału rozrządu

865 camshaft lobe
d Nockenerhebung *f*
f lobe *m* de came
n rond uitsteeksel van nokkenas
p wzniesienie krzywki

866 camshaft sensor
d Nockenwellensensor *m*
f sensor *m* d'arbre à cames
n nokkenassensor
p sensor wału rozrządu

867 can carrier
d Kanisterhalter *m*
f porteur *m* de jerrycan
n jerrycanhouder
p uchwyt na kanister

868 can combustion chamber
d Einzelbrennkammer *f*
f chambre *f* de combustion séparée
n enkelvoudige verbrandingskamer
p dzbanowa komora spalania

869 canister
d Benzinkanister *m*
f nourrice *f*
n jerrycan
p kanister

870 cantilever spring
d Auslegerfeder *f*
f ressort *m* cantilever
n cantileverveer
p wysięgnikowy resor półepileptyczny

871 canvas top
d Planenverdeck *n*
f bâche *f* en grosse toile
n canvas kap
p dach brezentowy

872 cap
d Kappe *f*
f capuchon *m*
n kap; deksel
p kapa; pokrywa

873 cap clamp spring
d Verteilerkappeverschlussfeder *f*
f fixation *f* de chapeau
n verdeelkapsluitveer
p zatrzask sprężynowy głowicy rozdzielacza

874 cape chisel
d Kreuzmeissel *m*
f bédane *f*
n ritsbeitel
p wycinak ślusarski prostokątny

875 cap nut
d Kapselmutter *f*; Hutmutter *f*
f écrou *m* à chapeau
n kapmoer
p nakrętka kołpakowa

876 cap screw
d Kopschraube *f*
f vis *f* à tête
n kopschroef
p wkręt z łbem; śruba z łbem

877 caravan
d Wohnanhänger *m*
f caravane *f*
n caravan
p przyczepa turystyczna mieszkalna

878 car body sheet
d Karosserieblech *n*
f tôle *f* de carrosserie
n carrosserieplaat
p blacha karoseryjna

879 carbon brush
d Kohlenbürste *f*
f balai *m* de carbonne
n koolborstel
p szczotka węglowa

880 carbon deposit
d Ölkohlenablagerung *f*
f dépot *m* charbonneux
n koolaanslag
p osad węglowy

881 carbon fibre
d Kohlenstoffaser *m*
f fibre *f* de carbone
n koolstofvezel
p włókno węglowe

882 carbon removal
d Entkohlung *f*
f décarburation *f*
n ontkoling
p odwęglanie

883 carburation
d Vergasung *f*
f carburation *f*
n vergassing
p karburacja

884 carburetter; carburettor
d Vergaser *m*
f carburateur *m*
n carburateur
p gaźnik; karburator

885 carburetter adapter
d Vergaserzwischenstück *n*
f tubulure *f* de raccord de carburateur
n tussenstuk van carburateur
p element pośredni gaźnika

886 carburetter adjustment
d Vergasereinstellung *f*; Vergaserabstimmung *f*
f réglage *m* du carburateur
n carburateurafstelling
p ustawianie gaźnika

887 **carburetter body**
d Vergasergehäuse n
f corps m de carburateur
n carburateurhuis
p kadłub gaźnika

888 **carburetter cover**
d Vergaserdeckel m
f couvercle m de carburateur
n carburateurdeksel
p pokrywa gaźnika

889 **carburetter drain cock**
d Vergaserablasshahn m
f robinet m de vidange de carburateur
n aftapkraan van carburateur
p kurek spustowy gaźnika

890 **carburetter engine**
d Vergasermotor m
f moteur m à carburateur
n motor met carburateur
p silnik gaźnikowy

891 **carburetter flange**
d Vergaseranschlussstutzen m
f bride f de carburateur
n carburateurvoet
p króciec podłączeniowy gaźnika

892 **carburetter float**
d Vergaserschwimmer m
f flotteur m de carburateur
n vlotter van carburateur
p pływak gaźnika

893 **carburetter jacket**
d Vergaserheizmantel m
f enveloppe f de réchauffage du carburateur
n warmvloeistofmantel om carburateur
p płaszcz grzewczy gaźnika

894 **carburetter jet**
d Vergaserdüse f
f gicleur m
n doseur van carburateur; sproeier van
  carburateur
p dysza gaźnika

895 **carburetter pipe wrench**
d Steckschlüssel m für Vergaser
f clé f à tube pour carburateur
n steeksleutel voor carburateur
p klucz nasadowy do gaźnika

896 **carburetter primer**
d Vergasertipper m

f poussoir m d'appel d'essence au carburateur
n carburateurstartklep
p igła gaźnika

* **carburettor → 884**

897 **carcass saw**
d grobe Rückensäge f
f scie f grosse à dos
n grove rugzaag
p piła grzbietnica

* **cardan joint → 5265**

* **cardan shaft → 1773**

898 **cardan spider**
d Kardankreuz n
f croisillon m de cardan
n kruisstuk van cardankoppeling
p krzyżulec Cardana

899 **cardan tube**
d Kardanrohr n
f tube m de cardan
n cardanpijp
p rura wału przegubowego; rura wału
  kardanowego

900 **cardboard panel**
d Papptafel m
f panneau m carton
n kartonnen plaat
p tablica tekturowa

901 **car heater**
d Wagenheizgerät n
f appareil m de chauffage pour voiture
n autoverwarming; autoverwarmingsapparaat
p ogrzewanie samochodowe

902 **car importer**
d Automobilimporter m
f importateur m d'automobiles
n auto-importeur
p importer samochodów

903 **car lift**
d Wagenhebebühne f
f pont m élévateur pour autos
n hefbrug
p dźwig samochodowy stały

904 **car line**
d Modellnummer f
f code f modèle
n modelcode
p kod modelu

905 car polish
d Wagenpflegemittel *n*
f produit *m* pour lustrer les carrosseries
n autopoetsmiddel
p środek do polerowania nadwozia samochodu

906 car radio
d Autoradio *n*
f poste *f* autoradio; radio *m* d'automobile
n autoradio
p radio samochodowe

907 carrier plate cap
d Lagerkapsel *f*
f cloche *f*
n flenskap
p kołpak

908   carrier rack
(metal rack for pullers)
d Stahlträger *m*
f stand *m* portatif
n stalen draagbak
p stojak nośny

* carrying axle → 445

909 car transporter
d Spezialautotransporter *m*
f semi-remorque *f* de transport de voitures
n autotransporter
p naczepa transporter-samochodów

910 car washing brush
d Wagenwaschbürste *f*
f brosse *f* à laver les voitures
n autowasborstel
p szczotka do mycia samochodu

911 car wash installation
d Autowaschanlage *f*
f installation *f* de lavage de voitures
n autowasinrichting
p instalacja do mycia samochodów

912 case hardened steel
d Einsatzstahl *m*
f acier *m* de cémentation
n gecementeerd staal
p stal do nawęglania

913 case hardening
d Einsatzhärten *n*
f cémentation *f*
n cementeren
p utwardzanie

914 casing
d Blockkasten *m*; Gehäuse *n*

f bac *m*
n ommanteling; huis
p obudowa; osłona

915 casing plies
d Gummizwischenlage *f*
f gomme *f* entre nappes
n rubberen tussenlaag
p guma międzywarstwowa

916 cassette stowage
d Cassettenlagerung *f*
f cache *m* pour cassette
n bergplaats voor cassettes
p miejsce na schowanie kaset

917 cast *v*
d giessen
f couler
n gieten
p odlewać

* castellated nut → 922

918 caster
d Vorlauf *m*; Nachlauf *m*
f chasse *f*
n naspoor
p wyprzedzenie sworznia zwrotnicy

919 caster angle
d Nachlaufwinkel *m*
f angle *m* de chasse
n naspoorhoek
p kąt wyprzedzenia

920 casting
d Gussstück *n*
f moulage *m*; pièce *f* de fonte
n gietstuk
p odlew

921 cast iron
d Giesseisen *n*
f fonte *f* moulée
n gegoten staal
p żeliwo

922 castle nut; castellated nut
d Kronenmutter *f*
f écrou *m* à créneaux
n kroonmoer
p nakrętka koronowa

923 castor oil
d Rizinusöl *n*
f huile *f* de ricin
n castorolie
p olej rycynowy

52

924 **cast steel**
d Gussstahl *m*
f acier *m* coulé
n gietstaal
p staliwo

925 **catalyst**
d Katalysator *m*
f catalyseur *m*
n katalysator
p katalizator

926 **catalytic converter substrate**
d Katalysatorträger *m*
f porte-catalyseur *m*
n katalysatordrager
p nośnik katalizatora

927 **catalytic cracking**
d katalytisches Spaltverfahren *n*
f cracking *m* catalytique
n katalytisch kraken
p krakowanie katalityczne

928 **catch**
d Sperrklinke *f*
f cliquet *m* d'arrêt
n grendel; pal; klink
p zatrzask; zaczep; zapadka

929 **catch plate**
d Mitnehmerscheibe *f*
f disque *m* d'entraînement de déclic
n meeneemplaat
p tarcza zabierakowa

930 **cathode**
d Kathode *f*
f cathode *f*
n kathode
p katoda

931 **cattle crossing**
d Achtung, Tiere
f passage *m* de bétail
n overstekend vee
p uwaga; zwierzęta

932 **cattle truck**
d Viehtransporter *m*
f camion *m* bétaillère
n veetransportauto; veewagen
p transporter zwierząt

* **C clamp** → 933

933 **C cramp; C clamp**
d Schraubzwinge *f*
f presse *f* à vis en C

n ketelklem
p zacisk śrubowy

934 **ceiling lamp; roof lamp**
d Deckenlampe *f*
f plafonnier *m*; lampe *f* de plafond
n plafondlamp
p plafoniera; oprawa przysufitowa; lampa sufitowa

935 **cell**
d Zelle *f*
f élément *m*
n cel; element
p ogniwo

936 **cell cover**
d Zellendeckel *m*
f couvercle *m* d'élément
n celdeksel
p pokrywa ogniwa

937 **cell filler plug**
d Zellenfüllöffnungsstopfen *m*
f bouchon *m* de remplissage de batterie
n vuldop van de cel
p korek wlewu pokrywy ogniwa akumulatora

938 **cell hanger**
d Zellenanhängezapfen *m*
f téton *m* de fixation d'élément
n bevestigingsbout van de cel
p czop zawieszenia ogniwa

939 **cell negative terminal**
d Zellenminuspol *m*
f tige *f* polaire négative
n minpool van de cel
p ujemna końcówka ogniwa

940 **cell separator**
d Zellenseparator *m*
f cloison *m* de séparation
n scheidingswand tussen de cellen
p gródź międzyogniwowa

* **cellular radiator** → 2693

941 **cellulose tip mallet**
d Plastikhammer *m*
f massette *f* embouts cellulose
n kunststof hamer
p młotek z plastiku

942 **cell voltage**
d Zellenspannung *f*
f tension *f* d'un élément
n celspanning
p napięcie ogniwa akumulatora

53

* **center bar** → 1756

943 center bearing
d Mittellager *n*
f palier *m* intermédiaire
n tussenlager
p łożysko pośrednie

* **center console** → 959

944 center cut
d axialer Schnitt *m*
f coupe *f* axiale
n middeninsnijding
p cięcie poosiowe

945 center electrode
d Mittelelektrode *f*
f électrode *f* centrale
n middenelektrode
p elektroda środkowa

946 center finding gauge
d Zentrierwinkel *m*
f équerre *f* de centrage
n centreerhaak
p kątownik z przekątną; środkownik

947 centering drilling machine
d Zentrierbohrmaschine *f*
f perceuse *f* à centrer
n centreerboormachine
p wiertarko-centrówka

948 centering pin
d Zentrierstift *m*
f goujon *m* de centrage
n centreerpen
p sworzeń ustalający

949 centering ribs
d Zentrierrippen *fpl*
f nervures *fpl* de centrage
n centreringsribben
p żeberka centrujące

950 centerless grinding
d spitzenloses Schleifen *n*
f rectification *f* sans pointes
n puntloos slijpen
p szlifowanie bezkłowe

951 center multiplate clutch
d mittlere Lamellenkupplung *f*
f embrayage *m* multidisque
n middelste lamellenkoppeling
p sprzęgło wielotarczowe

952 center of gravity
d Schwerpunkt *m*
f centre *m* de gravité
n zwaartepunt
p środek ciężkości

953 center of pressure
d Druckmittelpunkt *m*
f centre *m* de pression
n drukpunt
p środek parcia

954 center pillar
d Mittelpfosten *m*; Türsaule *f*
f pied *m* milieu; pilier *m* central
n middelste deurstijl
p słupek środkowy; słupek drzwiowy

955 center pillar facing
d Türsäulenbezug *m*
f garniture *f* de pied milieu
n bekleding van de deurstijl
p pokrycie słupka środkowego

956 center pillar shield
d Türsäulenschild *f*
f revêtement *m* de pied milieu
n bekleding van de middenstijlconsole
p nakładka dolna słupka środkowego

957 center punch
d Körner *m*
f pointeau *m*; pointe *f* du poinçon
n centerpons
p punktak

958 center sleeve; pilot sleeve
d Zentrierhülse *f*
f douille *f* de centrage
n centreerhuls
p tuleja środkująca

959 central console; center console
d Mittelkonsole *f*
f console *f* centrale
n middenconsole
p konsola środkowa

960 centralized lubrication
d Zentralschmierung *f*
f graissage *m* centralisé
n centrale smering
p smarowanie centralne

961 central locking switch
d Zentralvernagelungsschalter *m*
f commutateur *m* condamnation de porte électromagnétique

n schakelaar van centrale portiervergrendeling
p wyłącznik centralnej blokady drzwi

962 central passage; gangway
d Mittelgang *m*
f passage *m* du milieu; couloir *m* central
n gangpad
p przejście

* **central spring clutch** → **4509**

963 central tube frame
d Zentralträgerrahmen *m*
f cadre *m* à poutre centrale
n ruggegraatchassis
p rama centralna

964 central vent
d Zentralausströmer *m*
f aérateur *m* central
n ventilatierooster
p wylot środkowy

965 centrifugal clutch
d Fliehkraftkupplung *f*
f embrayage *m* centrifuge
n centrifugaalkoppeling
p sprzęgło odśrodkowe

966 centrifugal clutch vacuum actuator
d Fliehkraftkupplungsunterdruckversteller *m*
f servo *m* à dépression d'embrayage centrifuge
n onderdrukbekrachtiger van centrifugaalkoppeling
p siłownik podciśnieniowy sprzęgła odśrodkowego

967 centrifugal filter
d zentrifugaler Filter *m*
f filtre *m* centrifuge
n centrifugaalfilter
p filtr odśrodkowy

968 centrifugal force
d Fliehkracht *f*
f force *f* centrifuge
n centrifugaalkracht
p siła odśrodkowa

* **centrifugal governor** → **3293**

969 centrifugal pump
d Fliehkrachtpumpe *f*
f pompe *f* centrifuge
n centrifugaalpomp
p pompa odśrodkowa

970 centrifugal switch
d Fliehkraftschalter *m*
f interrupteur *m* centrifuge
n centrifugaalschakelaar
p wyłącznik odśrodkowy

971 ceramic substrate catalytic converter
d Keramikkatalysator *m*
f catalyseur *m* à céramique
n keramische katalysator
p katalizator ceramiczny

972 certificate
d Attest *n*; Zeugnis *n*
f attestation *f*
n bewijs
p atest

973 cetane booster
d Zündbeschleuniger *m*
f survolteur *m*
n ontstekingsversneller
p dodatek przyśpieszający zapłon

974 cetane number
d Cetanzahl *f*
f indice *m* cétane
n cetaangetal
p liczba cetanowa

975 CFR engine
d CFR-Research-Motor *m*
f moteur *m* CFR
n CFR-proefmotor
p silnik CFR; znormalizowany silnik do określania liczby oktanowej paliw

976 chain
d Kette *f*
f chaîne *f*
n ketting
p łańcuch

977 chain block
d Kettenrolle *f*
f poulie *f* à chaîne
n kettingtakelblok
p wciągnik łańcuchowy

978 chain coupling
d Kettenkupplung *f*
f accouplement *m* par chaîne
n kettingkoppeling
p sprzęgło łańcuchowe

979 chain drive
d Kettenantrieb *m*
f transmission *f* par chaîne

n kettingoverbrenging
p napęd łańcuchowy

980 chain hoist
d Kettenflaschenzug *m*
f palan *m* à chaîne
n takelketting
p wciągnik łańcuchowy

981 chain link
d Kettenglied *n*
f maillon *m* de chaîne
n kettingschakel
p ogniwo łańcucha

982 chain pin
d Kettenbolzen *m*
f tourillon *m* de chaîne
n pen van ketting
p sworznie łańcucha

983 chain pipe wrench
d Kettenrohrzange *f*
f serre-joint *m* à chaîne
n kettingpijptang
p klucz łańcuchowy do rur

984 chain pitch
d Kettenteilung *f*
f pas *m* de chaîne
n kettingsteek
p podziałka łańcucha

985 chain rivet
d Kettenniet *m*
f rivet *m* de chaîne
n kettingniet
p nit równoległy

986 chain roller
d Kettenrolle *f*
f rouleau *m* de chaîne
n kettingrol
p rolka łańcuchowa

987 chain tensioner
d Kettenspanner *m*
f tendeur *m* de chaîne
n kettingspanner
p napinacz łańcucha

988 chain wheel; sprocket wheel
d Kettenrad *n*
f roue *f* pour chaîne
n kettingwiel
p koło łańcuchowe

989 chain wrench
(tool used for suspension spheres, allowing

either loosening or tightening)
d Kettenrohrzange *f*
f clé *f* à chaîne
n kettingtang
p klucz łańcuchowy

990 chamber
d Kammer *f*; Raum *m*
f chambre *f*
n kamer; ruimte
p komora

991 chamber cover
d Gehäusedeckel *m*
f dessus *m* cuve
n huisdeksel
p pokrywa obudowy

992 chamfer
d Fase *f*
f chanfrein *m*
n afschuining; schuine kant
p skos; ścięcie

993 chamfering vice
d Reifkloben *m*
f étau *m* à chanfrein
n afschuinbankschroef
p imadełko skośne

994 change
d Veränderung *f*
f changement *m*
n verandering
p zmiana

995 change speed fork; striking fork
d Schaltgabel *f*
f fourchette *f* de changement de vitesse
n schakelvork
p widełki przełączania biegów

996 channel
d Kanal *m*; Leitung *f*
f conduit *m*; canal *m*; tuyau *m*
n kanaal
p kanał

997 channel section axle
d Achse *f* mit U-Querschnitt
f essieu *m* à profil en U
n U-profielas
p oś ceowa; oś o kształcie U

998 channel side member
d U-Längsträger *m*
f langeron *m* en U
n langsdraagbalk met U-profiel
p podłużnica ceowa; podłużnica o kształcie U

999 characteristic
d Charakteristik *f*
f caractéristique *f*
n karakteristiek
p charakterystyka

1000 characteristic curve
d Kennlinie *f*
f caractéristique *f*
n karakteristiek
p krzywa charakterystyczna

1001 charge cylinder for purging cooling circuits
d Lastzylinder *m* zur Entluftung der Kuhlkreise
f cylindre *m* de charge pour purger des circuits de refroidissement
n cilindervulling voor doorspoelen van koelkringloop
p cylinderek obciążenia do odpowietrzania obwodu chłodzenia

* charge indicator → 1003

1002 charge pump
d Ladepumpe *f*
f pompe *f* de suralimentation
n compressor
p pompa zasilająca

1003 charging control lamp; charge indicator
d Ladeanzeigeleuchte *f*; Ladekontrolleuchte *f*
f lampe *f* témoin de charge; indicateur *m* de charge
n laadstroomcontrolelicht
p lampka kontrolna ładowania akumulatora

1004 charging current
d Ladestrom *m*
f courant *m* de charge
n laadstroom
p prąd ładowania

1005 charging of the battery
d Ladung *f* des Akkumulators
f charge *f* d'accumulateur
n opladen van de accu
p ładowanie akumulatora

1006 charging plug
d Ladestöpsel *m*
f bouchon *m* de charge
n vuldop
p korek załadowania

1007 charging valve
d Füllventil *n*
f soupape *f* de charge
n vulklep
p zawór napełniania

1008 charging voltage
d Ladespannung *f*
f voltage *m* de charge
n laadspanning
p napięcie ładowania akumulatora

1009 chasing hammer
d Ziselierhammer *m*
f marteau *m* de ciseleur
n drijfhamer
p młotek cyzelerski

1010 chassis
d Chassis *n*
f châssis *m*
n chassis; onderstel
p podwozie

1011 chassis dynamometer
d Kraftwagenrollprüfstand *m*; Dynamometer *n*; Leistungsprüfstand *m*
f dynamomètre *m*
n vermogentestbank
p dynamometr; siłomierz

1012 chassis frame accessories
d Fahrgestellteile *npl*
f organes *mpl* pour de châssis
n chassisdelen
p części podwozia

1013 chassis frame height
d Rahmenhöhe *f*
f hauteur *f* de châssis
n chassishoogte
p wznios ramy

1014 chassis length
d Fahrgestellänge *f*
f longueur *f* de châssis
n chassislengte
p długość podwozia

1015 chassis number
d Fahrgestellnummer *f*
f numéro *m* de châssis
n chassisnummer
p numer podwozia

1016 chassis roller
d Rollenprüfstand *m*
f banc *m* d'essai à rouleaux
n rollentestbank
p rolkowe stanowisko kontrolne

1017 chassis unit
d Bodengruppe *f*

f ensemble *m* de châssis
n vloergroep
p zespół podwozia

1018 chassis weight
d Fahrgestellgewicht *n*
f poids *m* du châssis
n onderstelgewicht
p ciężar podwozia

* **checkered flag → 4749**

1019 checking plug
d Kontrollstopfen *m*
f bouchon *m* de visite
n controlestop
p czop kontrolny

1020 check out lock
d Kassenschlösser *m*
f serrure *f* de caisse
n controleslot voor toepassing
p zamek kasowy

* **check valve → 342**

1021 check valve
d Rückschlagventil *n*
f clapet *m* de recharge
n dubbelwerkende voetklep
p zawór zwrotny; zawór jednokierunkowy

1022 cheese head bolt; round head bolt
d rundkörpiger Bolzen *m*
f boulon *m* à tête ronde
n bout met ronde kop
p śruba z łbem kulistym

1023 chemical wear
d chemischer Verschleiss *m*
f usure *f* chemique
n chemisch verbruik
p zużycie chemiczne

1024 chest rolley
d Werkzeugkastenwagen *m*
f chariot *m* de manutention
n kastrolwagentje
p wózek narzędziowy

1025 child proof door lock
d Kindersicherung *f*
f sécurité *f* d'enfants
n kinderslot
p zabezpieczenie tylnych drzwi przed otwarciem
    przez dzieci

1026 child safety seat
d Sicherheitskindersitz *m*

f siège *m* de sécurité pour enfant
n veiligheidskinderzitje
p bezpieczne siedzenie dziecka

1027 chilled glass
d Sicherheitsglas *n*
f verre *m* trempé
n gehard glas
p szkło hartowane

1028 chipping knife
d Kratzmesser *m*
f bouloir *m* de maréchal
n schraapmes
p nóż kablowy

1029 chisel for exhaust system
d Auspufftrennmeissel *m*
f burin *m* pour l'échappement
n uitlaatbeitel
p dłuto do układu wylotowego

1030 chisel for rivets
d Nietenquetscher *m*
f tranche *f* à dériver
n beitel voor klinknagels
p dłuto do nitów

1031 chisels
d Beitels *mpl*
f ciseaux *mpl*
n beitels
p dłuta

1032 chlorinated rubber
d Chlorkautschuk *m*
f caoutchouc *m* chloré
n chloorrubber
p kauczuk chlorowany; chlorokauczuk

1033 choke
d Starterklappe *f*
f starter *m*
n choke
p przepustnica rozruchowa powietrza

1034 choke control
d Starterzug *m*
f commande *f* de starter
n chokekabel
p cięgło zsania; cięgło rozruchowe

1035 choke control cable handle
d Griff *m* des Startklappenbowdenzuges
f bouton *m* de commande de volet d'air
n chokeknop
p uchwyt cięgna zasysacza

* choke control light → 1040

1036 choke rod
  d Saugantriebsstange f
  f barre f d'étrangleur
  n chokestang
  p cięgno ssania

1037 choke thermostat
  d Starterthermostat m
  f thermostat m de starter
  n chokethermostaat
  p termostat rozrusznika

1038 choke tube
  d Mischrohr n
  f diffuseur m
  n mengbuis
  p gardziel

1039 choke valve
  d Starterdrosselklappe f
  f soupape f d'étranglement
  n chokeklep
  p przepustnica

1040 choke warning light; choke control light
  d Luftklappenwarnleuchte f;
     Starterklappenkontrolleuchte f
  f témoin m de starter; témoin m d'étrangleur
  n chokecontrolelicht
  p lampa kontrolna zasysacza

1041 choking passage
  d Drosseldurchlass m
  f orifice m passage
  n vernauwd doorvoerkanaal
  p przewężenie; gardziel

1042 chrome-vanadium jaws
  d Chromvanadiumbacken mpl
  f mâchoires fpl chrome-vanadium
  n chroom-vanadium klauwen
  p szczęki chromowo-wanadowe

1043 chrome protection
  d Chromschutzmittel n
  f protecteur m des chromes
  n chroombeschermmiddel
  p ochrona chromowa

1044 chromium plated
  d verchromt
  f chromé
  n verchroomd
  p chromowany

1045 chuck jaws
  d Bohrfutterbacken fpl
  f becs mpl de mandrin
  n boorhouderbekken
  p szczęki uchwytu wiertarskiego

1046 chuck tap wrench
  d Gewindebohrerhalter m
  f porte-tarauds m
  n tapkruk
  p oprawka do gwintownika

1047 chunked shoulder
  d Schulterausbruch m
  f épaulement m décolé
  n krachtige schouder
  p wyłom obsadzania

1048 cigar lighter
  d Zigarrenanzünder m
  f allume-cigare m
  n sigarettenaansteker
  p zapalniczka

1049 circlip; snap ring
  d Sprengring m; Sicherungsring m; Nutring m
  f anneau m de retenue; jonc m d'arrêt
  n borgveer; veerring; seegerring
  p pierścień sprężynujący zabezpieczający

1050 circlip for ball bearing
  d Sicherungsring m für Kugellager
  f segment m de roulement
  n seegerring voor kogellager
  p pierścień sprężynujący zabezpieczający
     łożyska

1051 circlip pliers
  d Seegerzange f; Seegerringzange f
  f pince f pour circlips; pince f à circlip
  n borgveertang
  p szczypce do pierścienia osadczego
     sprężynującego; szczypce do pierścienia
     Seegera

1052 circuit breaker
  d Unterbrecher m
  f disjoncteur m
  n stroomkringonderbreker
  p wyłącznik prądu

1053 circuit breaker insulating plate
  d Sicherungsisolierplatte f
  f plaque f isolante de coupe-circuit
  n stroomkringonderbreker-isolatieplaat
  p płytka izolacyjna wyłącznika prądu

1054 circuit tester; multimeter
(instrument to measure voltages, conductor
continuity, voltage polarity)
d Mehrfachmessgerät *n*
f multimètre *m*
n multimeter
p multimetr; woltoamperomierz; miernik
   uniwersalny

1055 circular cutting snip
d Universalblechschere *f* mit Hebelübersetzung
f cisaille *f* à tôle à chantourner universelle
n blikschaar met hefboomoverbrenging
p nożyce do blach z przekaźnią dźwigniową

1056 circular guideway
d ringförmige Führung *f*
f glissière *f* circulaire
n ringgeleiding
p prowadnica pierścieniowa

1057 circular lever
d Dosenlibelle *f*
f niveau *m* circulaire
n dooswaterpas; luchtbelwaterpas
p libela pudełkowa

1058 circular saw blade
d Kreissägeblatt *n*
f lame *f* de scie circulaire
n cirkelzaagblad
p piła tarczowa

1059 circular saw blade with straight teeth
d Kreissägeblatt *n* mit Normalverzahnung
f lame *f* de scie circulaire à denture droite
n cirkelzaagblad met rechte vertanding
p piła tarczowa z normalnym uzębieniem

1060 circular solid die
d geschlossenes Gewindeschneideisen *n*
f filière *f* ronde solide
n ronde snijplaat gesloten model
p narzynka do gwintów-model zamknięty

1061 circular split die
d geschlitztes Gewindeschneideisen *n*
f filière *f* ronde extensible
n ronde snijplaat gespleten model
p narzynka do gwintów-model rozcięty

1062 circulation
d Umlauf *m*; Kreizlauf *m*
f circulation *f*
n circulatie
p cyrkulacja

1063 circulation chart
d Arbeitsdiagramm *n*

f diagramme *m* de cycle
n cyclusschema
p wykres obiegu

1064 circumferential rib pattern
d Längsrippenprofil *n*
f profil *m* à nervures longitudinales
n profiel geribd in de lengte
p wzdłużny profil żeberkowy

1065 city map
d Stadtplan *m*
f plan *m* de la ville
n plattegrond
p plan miasta

1066 clamp; clip
d Klemme *f*; Schelle *f*; Bügel *m*
f bride *f* de serrage; console *f*; support *m*
n beugel; klem; steun
p zacisk

1067 clamp bearing
d Klemmlager *n*
f palier *m* de serrage
n klemlager
p łożysko zaciskowe

1068 clamping bolt
d Klemmbolzen *m*
f boulon *m* de serrage
n klembout
p sworzeń zaciskowy

*   clamping plate → 1620

1069 clamping power
d Spannkraft *f*
f puissance *f* de serrage
n spankracht
p siła zamocowania

1070 clamping shaft
d Spannwelle *f*
f arbre *m* de serrage
n spanas
p wałek zaciskowy

1071 clamping shoe
d Klemmschuh *m*
f sabot *m* de serrage
n klemschoen
p widełki zaciskowe

1072 clamping tool
d Klemmwerkzeug *n*
f outil *m* de serrage
n opspanblok
p narzędzie do zaciskania

1073 clamp ring
d Klemmring *m*
f anneau *m* de serrage
n klemring
p pierścień zaciskowy

1074 clamp vice
d Schraubstock *m* mit Klemmschraube
f étau *m* à agrafe
n klembankschroef
p imadło ze śrubą zaciskową

1075 clashing of gear; grating of gear
d Schaltgeräusch *n*
f grincement *m* des pignons
n "krakende tandwielen"
p koła zębate przesuwne

1076 clasp fastening set
d Heftmaschine *f*
f agrafeuse *f*
n nietmachine
p zszywarka

1077 claw collar
d Klauenmuffe *f*
f manchon *m* à griffes
n mof met klauw
p tarcza kłowa

1078 claw hammer
d Klauenhammer *m*
f marteau *m* de coffreur
n klauwhamer
p młotek do gwoździ

1079 cleaning
d Reinigung *f*
f nettoyage *m*
n reiniging
p oczyszczanie

1080 cleaning agent
d Reinigungsmittel *n*
f détergent *m*
n reinigingsmiddel
p środek czyszczący

1081 cleaning edges
d Abstreifscheiben *fpl*
f lames *fpl* racleurs
n afstrijklamellen
p płytki zbierające

1082 cleaning edges spindle
d Abstreifscheibenachse *f*
f axe *m* de support des lames fixes de décrassage

n as van afstrijklamellen
p trzpień płytek zbierających

1083 clearance
d Spiel *n*
f jeu *m*
n speling; speelruimte
p luz

1084 clearance filter
d Stabfilter *m*
f filtre *m* tige
n staaffilter
p filtr labiryntowy

1085 cleft cap
d Spitzteil *n*; versetzte Bürste *f*
f pied *m* de mouton
n puntstuk; bokkenpoot
p skośna szczotka

1086 clevis pin
d Splintbolzen *m*
f boulon *m* à goupille
n gaffelpen
p sworzeń z łbem płaskim i otworem na zawleczkę

1087 climbing capacity
d Steigvermögen *n*; Steigfähigkeit *f*
f performance *f* en côte; tenue *f* en côte
n klimvermogen
p zdolność pokonywania wzniesień

* **clinometer** → 2514

* **clip** → 1066

1088 clock
d Uhr *f*
f montre *f*
n klok
p zegar

* **closed end turn** → 4711

1089 closed guideways
d geschlossene Führungen *fpl*
f glissières *fpl* fermées
n gesloten geleidingen
p prowadnice zamknięte

1090 closing clip
d Spange *f* für Verschluss
f agrafe *f* de fermeture
n klemslot
p zacisk zamykający

1091 cloth disc
  d Leinenschleifscheibe f
  f disque m à toile
  n slijpschijf met schuurlinnen
  p lniana tarcza szlifierska

1092 clouded glass
  d Mattglas n
  f verre m terne
  n matglas
  p szkło matowane

1093 clover leaf crossing
  d Kleeblattkreuzung f
  f croisement m en feuille de trèfle
  n klaverbladkruising
  p skrzyżowanie w kształcie liścia koniczyny

1094 club hammer; lump hammer
  d Fäustel m
  f masse f
  n vuistmoker
  p pobijak ręczny; kowadełko blacharskie

1095 cluster gear
  d Block m von Zahnrädern
  f ensemble m de pignons et un arbre
  n tandwielgroep
  p blok kół zębatych

1096 clutch
  d Kupplung f
  f embrayage m
  n koppeling
  p sprzęgło

  * clutch casing → 1100

1097 clutch cone
  d Kupplungskegel m
  f cône m de la friction; cône m d'embrayage
  n koppelingsconus
  p stożek sprzęgła

1098 clutch cover
  d Kupplungsdeckel m
  f plateau m de fermeture
  n koppelingsdeksel
  p osłona sprzęgła

  * clutch disc → 1111

1099 clutch facing
  d Kupplungsbelag m
  f garniture f de disque d'embrayage
  n koppelingsvoering
  p okładzina sprzęgła

1100 clutch housing; clutch casing
  d Kupplungsgehäuse n
  f carter m d'embrayage
  n koppelingshuis
  p obudowa sprzęgła

1101 clutch housing cover
  d Kupplungsgehäusedeckel m
  f couvercle m d'embrayage
  n koppelingshuisdeksel
  p pokrywa obudowy sprzęgła

1102 clutch housing flange
  d Flansch m des Kupplungsgehäuses
  f bride f de carter d'embrayage
  n koppelingshuisflens
  p kołnierz obudowy sprzęgła

1103 clutch housing gasket
  d Kupplungsgehäusedichtung f
  f joint m de carter d'embrayage
  n koppelingshuispakking
  p uszczelnienie obudowy sprzęgła

1104 clutch hub
  d Kupplungsnabe f
  f moyeu m d'embrayage
  n schakelnaaf; naaf van koppelingsplaat
  p piasta sprzęgła

1105 clutch operating fork; clutch release fork
  d Ausrückgabel f
  f fourchette f d'embrayage
  n ontkoppelingsvork
  p widełki wyłączające sprzęgła

1106 clutch pedal
  d Kupplungspedal n
  f pédale f d'embrayage
  n koppelingspedaal
  p pedał sprzęgła

1107 clutch pedal lever
  d Kupplungspedalhebel m
  f levier m de pédale d'embrayage
  n koppelingspedaalhefboom
  p dźwignia pedału sprzęgła

1108 clutch pedal mounting bracket
  d Kupplungspedalstütze f
  f support m de pédale d'embrayage
  n ophangsteun voor koppelingspedaal
  p wspornik dźwigni pedału sprzęgła

1109 clutch pedal pad
  d Kupplungspedalauflage f
  f patin m de pédale d'embrayage

n koppelingspedaalrubber; opvulling van
koppelingspedaal
p stopka pedału sprzęgła

1110 clutch pedal shaft
d Kupplungspedalwelle *f*
f axe *m* de pédale d'embrayage
n as van koppelingspedaal
p wałek pedału sprzęgła

1111 clutch plate; clutch disc
d Kupplungsscheibe *f*
f disque *m* d'embrayage
n koppelingsplaat
p tarcza sprzęgła

1112 clutch plate hub
d Kupplungsscheibennabe *f*
f moyeu *m* de disque d'embrayage
n naaf van koppelingsschijf
p piasta tarczy sprzęgła

1113 clutch plate lining
d Kupplungsscheibenreibbelag *m*
f garniture *f* de disque d'embrayage
n voering van koppelingsschijf
p okładzina cierna tarczy sprzęgła

1114 clutch play
d Kupplungsspiel *n*
f jeu *m* d'embrayage
n koppelingsspeling
p luz sprzęgła

1115 clutch release cable
d Kupplungsseil *n*
f câble *m* d'embrayage
n koppelingskabel
p linka wyprzęgająca; cięgno wyprzedzające

* **clutch release fork** → 1105

1116 clutch release lever axle
d Ausrückhebelachse *f*
f axe *m* de levier de débrayage
n as van loshefboom
p ośka dźwigni wyłączającej

1117 clutch release master cylinder
d Ausrückpumpe *f*
f maître-cylindre *m* de débrayage
n hoofdcilinder voor ontkoppeling
p pompa wyprzęgnika

1118 clutch release shaft
d Zugstangenhebelwelle *f*
f arbre *m* de relais avec leviers

n hefboomas
p wałek dźwigni cięgien

1119 clutch remover
d Kupplungsabzieher *m*
f extracteur *m* d'embrayage
n koppelingstrekker
p przyrząd do demontażu sprzęgła

1120 clutch shaft
d Kupplungswelle *f*
f arbre *m* d'embrayage
n koppelingsas
p wał sprzęgłowy

1121 clutch shaft bearing
d Kupplungswellenlager *n*
f roulement *m* d'arbre d'embrayage
n koppelingsaslager
p łożysko wału sprzęgłowego

1122 clutch shaft sleeve
d Kupplungswellenbuchse *f*
f manchon *m* d'arbre d'embrayage
n geleidbus voor koppelingsas
p obejma wału sprzęgłowego

1123 clutch slip
d Kupplungsschlupf *m*
f glissement *m* de l'embrayage
n slippen van de koppeling
p poślizg sprzęgła

* **coarse file** → 4224

1124 coast *v*
d ausrollen
f rouler sur l'erre
n zonder aandrijving voortrollen
p poruszanie się ruchem bezwładnym

1125 coated tweezers
d isolierte Pinzette *f*
f brucelle *f* gainée
n geïsoleerd pincet
p szczypce izolowane

1126 coated wire stripping pliers
d Lackabziehpinzette *f*
f pince *f* gratte-laque
n isolatiesnijtang
p szczypce do cięcia izolacji

1127 coat hanger hook
d Kleiderhaken *m*
f portemanteau *m*
n kledinghaak
p wieszak

1128 **coat of paint**
 d Farbanstrich *m*
 f couche *f* de peinture
 n laklaag
 p warstwa lakieru

1129 **cock**
 d Hahn *m*
 f robinet *m*
 n kraan
 p kurek

1130 **cocking lever**
 d Spannhebel *m*
 f levier *m* poussoir
 n hefboom van kleminrichting
 p dźwignia zabezpieczająca

1131 **coded**
 d verschlüsselt
 f chifré par codes
 n gecodeerd
 p zakodowany

1132 **coefficient of friction**
 d Reibungszahl *f*
 f coefficient *m* de frottement
 n wrijvingscoëfficiënt
 p współczynnik tarcia

1133 **cogging iron**
 d Stemmeisen *n*
 f ciseau *m* de matage
 n aanzetijzer
 p uszczelniak; dłuto dziobak

1134 **coil**
 d Spule *f*
 f bobine *f*
 n spoel
 p cewka

1135 **coil bracket**
 d Zündstütze *f*
 f support *m* de bobine
 n bobinesteun
 p wspornik cewki

1136 **coil case**
 d Zündspulengehäuse *n*
 f boîtier *m* de bobine
 n bobinehuis
 p obudowa cewki

1137 **coil spring; cylindrical spring**
 d Schraubenfeder *f*
 f ressort *m* hélicoïdal
 n schroefveer; spiraalveer
 p śrubowa sprężyna resorowa

1138 **cold chisel**
 d Kaltmeissel *m*
 f ciseau *m* à froid
 n koudbeitel
 p przecinak ślusarski

1139 **cold plug; hard plug**
 d kalte Zündkerze *f*
 f bougie *f* froide
 n "koude bougie"
 p świeca "zimna"

1140 **cold roller steel**
 d kalt gewalzter Stahl *m*
 f acier *m* laminé à froid
 n koudgewalst staal
 p stal cienka walcowana na zimno

1141 **cold starting ability**
 d Kaltstartvermögen *n*
 f capacité *f* de démarrage à froid
 n koudestartvermogen
 p zdolność rozruchu na zimno

1142 **cold starting cam**
 d Kaltstartnocken *m*
 f came *f* de départ à froid
 n koudestartnok
 p krzywka rozruchu zimnego silnika

1143 **cold start injector**
 d Kaltstartventil *n*
 f injecteur *m* de démarrage à froid
 n koudestartverstuiver
 p wtryskiwacz rozruchowy

1144 **cold start unit cover**
 d Deckel *m* der Starterinrichtung
 f couvercle *m* de dispositif de départ à froid
 n deksel van startinrichting
 p pokrywa urządzenia rozruchowego

1145 **cold start unit stem**
 d Verstellachse *f* der Starterinrichtung
 f axe *m* de dispositif de départ à froid
 n as van startinrichting
 p ośka urządzenia rozruchowego

1146 **collapsible hood**
 d Plane *f*
 f prélant *m*; bâche *f*
 n dekzeil
 p plandeka

1147 **collar bush**
 d Flanschbuchse *f*
 f bague *f* à épaulement

n kraagbus
p tulejka kołnierzowa

1148 collar journal
d Kammzapfen *m*
f tourillon *m* cannelé
n kraagtap
p czop główny grzebieniowy

1149 collar nut
d Bundmutter *f*
f écrou *m* à bride
n kraagmoer
p nakrętka wieńcowa

1150 collar screw
d Bundschraube *f*
f boulon *m* à collet
n kraagbout
p śruba z kołnierzem

1151 collar thrust bearing
d Kammlager *n*
f palier *m* à cannelures
n kraaglager
p łożysko kołnierzowe

1152 collector
d Kollektor *m*
f collecteur *m*
n collector
p komutator

1153 collet
d Klemmbüchse *f*; Spannbüchse *f*
f douille *f* de serrage
n klemring; klemhuls
p tuleja zaciskowa

1154 collet chuck
d Büchsenfutter *m*
f mandrin *m* de serrage
n collethouder
p uchwyt tulei zaciskowej

1155 collet clamp
d Ventiltellerdrücker *m*
f presse *f* coupelles
n klepschoteldrukker
p przyrząd do wyciskania talerza zaworu

1156 collision
d Zusammenstoss *m*
f collision *f*
n botsing; aanrijding
p kolizja

1157 colour code of the varnish
d Farbenlackkode *f*
f code *m* de couleur
n kleurcode van de lak
p kod koloru lakieru

1158 colour combination
d Farbkombination *f*
f combinaison *f* de couleurs
n kleurcombinatie
p kombinacja barw

1159 colour disc
d Farbenscheibe *f*
f disque *m* en couleur
n kleurschijf
p tarcza kolorowa

1160 coloured spectrum
d Farbenspektrum *n*
f spectre *m* de coloré
n kleurspectrum
p spektrum koloru

1161 colour fast
d farbecht
f à couleur durable
n kleurecht
p barwa trwała

1162 colour key
d Farbskala *f*
f gamme *f* des couleurs
n kleurenscala
p skala barw

1163 colour temperature
d Farbtemperatur *f*
f température *f* de couleur
n kleurtemperatuur
p temperatura barwy

1164 colour tint
d Farbton *m*; Farbstich *m*
f tinte *f* de couleur
n kleurtint
p odcień koloru; odcień barwy

1165 column
d Säule *f*
f colonne *f*
n kolom
p kolumna

1166 combination saw blade
d Sägeblatt *n* mit kombinierter Zahnung
f lame *f* de scie à dentures combinées

n combinatiezaagblad
p piła z kombinowanym uzębieniem

1167 combination square
d Universalwinkelmesser *m*
f équerre *f* universelle
n combinatiehaak
p kątomierz uniwersalny

1168 combined transport
d kombinierter Verkehr *m*
f transport *m* combiné
n gecombineerd transport
p transport kombinowany

1169 combustible mixture
d brennbares Gemisch *n*
f mélange *m* combustible
n brandbaar mengsel
p mieszanka paliwa

1170 combustion
d Verbrennung *f*
f combustion *f*
n verbranding
p spalanie

* **combustion chamber volume** → 1194

1171 combustion pressure; firing pressure
d Verbrennungsdruck *m*
f pression *f* de combustion
n verbrandingsdruk
p ciśnienie spalania

1172 commercial vehicle
d Nutzfahrzeug *n*
f véhicule *m* industriel
n bedrijfsauto
p samochód firmy

1173 commutator bearing bush
d Kollektorlagerbuchse *f*
f bague *f* palier collecteur
n collector-lagerschaal
p tuleja łożyska kolektora

1174 compact push jack
d kurze Schraubenwinde *f*
f vérin *m* pousseur court
n korte vijzel
p podnośnik przyciskowy kompaktowy

1175 companion flange
d Anschlussflansch *m*
f bride *f* de raccordement
n aandrijfflens
p tarcza sprzęgająca

1176 comparator
d Vorgleicher *m*
f comparateur *m*
n pulsomvormer
p komparator

1177 compartment
d Abteil *n*
f compartiment *m*
n compartiment
p pomieszczenie; przedział

1178 compartment dimensions
d Innenabmessungen *fpl*
f dimensions *fpl* de l'intérieur
n interieurafmetingen
p wymiary wnętrza

1179 compass saw
d Stichsäge *f*; Lochsäge *f*
f scie *f* à guichet
n schrobzaag
p otwornica; lisica

1180 compensating jet
d Ausgleichsdüse *f*; Zusatzdüse *f*
f gicleur *m* de compensation
n compensatiesproeier
p dysza kompensacyjna

1181 compensating reservoir
d Ausgleichbehälter *m*
f réservoir *m* de compensation
n expansievat
p zbiornik wyrównawczy

1182 compensating resistor
d Temperaturangleichwiderstand *m*
f résistance *f* compensatrice
n compensatieweerstand
p rezystor kompensacji temperatury

1183 compensating ring
d Ausgleichsfeder *f*
f ressort *m* compensateur
n compensatieveer
p sprężyna kompensacyjna

1184 compensatory system
d Kompensationsschaltung *f*
f système *m* de compensation
n compensatiesysteem
p układ kompensacyjny

1185 complete combustion
d volkommene Verbrennung *f*
f combustion *f* complète

 n volledige verbranding
 p spalanie zupełne

1186 component
 d Bestandteil *m*
 f composant *m*
 n component
 p składnik; część składowa

1187 component of force
 d Teilkraft *f*
 f composant *m* de la force
 n component van een kracht
 p składowa siły

1188 compound oil
 d Compoundöl *n*; Mischöl *n*
 f huile *f* composée
 n compound-olie
 p olej kompandowany; olej mieszany

1189 comprehensive insurance
 d Kaskoversicherung *f*
 f assurance *f* tous risques
 n all-riskverzekering
 p ubezpieczenie casco

1190 compressed air
 d Druckluft *f*
 f air *m* comprimé
 n druklucht; perslucht
 p powietrze sprężone

1191 compressed air tank
 d Druckluftbehälter *m*
 f réservoir *m* d'air comprimé
 n persluchttank
 p zbiornik sprężonego powietrza

1192 compressed gas
 d Druckgas *n*
 f gaz *m* comprimé
 n gecomprimeerd gas
 p gaz sprężony

1193 compression
 d Verdichtung *f*
 f compression *f*
 n compressie
 p sprężanie

1194 compression chamber volume; combustion chamber volume
 d Verbrennungsraum *m*
 f volume *m* de compression
 n verbrandingsruimte
 p przestrzeń spalania; komora spalania

1195 compression coil spring
 d Schraubendruckfeder *f*
 f ressort *m* hélicoïdal de compression
 n drukveer
 p ściskana sprężyna śrubowa

1196 compression gauge
 d Kompressormesser *m*
 f indicateur *m* de compression
 n compressiemeter
 p ciśnieniomierz do pomiaru sprężania

1197 compression grease cup
 d Staufferbuchse *f*
 f graisseur *m*
 n stauffer-vetpot
 p smarownica kapturowa; smarownica Stauffera

1198 compression ignition engine
 d Eigenzündungsmotor *m*; Dieselmotor *m*
 f moteur *m* à allumage par compression; diesel *m*
 n dieselmotor
 p silnik Diesla; silnik wysokoprężny; silnik z zapłonem samoczynnym Diesla

1199 compression pressure
 d Verdichtungsendspannung *f*
 f pression *f* finale de compression
 n compressiedruk
 p ciśnienie sprężania

1200 compression ring
 d Verdichtungsring *m*
 f segment *m* de compression
 n compressieveer
 p uszczelniający pierścień tłokowy

1201 compression spring
 (chassis-steering linkage ball joint)
 d Druckfeder *f*
 f ressort *m* de compression
 n drukveer
 p sprężyna dociskowa; sprężyna oporowa

1202 compression stroke
 d Verdichtungstakt *m*
 f temps *m* de compression
 n compressieslag
 p suw sprężania

1203 compression test
 d Kompressionsdruckmessung *f*
 f contrôle *m* de compression
 n compressietest
 p pomiar ciśnienia sprężania

1204 compression tester
 d Kompressionsprüfer *m*; Kompressionsmeter *n*

f   indicateur *m* de compression
n   compressiemeter
p   manometr do pomiaru ciśnienia sprężania

*   **compression valve** → 790

1205 compressive force
d   Druckkraft *f*
f   force *f* de compression
n   drukkracht
p   siła nacisku; siła ściskająca

1206 compressive strength
d   Druckfestigkeit *f*
f   résistance *f* à la compression
n   drukvermogen
p   wytrzymałość na ściskanie

1207 compressive stress
d   Druckbeanspruchung *f*
f   effort *m* de compression
n   drukbelasting; drukspanning
p   naprężenie ściskające

1208 compressor
d   Verdichter *m*
f   compresseur *m*
n   compressor
p   sprężarka

1209 compressor horn
d   Kompressorhorn *m*
f   avertisseur *m* à compresseur
n   compressorhoorn
p   sygnał dźwiękowy sprężarki

1210 computer controlled machine
d   rechnergesteuerte Maschine *f*
f   machine *f* à commande numérique
n   computergestuurde machine
p   maszyna sterowana komputerem

1211 concave mirror
d   Hohlspiegel *m*
f   miroir *m* concave
n   concave spiegel
p   lustro wklęsłe

1212 concrete mixer truck; truck mixer
d   Transportbetonmischer *m*
f   malaxeur *m* sur camion
n   betonmixer
p   betoniarka

1213 condense *v*
d   sich niederschlagen
f   condenser

n   condenseren
p   kondensować; zagęszczać

1214 condenser
d   Kondensator *m*
f   condensateur *m*
n   condensator; verdichter
p   kondensator

1215 condenser capacity
d   Kondensatorkapazität *f*
f   capacité *f* de condensateur
n   condensatorcapaciteit
p   pojemność kondensatora

1216 condenser ignition
d   Kondensatorzündung *f*
f   allumage *m* par condensateur
n   condensatorontsteking
p   zapłon kondensatorowy

1217 conducting wire
d   Leitungskabel *n*
f   fil *m* conducteur
n   stroomdraad
p   drut przewodzący

1218 conduction of heat
d   Wärmeleitung *f*
f   conductibilité *f* calorifique
n   warmtegeleiding
p   przewodzenie ciepła

1219 conductor
d   Leiter *m*
f   conducteur *m*
n   geleider
p   przewodnik

1220 conduit; duct; lead
d   Leitung *f*
f   conduit *m*; tuyau *m*
n   bus; leiding
p   przewód

1221 cone
d   Kegel *m*
f   cône *m*
n   kegel
p   stożek

1222 cone bits
d   Aufreissbohrer *mpl*
f   forets *mpl* coniques
n   conische boren
p   wiertła stożkowe

1223 cone clutch
d   Kegelkupplung *f*

f embrayage *m* par cônes de friction
n kegelkoppeling
p sprzęgło stożkowe (cierne)

1224 cone head rivet
d Kegelstumpfniet *m*
f rivet *m* à tête plate
n klinknagel met vlakke kop
p nit stożkowy ścięty

* **conical bearing** → 4955

1225 conical reamer
d Kegelreibahle *f*
f alésoir *m* conique
n conische ruimer
p rozwiertak stożkowy

* **conical roller** → 4950

1226 conical separator; sediment element
d Kegelabscheider *m*
f séparateur *m* conique
n afscheidingskegel
p wytrącacz

1227 conical spring
d Kegelfeder *f*
f ressort *m* conique à compression
n kegelveer
p sprężyna stożkowa

1228 conical spring clutch
d Kegelfederkupplung *f*
f embrayage *m* à ressort conique
n koppeling met conische veer
p sprzęgło o sprężynie stożkowej

1229 connecting adapter
d Gelenkklemme *f*
f attache *f* d'articulation
n draaibare aansluitklem
p łącznik przegubu

1230 connecting bolt
d Gehäusedeckelmutter *f*
f écrou *m* de serrage
n kapmoer
p nakrętka dociskowa

1231 connecting clamp
d Anschlussstück *n*
f raccord *m*
n aansluitstuk
p złączka; łącznik rurowy

1232 connecting lever rod
d Gestänge *f*

f tige *f* de levier
n gasklephevelstang
p sztywne cięgło dźwigni

1233 connecting line
d Verbindungsleitung *f*
f conduite *f* de raccordement
n verbindingsleiding
p przewód łączący

1234 connecting link
d Verbindungsglied *n*
f maillon *m* de liaison
n verbindingselement
p ogniwo złączne

1235 connecting plug
d Kontaktstöpsel *m*
f fiche *f* de connection
n contactstop
p wtyczka

1236 connecting rod
d Pleuelstange *f*; Verbindungsstange *f*
f bielle *f*
n drijfstang
p korbowód

1237 connecting rod bearing
d Pleuellager *n*
f coussinet *m* de tête de bielle
n drijfstanglager
p łożysko korbowe

1238 connecting rod big end
d Pleuelstangenfuss *m*
f tête *f* de bielle
n "big end" drijfstangvoet
p łeb korbowy korbowodu; stopa korbowodu

1239 connecting rod bolt
d Pleuelschraube *f*
f boulon *m* de bielle
n drijfstangbout
p śruba korbowodu

1240 connecting rod bushing
d Pleuelbuchse *f*; Pleuelstangenlager *n*
f bague *f* de pied de bielle
n zuigerpenlagerbus
p tulejka główki korbowodu

1241 connecting rod cap
d Pleuelstangendeckel *m*
f chapeau *m* de tête de bielle
n drijfstangkap
p pokrywa łba korbowodu

1242 **connecting rod small end**
 d Pleuerstangenkopf *m*
 f petite tête *f* de bielle
 n drijfstangkop "small end"
 p łeb korbowodu

1243 **connecting sleeve**
 d Verbindungsmuffe *f*
 f manchon *m* de jonction
 n verbindingsmof
 p mufka łącząca

1244 **connecting support**
 d Träger *f* für Verbindung
 f support *m* de liaison
 n verbindingsplaatje
 p płytka do połączenia

1245 **connection**
 d Anschluss *m*
 f accouplement *m*; raccord *m*
 n aansluiting
 p połączenie

1246 **connector**
 d Verbinder *m*; Verbindungsstück *n*
 f joint *m*; articulation *f*
 n aansluitnippel
 p łącznik

1247 **consistent grease**
 d konsistentes Fett *n*
 f graisse *f* consistante
 n consistentvet
 p smar stały

1248 **constant depression carburetter; variable**
       choke carburetter
 d Vergaser *m* mit veränderlichem Luftdurchlass
 f carburateur *m* à diffuseur variable
 n constantvacuümcarburateur
 p gaźnik o nastawnej gardzieli

1249 **constant mesh transmission**
 d Getriebe *n* mit Dauereingriff
 f engrenage *m* constamment en prise
 n constant-mesh-versnellingsbak
 p skrzynka przekładniowa z kołami w stałym
       zazębieniu

1250 **constant profile ribbed chisel**
 d Flachmeissel *m* mit Hohlprofil
 f burin *m* nervuré à profil constant
 n geprofileerde koudbeitel
 p dłuto udarowe proste o profilu wydrążonym

1251 **constant velocity joint**
 d homokinetisches Gelenk *n*
 f joint *m* homocinétique

 n homokinetische koppeling
 p przegub równobieżny

1252 **constructional steel**
 d Konstruktionsstahl *m*
 f acier *m* de construction
 n constructiestaal
 p stal konstrukcyjna

1253 **construction wrench**
 d Monteurgabelschlüssel *m* mit Dorn
 f clé *f* à fourche de monteurs
 n montagesleutel
 p klucz montażowy do konstrukcji stalowych

 * **consumption** → **5411**

1254 **contact**
 d Kontakt *m*
 f contact *m*
 n contact
 p styk

1255 **contact bracket**
 d Kontaktträger *m*
 f porte-contacts *m*
 n contactpuntdrager
 p podstawa styków

1256 **contact breaker**
 d Unterbrecher *m*
 f rupteur *m*
 n onderbreker
 p przerywacz

1257 **contact cleaner**
      (cleaner used for effective cleaning of
      electrical contacts and sensitive electronic
      components; removes grease and oil from
      metallic contact surfaces)
 d Kontaktreiniger *m*
 f nettoyant *m* de contact
 n contactreiniger
 p środek do czyszczenia styków

1258 **contact file**
 d Kontaktfeile *f*
 f lime *f* contact
 n contactpuntvijl
 p pilnik do czyszczenia styków przerywacza

1259 **contactor**
 d Schaltschütz *n*
 f contacteur *m*
 n contactgever
 p stycznik

1260 **contact rail**
 d Stromschiene *f*

f barrette *f* de connexion
n contactrail
p szynoprzewód

1261 contact seal
d Berührungsdichtung *f*
f garniture *f* fermée
n contactafdichting
p uszczelnienie stykowe

1262 contact spray
(spray which eliminates faults and breakdown
of the electrical system caused by damp)
d Kontaktsprüher *m*
f contact-spray *m*
n contactspray
p spray do złącz

1263 contact spring
d Kontaktfeder *f*
f lame *f* de contact
n contactveer
p sprężyna stykowa

1264 contact surface
d Berührungsfläche *f*
f surface *f* de contact
n contactvlak
p powierzchnia stykowa

1265 contact washer
d Unterlegscheibe *f*
f rondelle *f* contact
n contactring
p podkładka

1266 container
d Kontainer *m*
f conteneur *m*
n container
p kontener

1267 contents
d Inhalt *m*
f contenu *m*
n inhoud
p zawartość

1268 continuous fuel injection
d kontinuierliche Benzineinspritzung *f*
f injection *f* continue d'essence
n ononderbroken benzine-inspuiting
p ciągłe zasilanie wtryskowe benzyny

1269 continuous hinge
d Bandscharnier *m*
f charnière *m* continue
n bandscharnier
p zawiasa pasowa

1270 contour; outline
d Umriss *m*
f contour *m*
n profiel
p obrys; kontur

1271 contraction rule
d Schwindmassstab *m*
f mètre *m* à retrait
n krimpduimstok
p miara skurczowa; skurczówka

1272 control arm
d Schaltarm *m*
f bras *m* de commande
n bedieningsarm
p ramię sterownicze

1273 control button
d Betätigungsknopf *m*
f bouton *m* de commande
n bedieningsknop
p gałka sterowania

1274 control cable
d Schaltzug *m*; Betätigungszug *m*
f câble *m* de commande
n bedieningskabel; controleerkabel
p kabel sterowniczy

1275 control device
d Prüfgerät *n*
f appareil *m* de contrôle
n testapparaat
p przyrząd kontrolny

1276 control grid
d Steuergitter *n*
f grille *f* de commande
n stuurrooster
p siatka sterująca; siatka kontrolna

1277 controller
d Steuerer *m*
f soupape *f* de commande
n bedieningsklep
p sterownik

1278 control lever
d Verstellhebel *m*; Schalthebel *m*
f levier *m* de commande
n bedieningshendel
p dźwignia nastawcza

1279 control lever spring
d Verstellhebelfeder *f*
f ressort *m* de rappel de levier de commande

n bedieningshendelveer
p sprężyna dźwigni nastawczej

1280 control linkage
d Übertragungsgestänge *f*
f renvoi *m* de commande
n overbrengingsstang
p zespół dźwigni lub drążków kontrolnych

1281 control mechanism
d Steuermechanismus *m*
f mécanisme *m* de commande
n bedieningsmechanisme
p mechanizm sterujący

1282 control module
d Steuerungsbaueinheit *f*
f module *m* de commande
n regeleenheid; stuureenheid
p sterownik

1283 control quadrant
d Zahnrad *n*
f pignon *m*
n tandkwadrant
p kółko zębate

1284 control range
d Verstellbereich *m*
f étendue *f* de réglage
n regelbereik
p zakres regulacji

1285 control rod
d Steuerstange *f*; Regelstange *f*
f tige *f* de commande
n bedieningsstang; regelstang
p drążek sterujący

1286 control rod bush
d Führunghülse *f* für Stange
f bague *f* de tige
n geleiding van bedieningsstang
p tuleja pręta regulacyjnego

1287 control rod link
d Regelstange *f*
f tige *f* de crémaillère
n regelstang
p cięgno zębatki sterowniczej

* **control sleeve** → **86**

1288 control spindle
d Schaltstange *f*
f arbre *m* de changement de vitesse
n schakelas; schakelstang
p wodzik widełek zmiany biegów

1289 control stand
d Prüfstand *m*
f poste *f* de contrôle
n proefbank
p stanowisko kontrolne

1290 control system
d Regelungssystem *n*
f système *m* de réglage
n regelsysteem
p układ regulacyjny

1291 control valve
d Steuerventil *n*
f soupape *f* de commande
n stuurklep; regelklep
p zawór sterujący

1292 control weight
d Fliehkraftsteuergewicht *n*
f masselotte *f* centrifuge de commande
n controlegewicht
p regulacyjny odciążnik wirujący

1293 convection of heat
d Konvektion *f*
f échange *m* convectif de chaleur
n convectie
p konwekcja; unoszenie ciepła

1294 converter
d Wandler *m*
f convertisseur *m*
n omvormer
p przetwornica; konwertor

1295 converter housing
d Wandlergehäuse *n*
f carter *m* fixe
n huis van koppelomvormer
p obudowa kierownicy

1296 convex hand dolly
d Universalhandstock *m*
f table *f* genre universel
n bolle handtegenhouder
p chwyt ręczny uniwersalny

1297 conveyor belt
d Förderband *n*
f courroie *f* de transport
n transportband
p taśma przenośnikowa

1298 conveyor system of assembly
d Bandmontage *f*
f montage *m* en chaîne
n montageband; lopende band
p instalacja taśmowa

1299 coolant
  d Kühlmittel *n*
  f agent *m* de refroidissement
  n koelmiddel; koelvloeistof
  p ciecz chłodząca

1300 coolant outlet pipe
  d Kühlflüssigkeitsauslassstutzen *m*
  f tubulure *f* de sortie de fluide de refroidissement
  n koelvloeistofuitlaat
  p króciec odpływu cieczy chłodzącej

1301 coolant pressure
  d Kühlflüssigkeitsdruck *m*
  f pression *f* de liquide de refroidissement
  n koelvloeistofdruk
  p ciśnienie płynu chłodzącego

1302 coolant tank
  d Kühlflüssigkeitsbehälter *m*
  f réservoir *m* de liquide de refroidissement
  n koelvloeistofreservoir
  p zbiornik cieczy chłodzącej

  * **coolant temperature indicator** → 5407

1303 coolant temperature sensor
  d Messgeber *m* für Kühlwassertemperatur
  f témoin *m* de température d'eau de refroidissement
  n sensor voor koelvloeistoftemperatuur
  p czujnik temperatury cieczy chłodzącej

1304 cooling fin
  d Kühlrippe *f*
  f ailette *f* de refroidissement
  n koelribbe
  p żebro chłodzące

1305 cooling fluid delivery
  d Kühlmittelförderung *f*
  f débit *m* de fluide de refroidissement
  n koelmiddelverbruik
  p zużycie cieczy chłodzącej

1306 cooling system
  d Kühlsystem *n*
  f système *m* de refroidissement
  n koelsysteem
  p układ chłodzenia

1307 cooling system tester
  (tool to check cooling system tightness under pressure)
  d Gerät *n* zum Prüfung des Kuhlsystems
  f contrôleur *m* du circuit de refroidissement

  n koelvloeistoftestapparaat
  p tester układu chłodzenia

1308 cooling water temperature
  d Kühlwassertemperatur *f*
  f température *f* d'eau de refroidissement
  n koelvloeistoftemperatuur
  p temperatura cieczy chłodzącej

1309 cooling water thermostat
  d Kühlwasserthermostat *m*
  f thermostat *m* d'eau de refroidissement
  n koelvloeistofthermostaat
  p termostat cieczy chłodzącej

1310 cooper's knife
  d Fasszieher *m*; Krummeisen *n*
  f débordoir *m*; cochoir *m*
  n haalmes; steekmes
  p dłuto stolarskie wygięte

1311 coordinates drilling machine
  d Koordinatenbohrmaschine *f*
  f perceuse *f* à coordonnées
  n coördinatenboormachine
  p wiertarka współrzędnościowa

1312 coping saw
  d Dekupiersäge *f*
  f scie *f* à découper
  n decoupeerzaag; kromzaag
  p piła krzywica

1313 copper
  d Kupfer *m*
  f cuivre *m*
  n koper
  p miedź

1314 copper asbestos packing
  d Kupfer-Asbestosdichtung *f*
  f garniture *f* en cuivre-amiante
  n koper-asbestpakking
  p uszczelka miedziano-azbestowa

1315 copper gasket
  d Kupferdichtung *f*
  f joint *m* cuivre
  n koperpakking
  p uszczelka miedziana

1316 copper pipe
  d Kupferrohr *n*
  f tuyau *m* en cuivre
  n koperpijp
  p rurka miedziana

1317 coppersmith's hammer
  d Kupferschmiedehammer *m*

f marteau *m* de chaudronnier
n koperslagershamer
p młotek miedziany kowalski

1318 cord
d Kord *m*
f câble *m*
n koord; snoer; touw
p sznur; linka

1319 core
d Kern *m*
f noyau *m*
n kern
p rdzeń

1320 core drill
d Kernbohrer *m*
f trépan *m* de soudage
n kernboor; buisboor
p wiertło rurowe

1321 core plug
d Kernstopfen *m*
f tampon *m* de noyau
n helicol
p zaślepka rdzeniowa

1322 cork gasket
d Presskorkdichtung *f*
f joint *m* en liège comprimé
n kurkpakking
p uszczelka korkowa

1323 corner clamp
d Kantenzwinge *f*
f serre-joint *m* à coins
n kantklemschroef
p docisk krawędziowy

1324 corner drilling machine
d Eckbohrer *m*
f perceuse *f* à coins
n hoekenboormachine
p wiertarka do wiercenia pod kątem

1325 cornering stability
d Kurvenstabilität *f*
f tenue *f* de route en virage
n bochtvastheid
p stateczność zakręcania

1326 corner joint
d Eckstoss *m*
f assemblage *m* en angle
n hoeknaad
p złącze narożne

1327 corner spoiler
d Winkelspoiler *m*
f bacquet *m* d'angle
n hoekspoiler
p spoiler kątowy

1328 corner tool
d Winkelpolierknopf *m*
f lissoir *m* angulaire
n haakse slikker
p gładzik kątowy

1329 corner weld
d Kehlnaht *f*
f soudure *f* d'angle
n hoeklas
p spoina pachwinowa zewnętrzna w złączu kątowym

1330 corrosion
d Korrosion *f*
f corrosion *f*
n corrosie; roest
p korozja

1331 corrosion inhibitor
d Antikorrosionszusatz *m*
f inhibiteur *m* de corrosion
n antiroestmiddel
p dodatek przeciwkorozyjny

1332 corrosion resistance
d Korrosionsfestigkeit *f*;
   Korrosionsbeständigkeit *f*
f résistance *f* à la corrosion
n roestbestendigheid
p odporność na korozję

1333 corrugated sheet
d Wellblech *n*
f tôle *f* ondulée
n golfplaat
p blacha falista

1334 corrugated sheet cutting
d Ausschneiden *n* von Wellblech
f découpage *m* de feuilles ondulées
n uitsnijden van golfplaat
p wycinanie z blachy falistej

* **cotter pin** → **4654**

1335 counterbalance *v*
d auswuchten
f balancer
n uitbalanceren
p wyważać

1336 counter electromotive force
d elektromotorische Gegenkraft *f*
f force *f* contre-électromotrice
n elektromotorische tegenkracht
p siła przeciwelektromotoryczna

1337 counter nut
d Gegenmutter *f*; Kontermutter *f*
f contre-écrou *m*
n contramoer
p przeciwnakrętka

1338 counter pressure
d Gegendruck *m*
f contre-pression *f*
n tegendruk
p przeciwciśnienie

1339 countershaft cover
d Vorgelegewellendeckel *m*
f couvercle *m* d'arbre de renvoi
n deksel van secondaire as
p pokrywa wału pośredniczącego skrzynki
   przekładniowej

1340 countershaft drive gear
d Vorgelegerad *n*
f pignon *m* de l'arbre intermédiaire
n tandwiel van secundaire as van
   versnellingsbak
p koło zębate wału pośredniego

1341 countersink bit
d Holzversenker *m*
f foret *m* à fraiser
n verzinkboor
p pogłębiacz czołowy; pogłębiacz walcowy

1342 countersunk bolt
d versenkter Bolzen *m*
f boulon *m* noyé
n verzonken bout; platkopbout
p śruba stożkowa płaska

1343 coupling length of trailer
d Kuppellänge *f* des Anhängers
f longueur *f* d'accrochage de remorque
n totale koppellengte
p długość sprzęgowa przyczepy

1344 coupling nut
d Überwurfmutter *f*
f écrou *m* à raccord
n wartel
p nakrętka złączkowa

1345 coupling timing valve
d Kupplungsschaltabstimmungsventil *n*

f soupape *f* temporisateur d'embrayage
n klep van koppelingstijdschakeling
p zawór włączania sprzęgła ciernego

1346 cover
d Deckel *m*
f couvercle *m*
n deksel
p pokrywa

1347 cover gasket
d Deckeldichtung *f*
f joint *m* de couvercle
n dekselpakking
p uszczelka pokrywy

1348 covering
d Überzug *m*
f housse *f*
n bedekking; hoes
p pokrycie

1349 cowl light
d Motorraumleuchte *f*
f éclairage *m* de compartiment moteur
n motorruimteverlichting
p lampa oświetlenia silnika

1350 cowl support
d Stirnwandstütze *f*
f support *m* de tablier
n scheidingswandsteun
p podpora ściany przedniej

1351 crack; fissure
d Bruch *m*; Sprung *m*
f fissure *f*
n scheur
p pęknięcie; zerwanie

1352 cramp heads
d Klemmbacken *fpl*
f mâchoires *fpl* de serrage
n klemkoppen; klembekken
p szczęki zaciskowe

1353 crankcase
d Kurbelgehäuse *n*
f carter *m* moteur
n carter; krukascarter
p skrzynia korbowa

1354 crankcase bleed pipe
d Kurbelgehäuseentlüftungstutzen *m*
f tubulure *f* de ventilation de carter
n aansluiting voor de carterventilatie
p króciec odpowietrzania skrzyni korbowej

1355 crankcase explosion
d Kurbelkastenexplosion f
f explosion f dans le carter moteur
n explosie in krukkast
p wybuch w skrzyni korbowej

1356 crank handle
d Handkurbel f
f poignée f de manivelle
n slinger
p korba ręczna

1357 crankpin
d Kurbelzapfen m; Kurbelbolzen m
f maneton m de vilebrequin
n krukpen
p czop korbowy

1358 crankpin diameter
d Durchmesser m der Kurbelbolzen
f alésage m des manetons
n krukpendiameter
p średnica czopu korbowego

1359 crank radius
d Kurbelradius m
f rayon m de manivelle
n krukstraal; kruklengte
p promień korby

1360 crank set
d Kurbeltrieb m
f embiellage m
n krukasmechanisme
p zespół korbowy

1361 crankshaft
d Kurbelwelle f
f vilebrequin m
n krukas
p wał korbowy

1362 crankshaft bearing
d Kurbelwellenhauptlager n
f palier m de vilebrequin; roulement m de vilebrequin
n krukaslager
p łożysko główne wału korbowego

1363 crankshaft bearing cap
d Kurbenwellenlagerdeckel m
f chapeau m de palier de vilebrequin
n krukaslagerdeksel
p pokrywa łożyska wału korbowego

1364 crankshaft bearing shell
d Kurbelwellenlagerschale f
f coussinet m de palier de vilebrequin
n krukaslagerschaal
p panewka łożyska wału korbowego

1365 crankshaft collar
d Kurbelwellenflansch m
f bride f de vilebrequin
n krukasflens
p kołnierz wału korbowego

1366 crankshaft cross degree
d Kurbelwellengrad m
f degré m de vilebrequin
n krukasgraad
p współczynnik wału korbowego

1367 crankshaft drive
d Kurbeltrieb m
f mécanisme m d'embiellage
n krukasmechanisme
p mechanizm korbowy

1368 crankshaft front end
d Vorderzapfen m der Kurbelwelle
f bout m avant de vilebrequin
n voorste gedeelte van de krukas
p przedni koniec wału korbowego

1369 crankshaft grinder
d Kurbelwellenschleifmaschine f
f rectifieuse f de villebrequin
n krukasslijpbank
p szlifierka do wałów korbowych

1370 crankshaft position sensor
d Kurbelwinkelsensor m
f sensor m d'angle de vilebrequin
n krukashoeksensor
p czujnik pozycji wału korbowego

1371 crankshaft pulley; driving pulley
d Antriebsriemenscheibe f
f poulie f d'entraînement
n krukaspoelie
p koło pasowe wału korbowego

1372 crankshaft sprocket
d Kurbelwellenkettenrad n
f pignon m à chaîne de distribution
n krukaskettingwiel
p pędne koło łańcuchowe

1373 crankshaft thrust half ring
d Druckhalbring m des Begrenzungslagers
f semi-flasque m de butée latérale
n druklagerring tegen axiale speling
p półpierścień łożyska oporowego

1374 crankshaft timing gear
 d Kurbelwellenrad *n*; Kurbelwellenritzel *n*;
   Kurbelwellenzahnrad *n*
 f pignon *m* du vilebrequin
 n krukasdistributietandwiel; krukastandwiel
 p pędne koło zębate

1375 crank web
 d Kurbelwange *f*
 f bras *m* de manivelle
 n krukwang
 p ramię korby

1376 crash sensor
 d Aufprallsensor *m*
 f capteur *m* de choc
 n botssignaleringssensor
 p czujnik zderzenia

1377 crawler tractor; tracted tractor
 d Raupenschlepper *m*; Kettenschlepper *m*
 f tracteur *m* à chenilles
 n rupstrekker
 p ciągnik gąsienicowy

1378 cross chisel
 d Meissel *m* mit kreuzförmiger Fase;
   Schlossermeissel *m*
 f ciseau *m* à biseau cruciforme
 n beitel met kruisvormige snede
 p dłuto ślusarskie

1379 cross cut saw
 d Ablängsäge *f*
 f scie *f* à tronçonner
 n afkortzaag
 p piła poprzeczna

1380 crossed belt drive
 d gekreuzter Riementrieb *m*
 f transmission *f* par courroie croisée
 n gekruiste riemoverbrenging
 p przekładnia pasowa skrzyżowana

1381 cross filing
 d kreuzweises Feilen *n*
 f limage *m* croisé
 n kruiselings vijlen
 p pilnikowanie krzyżowe

1382 cross fitting
 d Kreuzstück *n*
 f raccord *m* en croix
 n kruisstuk
 p krzyżak

1383 crosshead bush
 d Kreuzstück *n*

 f croisillon *m*
 n kruisstuk
 p element krzyżowy

1384 crosshead lever
 d Kreuzkopfhebel *m*
 f levier *m* de goujon
 n kruiskophefboom
 p dźwignia wodzika krzyżulca

1385 crossing
 d Kreuzung *f*
 f croisement *m*; intersection *f*
 n kruising
 p skrzyżowanie

1386 cross joint
 d Kreuzverbindung *f*
 f assemblage *m* en croix
 n kruisverbinding
 p połączenie krzyżowe

1387 cross key
 d Kreuzschlüssel *m*
 f clé *f* à croix
 n kruissleutel
 p klucz krzyżowy

1388 cross member
 d Querträger *m*
 f traverse *f* en U
 n dwarsdraagbalk met U-profiel
 p poprzeczka ceowa

1389 cross ply patch
 d Kreuzpflaster *n*
 f emplâtre *m* croisé
 n pleister voor radiaalbanden
 p plaster krzyżowy

1390 cross ply tyre
 d Diagonalreifen *m*
 f pneumatique *m* à carcasse diagonale
 n diagonaalband
 p opona diagonalna

 * cross recessed head screw → 3729

1391 crossroad
 d Querstrasse *f*
 f traverse *f*
 n zijweg
 p przecznica

1392 cross shaft
 d Querwelle *f*
 f arbre *m* transversal

n dwarsas
p wał poprzeczny

1393 crowbar
d Brechstange *f*
f pince *f*
n koevoet
p łom

1394 crown circle; crown line
d Kopfkreis *m*
f cercle *m* de tête
n buitenomtrek van tandwiel
p koło wierzchołkowe

* **crown line → 1394**

1395 crown pressure spring
d Welltellerfeder *f*
f ressort *m* à diaphragme ondulé
n schoteldrukveer
p falista sprężyna talerzowa

* **crown wheel → 4158**

1396 crown wheel screw
d Schraube *f* für Zahnkranz
f vis *f* de couronne
n kroonwielschroef
p śruba koła zębatego tarczowego

1397 crucible
d Tiegel *m*
f creuset *m*
n gietkroes
p tygiel

1398 crucible tongs
d Tiegelzange *f*
f pince *f* à creuset
n kroestang
p szczypce do tygli

1399 cruciform frame
d X-Rahmen *m*
f cadre *m* en X
n X-vormige versteviging; kruisbalk
p rama krzyżowa

1400 cruising speed
d Reisegeschwindigkeit *f*
f vitesse *f* de croisière
n kruissnelheid
p prędkość podróżna

* **cup point set screw → 2622**

1401 cup seal
d Tropfmanchette *f*; Napfmanchette *f*
f cuvette *f* d'étanchéité
n afdichtingsmanchet
p pierścień uszczelniający

1402 curbstone
d Randstein *m*
f bordure *f* de route
n kantsteen
p krawężnik kamienny

1403 curb weight
d Gewicht *n* des fahrbereiten Wagens
f poids *m* de la voiture en ordre de marche
n eigengewicht
p ciężar własny samochodu gotowego do jazdy

1404 current
d Strom *m*
f courant *m*
n stroom
p prąd

1405 current consumption
d Stromverbrauch *m*
f consommation *f* de courant
n stroomverbruik
p zużycie prądu

1406 current division
d Stromverteilung *f*
f distribution *f* d'allumage
n stroomverdeling
p rozdział prądu

1407 current regulator
d Strombegrenzer *m*; Stromregler *m*
f limiteur *m* de débit
n stroombegrenzer; stroomregelaar
p ogranicznik prądu

* **curved rim → 497**

1408 curved scissors
d gebogene Schere *f*
f criseaux *mpl* courbes
n gebogen schaar
p nożyce wygięte

1409 curving blade saw
d Bauchsäge *f*
f scie *f* ventrée
n buikzaag
p piła poprzeczna

1410 cushion rubber
d Polstergummi *m*

f gomme *f* de liaison
n rubberen ring
p guma przeciwstrząsowa; guma amortyzacyjna

1411 cut off cock; spigot valve
d Absperrhahn *m*
f robinet *m* d'isolement
n stopkraan
p kurek odcinający

1412 cut off wheel
d Trennscheibe *f*
f meule *f* de tronçonnage
n doorslijpschijf
p przecinak ścierny

1413 cut out switch
d Rückstromschalter *m*
f disjoncteur *m*
n terugstroomautomaat van gelijkstroomdynamo
p wyłącznik prądu zwrotnego

1414 cut out terminal
d Ladeschalterklemme *f*
f borne *f* de disjoncteur
n laadstroomklem
p zacisk samoczynnego wyłącznika

1415 cutter arbor
d Frässpindel *f*
f mandrin *m* portefraises
n freesspil
p wrzeciono freza

1416 cutter head
d Fräskopf *m*
f tête *f* de moulurage
n freeskop; meskopfrees
p głowica frezowa

1417 cutter head knife
d Profilfräser *m*
f fraise *f* à profiler
n profielfrees
p frez kształtowy

1418 cutting die
d Schneideisen *n*
f filière *f* à tarauder
n snijijzer
p narzynka

1419 cutting edge
d Schneide *f*
f tranchant *m*
n snede
p krawędź skrawająca

1420 cutting ring
d Schneidring *m*
f bicône *f*
n snijring
p pierścień skrawający

1421 cutting table; milling table
d Frästisch *m*
f table *f* de fraisage
n freestafel
p stół frezarki

1422 cutting torch
d Schneidbrenner *m*
f chalumeau *m* à découper
n snijbrander
p palnik do cięcia; przecinak gazowy

1423 cylinder barrel; cylinder casing
d Zylinderoberfläche *f*
f fût *m* de cylindre
n cilinderoppervlak
p powierzchnia cylindra

1424 cylinder bearing surface
d Zylinderlauffläche *f*
f surface *f* de portée de cylindre
n cilinderdraagvlak
p gładź cylindra

1425 cylinder block
d Zylinderblock *m*
f bloc-moteur *m*
n cilinderblok
p blok cylindrów

1426 cylinder bolt
d Zylinderschraube *f*
f vis *f* de cylindre
n cilinderschroef
p śruba cylindra

1427 cylinder bore
d Zylinderbohrung *f*
f alésage *m* du cylindre
n cilinderboring
p średnica cylindra

* **cylinder casing** → **1423**

1428 cylinder filling ratio
d Zylinderfüllungsgrad *m*
f taux *m* de remplissage du cylindre
n cilindervullingsgraad
p stopień napełniania cylindra

1429 cylinder fin
d Kühlrippe *f* des Zylinders

f ailette *f* de cylindre
n koelribbe van cilinder
p żebro cylindra

1430 cylinder grinding
d Zylinderschleifen *n*
f rectification *f* de cylindre
n uitslijpen van cilinder
p szlifowanie cylindra

1431 cylinder head
d Zylinderkopf *m*
f culasse *f*
n cilinderkop
p głowica cylindra

1432 cylinder head bolt
d Zylinderkopfschraube *f*
f boulon *m* de culasse
n cilinderkopschroef
p śruba głowicy cylindra

1433 cylinder head cover
d Zylinderkopfdeckel *m*
f couvercle *m* de culasse
n cilinderkopdeksel
p pokrywa głowicy cylindra

1434 cylinder head cover gasket
d Zylinderkopfdeckeldichtung *f*
f joint *m* de couvre-culasse
n cilinderkopdekselpakking
p uszczelniacz pokrywy głowicy cylindra

1435 cylinder head fin
d Kühlrippe *f* des Zylinderkopfes
f ailette *f* de culasse de cylindre
n koelribbe van cilinderkop
p żebro głowicy cylindra

1436 cylinder head rest
d Motorenbock *m*
f support *m* de culasse
n cilinderkopsteun
p wspornik głowicy cylindra

1437 cylinder honing
d Zylinderhonen *n*
f rodage *m* du cylindre
n honen van cilinder
p gładź cylindrowa

1438 cylinder liner
d Zylinderlaufbüchse *f*
f chemise *f*
n cilindervoering
p tuleja cylindrowa

1439 cylinder rinsing
d Zylinderspülung *f*
f balayage *m*
n cilinderspoeling
p przepłukiwanie cylindra

1440 cylinder security insert
d Einbausicherung *f*
f dispositif *m* de sécurité
n veiligheidsvoorziening
p zabezpieczenie wmontowane

1441 cylinder spacing
d Abstand *m* der Zylinderachsen
f entre-axe *m* des cylindres
n cilinder-hartafstand
p rozstaw osi cylindrów

1442 cylinder wall
d Zylinderwand *f*
f paroi *f* de cylindre
n cilinderwand
p ścianka cylindra

1443 cylinder wear
d Zylinderverschleiss *m*
f usure *f* du cylindre
n cilinderslijtage
p zużycie cylindra

1444 cylindrical grinding
d Zylindrischschleifen *n*
f meulage *m* cylindrique
n cilindrisch slijpen
p szlifowanie na okrągło

1445 cylindrical joint
d zylindrische Verbindung *f*
f assemblage *m* cylindrique
n cilinderverbinding
p połączenie cylindryczne; złącze cylindryczne

1446 cylindrical roller
d Zylinderrolle *f*
f rouleau *m* cylindrique
n cilindervormige rol
p wałeczek walcowy

1447 cylindrical roller cage
d Zylinderrollenkäfig *m*
f cage *f* à galets cylindriques
n cilindrische rolkooi
p złożenie walcowe

* **cylindrical spring** → 1137

1448 cylindrical thread
d zylindrisches Gewinde *n*

   f filet *m* cylindrique
   n cilindrische schroefdraad
   p gwint walcowy

1449 cylindrical worm
   d Zylinderschnecke *f*
   f vis *f* sans fin cylindrique
   n cilindervormige worm; cilinderworm
   p ślimak walcowy; ślimak cylindryczny

D

1450 damage
d Beschädigung f
f endommagement m
n beschadiging; schade
p uszkodzenie; szkoda

1451 damage of lacquer; damage of paintwork
d Lackbeschädigung f
f dommage m à la peinture
n lakbeschadiging; lakschade
p uszkodzenie lakieru

* **damage of paintwork** → 1451

1452 damped oscillation
d gedämpfte Schwingung f
f oscillation f amortie
n gedempte trilling
p drgania tłumione

* **damper** → 4444

1453 damper valve
d Dämpfungsventil n
f soupape f d'amortissement
n dempklep van hydropneumatisch veersysteem
p zawór amortyzacyjny; zawór wygaszający

* **damping** → 4451

1454 damping coefficient
d Dämpfungskoeffizient m
f coefficient m d'amortissement
n dempingscoëfficiënt
p współczynnik tłumienia

1455 damping disc
d Dämpfungsscheibe f; Dämpferscheibe f
f disque m amortisseur de vibrations
n trillingsdemperschijf
p tarcza tłumika

1456 damping leaf
d Dämpfungsfederblatt n
f contre-lame f de ressort
n dempveerblad
p pióro sprężyste amortyzatora

1457 damping power
d Dämpfungsvermögen n
f capacité f d'absorption
n dempingsvermogen
p zdolność tłumienia

1458 danger
d Gefahr m
f danger m
n gevaar
p niebezpieczeństwo

1459 dangerous turn
d gefährliche Kurve f
f virage m dangereux
n gevaarlijke bocht
p niebezpieczny zakręt

1460 dashboard; facia
d Instrumententafel f; Instrumentenbrett n
f tableau m de bord; panneau m d'instruments
n dashboard
p tablica rozdzielcza

1461 dashboard cowl
d Stirnwand f; Spritzwand f
f planche f de tableau de bord
n schutbord
p przegroda czołowa

1462 dashboard instruments
d Armaturenbrettinstrumenten npl
f cadrans mpl de bord
n dashboardinstrumenten
p przyrządy tablicy rozdzielczej

1463 dashboard lamp
d Instrumentenleuchte f
f lampe f d'éclairage de tableau de bord
n dashboardverlichting
p oświetlenie tablicy rozdzielczej

1464 dash pan
d Vorderwand f
f cloison m avant
n schutbord
p ścianka przednia

1465 dash pot
d Vibrationsdämpfer m
f amortisseur m de vibration
n trillingsdemper
p tłumik drgań

1466 data plate
d Typenschild n
f plaque f signalétique
n typeplaatje; identificatieplaatje
p tabliczka znamionowa

* dazzle v → 2493

1467 **dazzling light intensity**
d Blendlichtstärke f
f intensité f de la lumière d'éblouissement
n verblindinglichtsterkte
p intensywność światła oślepiania

1468 **de-icer**
d Entfrostungsmittel n
f moyen m de dégivrage
n ontdooimiddel
p substancja przeciwoblodzeniowa

1469 **de-icing paste**
d Frostschutzpaste f
f pâte f antigivre
n antivriespasta
p pasta przeciw zamarzaniu

1470 **dead axle**
d feste Achse f
f essieu m fixe
n niet-aangedreven as
p oś stała; oś nienapędowa

1471 **dead center stop**
d Anschlag m für Neutralstellung
f butée f de point mort
n dood punt
p martwy punkt

1472 **dead end street**
d Sackstrasse f
f roue f cul-de-sac
n doodlopende weg
p ulica ślepa

* dead load → 4963

1473 **dealer**
d Händler m
f merchand m
n dealer; agent
p handlowiec; sprzedawca

1474 **deburring wheel**
d Schruppscheibe f
f meule f d'ébarbage
n afbraamschijf
p ściernica do zdzierania

1475 **decantation**
d Dekantierung f
f décantation f
n decantering
p dekantacja

1476 **deceleration**
d Verzögerung f
f retard m
n vertraging
p opóźnienie

1477 **deceleration limit**
d Beschleunigungsgrenze f
f limite f de décéleration
n acceleratielimiet
p granica przyśpieszenia

1478 **decelerometer**
d Bremsverzögerungsmesser m
f décéléromètre m
n vertragingsmeter
p opóźnieniomierz

1479 **decompression cam**
d Nocken m für Kompressionsverminderung
f came f de décompression
n decompressienok
p krzywka dekompresyjna

1480 **decompression lever**
d Kompressionshebel m
f levier m de décompression
n decompressiehefboom
p dźwignia ciśnienia sprężania

1481 **decorative tape**
d Klebeschablone f
f gabarit m de collage
n zelfklevende sjabloon
p nalepka

1482 **decrimping pliers**
d Demontagezange f für Stahlblech
f pince f à désassembler les tôles
n demontagetang voor staalplaat
p szczypce do rozdzielania blach stalowych

1483 **deep groove ball bearing**
d Rillenkugellager n
f roulement m rainuré à billes
n oliegroefkogellager
p łożysko kulkowe zwykłe

* defect → 729

1484 **defective mounting**
d Montagefehler m
f erreur f de montage
n montagefout
p błąd montażowy; montaż nieprawidłowy

1485 **deflection**
d Durchbiegung f

f flèche f
n doorbuiging
p zginanie; zgięcie

1486 deflector guide
d Ablenkfläche f
f déflecteur m
n deflector
p powierzchnia odchylająca

1487 deflector knob
d Leiteinsatzstellschraube f
f bouton m de commande de filtre
n arreteringsknop
p uchwyt przesłony powietrza

1488 deflectorless piston
d Flachkolben m
f piston m plat
n zuiger met platte bodem
p tłok z płasim dnem

1489 deflector of piston
d Kolbennase f
f déflecteur m de piston
n afbuigplaat
p deflektor

1490 deflector pipe
d Ableitrohr n um Wärmetauscher
f tube m déflecteur monté de l'échangeur
n afbuigpijp
p rurka odpływowa przy wymienniku ciepła

1491 deflector piston
d Ablenkerkolben m
f piston m déflecteur
n kantzuiger
p tłok odchylający

1492 deflector plate
d Ablenkplatte f
f déflecteur m
n afbuigplaat
p płytka odchylająca

1493 deformability
d Verformbarkeit f
f déformabilité f
n vervormbaarheid
p odkształcalność

1494 deformation
d Formänderung f
f déformation f
n vervorming
p odkształcenie; deformacja

1495 defroster
d Scheibenentfroster m
f dégivreur m
n ruitontdooier
p odszraniacz szyb

1496 defroster nozzle
d Entfrosterdüse f
f buse f de dégivreur
n ruitontdooi-inrichtingsproeier
p dysza odmrażacza

1497 defrosting
d Entfrostung f
f dégivrage m
n ontdooiing
p odszranianie

1498 degreasing
d Entfettung f
f dégraissage m
n ontvetten
p odtłuszczanie

1499 degreasing unit
d Entfettungsanlage f
f installation f de dégraissage
n ontvettingsinstallatie
p urządzenie odtłuszczające

1500 degree of dilution
d Verdünnungsgräd m
f degré m de dilution
n verdunningsgraad
p stopień rozcieńczenia

1501 degree of freedom
d Freiheitsgrad m
f degré m de liberté
n vrijheidsgraad
p stopień swobody

1502 degree of purity
d Reinheitsgrad m
f degré m de pureté
n graad van zuiverheid
p stopień czystości

1503 degree of reaction
d Reaktionsgrad m
f degré m de réaction
n reactiegraad
p stopień reakcyjności

1504 degree of uniformity
d Gleichförmigkeitsgrad m
f degré m d'uniformité

n gelijkvormigheidsgraad
p stopień równomierności

1505 delay device
d Bewegungsbremse *f*
f temporisateur *m*
n bewegingsmembraan
p przekaźnik czasowy

1506 delayed explosion
d verzögerte Explosion *f*
f explosion *f* retardée
n naontploffing
p wybuch opóźniony

* **delayed firing** → 1507

1507 delayed ignition; delayed firing
d Spätzündung *f*
f allumage *m* retardé
n naontsteking; vertraagde ontsteking
p zapłon opóźniony

1508 delivery
d Auslieferung *f*
f livraison *f*
n levering
p dostawa

1509 delivery stroke
d Auspuffhub *m*
f temps *m* d'échappement
n persslag
p suw tłoka

* **delivery valve** → 3868

1510 delivery van
d Lieferwagen *m*
f fourgon *m*
n bestelauto; bestelwagen
p samochód dostawczy; dostawczy samochód furgon

1511 delta connection
d Dreieckschaltung *f*
f couplage *m* en triangle
n driehoeksschakeling
p połączenie trójkątowe

1512 delta metal
d Deltametall *n*
f métal *m* delta
n deltametaal
p metal delta

1513 demagnetizer
d Entmagnetisierungsapparat *n*

f désaimanteur *m*
n demagnetiseerapparaat
p odmagnesowywacz

1514 demountable rim
d abnehmbare Felge *f*
f jante *f* amovible
n demonteerbare velg
p zdejmowalna obręcz

1515 densimeter; hydrometer
(apparatus to measure the density of acid)
d Dichtemesser *m*; Säureprüfer *m*
f aréomètre *m*; densimètre *m*
n zuurweger
p gęstościomierz; areometr

1516 density
d Dichte *f*
f densité *f*
n dichtheid
p gęstość; zagęszczenie

* **dent** *v* → 2835

1517 depolarize *v*
d depolarisieren
f dépolariser
n depolariseren
p depolaryzować

1518 depth gauge
d Tiefenlehre *f*; Tiefenmesser *m*
f pied *m* à profondeur
n dieptemaat
p głębokościomierz

1519 depth of cut
d Schnittiefe *f*
f profondeur *m* de pénétration
n snijdiepte; sneded
p głębokość skrawania

1520 derusting
d Entrosten *n*
f dérouillement *m*
n ontroesten
p odrdzewiać

1521 descaling
d Kesselsteinentfernung *f*
f détartrage *m*
n verwijdering van ketelsteen
p usuwanie kamienia kotłowego

1522 description
d Beschreibung *f*

f description *f*
n beschrijving
p opis

1523 design *v*
d entwerfen
f projeter
n ontwerpen
p projektować

1524 detachable chain
d zerlegbare Gelenkkette *f*
f chaîne *f* à crochets
n haakketting
p łańcuch haczykowy

1525 detachable flange
d abnehmbarer Flansch *m*
f rebord *m* détachable
n afneembare flens
p kołnierz zdejmowalny

1526 detachable joint
d lösbare Verbindung *f*
f assemblage *m* démontable
n losneembare verbinding
p połączenie odłączalne

1527 detachable wheel
d abnehmbares Rad *n*
f roue *f* démontable
n afneembaar wiel
p koło zdejmowalne

1528 detector
d Detektor *m*
f détecteur *m*
n detector
p detektor

1529 detent pin
d Verriegelungszapfen *m*
f doigt *m* de verrouillage
n vergrendelpen
p palec ryglujący; czop ryglujący

1530 detergent
d Waschmittel *n*
f produit *m* de lavage
n reinigingsmiddel
p detergent; środek myjący

1531 detonation; knocking
d Klopfen *n*
f détonation *f*
n detonatie
p detonacja

1532 device; instrument
d Gerät *n*; Vorrichtung *f*
f appareil *m*
n instrument; apparaat
p przyrząd; aparat

1533 diagnosis test set
d Motorprüfgerät *n*
f appareil *m* d'essai de moteur
n motorentestapparaat
p przyrząd do badania silnika

1534 diagonal reaction rod
d Schrägschubstrebe *f*
f barre *f* de réaction latérale
n diagonaal gemonteerde reactiestang
p skośny drążek reakcyjny

*   **diagonal seat belt** → 4459

1535 diagram
d Schema *n*; Diagramm *n*
f schéma *m*; diagramme *m*
n grafiek; schema
p diagram; wykres; schemat

1536 dial plate
d Zifferblatt *n*; Skalenscheibe *f*
f cadran *m*
n wijzerplaat
p tarcza wskaźnika

1537 diamond drill; drill bit diamond
d Diamantbohrer *m*
f sonde *f* à pointe diamantée; perceuse *f* à
    diamant
n diamantboor
p wiertło diamentowe

1538 diamond point chisel
d Rautenstichel *m*
f grain *m* d'orge
n facetbeitel
p żłobik kątowy

1539 diamond saw
d Diamantsäge *f*
f scie *f* diamantée
n diamantzaag
p piła diamentowa

1540 diaphragm; membrane
d Membrane *f*
f membrane *f*
n membraan
p membrana

1541 diaphragm carburetter
d Membranvergaser *m*

f carburateur *m* à membrane
n membraancarburateur
p gaźnik przeponowy

1542 diaphragm control unit
d Unterdruckspritzmengeversteller *m*
f commande *f* à dépression
n onderdrukregulateur
p wybierak przeponowy

1543 diaphragm hydraulic actuator
d Membranzylinder *m*
f cylindre *m* hydraulique en membrane
n membraancilinder
p siłownik przeponowy

1544 diaphragm pump
d Membranpumpe *f*
f pompe *f* à membrane
n membraanpomp
p pompa przeponowa; pompa membranowa

1545 diaphragm spring
d Membranfeder *f*
f ressort *m* de membrane
n diafragmaveer
p sprężyna membranowa; sprężyna przeponowa

1546 diaphragm spring clutch
d Membranfederkupplung *f*
f embrayage *m* à ressort diaphragme
n diafragmakoppeling
p sprzęgło o sprężynie membranowej

1547 diaphragm spring sector
d Membranfederlappen *m*
f secteur *m* de diaphragme
n diafragmaveersector
p wycinek sprężyny membranowej

1548 diaphragm switch
d Membranschalter *m*
f contacteur *m* à membrane
n membraanschakelaar
p wyłącznik przeponowy

1549 diaphragm valve
d Membranventil *n*
f soupape *f* à diaphragme
n membraanafsluiter
p zawór przeponowy

1550 die casting machine
d Pressgussmaschine *f*
f machine *f* à mouler sous pression
n spuitgietmachine
p maszyna odlewnicza ciśnieniowa

1551 Diesel adapter
d Dieselhilfsteil *n*
f adaptateur *m* diesel
n dieselhulpstuk
p łącznik Diesla

1552 Diesel circulation
d Dieselprozess *m*
f cycle *m* diesel
n dieselcirculatie
p obieg Diesla

1553 die stock
d Schneideisenhalter *m*
f porte-filière *m*
n snijplaathouder
p oprawka do narzynek

1554 differential brake
d Differentialbremse *f*
f frein *m* de différentiel
n differentieelrem
p hamulec różnicowy

1555 differential case
d Ausgleichgehäuse *n*
f carter *m* de différentiel
n differentieelhuis
p obudowa mechanizmu różnicowego

1556 differential gear
d Ausgleichgetriebe *n*
f différentiel *m*
n differentieel; compensatiedrijfwerk
p mechanizm różnicowy

1557 differential interlock
d Ausgleichgetriebesperre *f*
f blocage *m* de différentiel
n blokkeermof van differentieel
p ryglownik mechanizmu różnicowego

1558 differential lock
d Ausgleichgetriebeschloss *n*
f verrouillage *m* de différentiel
n differentieelslot
p blokada mechanizmu różnicowego

1559 differential organ
d Differenzorgan *n*
f organe *m* différentiel
n differentieelorgaan
p mechanizm różnicowy

1560 differential pinion shaft
d Achse *f* der Ausgleichskegelrades
f porte-satellite *m*

n pignonas van differentieel
p ramię przekładni różnicowej

1561 differential piston
d Differentialkolben *m*
f piston *m* différentiel
n differentieelzuiger
p tłok różnicowy

1562 differential side gear
d Differentialkegelrad *n*
f planétaire *m* de différentiel
n planeetwiel
p koło obiegowe przekładni różnicowej

1563 differential stepped wheel cylinder
d Differentialradbremszylinder *m*
f cylindre *m* de roue à action différentielle
n getrapte differentieelwielcilinder
p cylinder hamulca koła mechanizmu
   różnicowego

1564 diffused light
d gestreutes Licht *n*
f lumière *f* diffuse
n verstrooide verlichting
p oświetlenie rozproszone

1565 digital display multimeter
d Multifunktionsprüfgerät *n* mit Digitalanzeige
f multimètre *m* à affichage digital
n multimeter met digitale aflezing
p miernik z wyświetlaczem cyfrowym

1566 digital engine control
d digitale Motorsteuerung *f*
f commande *f* digitale du moteur
n digitale motorbesturing
p programowane sterowanie silnika

1567 dilute acid
d verdünnte Säure *f*
f acide *m* dilué
n verdund zuur
p kwas rozcieńczony

1568 diluting agent
d Verdünnung *f*
f diluant *m*
n verdunning
p rozcieńczalnik

1569 dimensions
d Abmessungen *fpl*
f dimensions *fpl*
n afmetingen
p wymiary

1570 diminishing nipple
d Übergangsnippel *m*
f douille *f* de réduction
n verloopnippel
p trzpionek redukcyjny

* **dimmer switch → 238**

1571 dimple
d Beule *f*
f cabosse *f*
n deuk
p wgniecenie

1572 dinging hammer
d Ausbeulhammer *m*
f marteau *m* à garnir
n plaatwerkhamer
p młotek do wgłębień

1573 diode
d Diode *f*
f diode *f*
n diode
p dioda

1574 diode mounting plate
d Diodenplatte *f*
f support *m* de diodes
n diodenbrug
p płytka osadcza diod

1575 dip soldering
d Tauchlöten *f*
f soudure *f* par immersion
n dompelsolderen
p lutowanie zanurzeniowe

1576 dipstick guide
d Führung *f* für Ölmessstab
f guide *m* jauge
n oliemeetstaafgeleider; oliepeilstokgeleider
p prowadnica prętowego wskaźnika poziomu
   oleju

1577 dip switch
d Abblendschalter *m*
f interrupteur *m* code
n dimlichtschakelaar
p wyłącznik świateł mijania

1578 direct current motor
d Gleichstrommotor *m*
f électromoteur *m* à courant continu
n gelijkstroommotor
p silnik prądu stałego

1579 direct current terminal
 d Gleichstromklemme *f*
 f prise *f* de courant continu
 n gelijkstroomklem
 p zacisk prądu wyprostowanego

1580 direct drive clutch
 d Kupplung *f* des direkten Ganges
 f embrayage *m* de prise directe
 n prise directe
 p sprzęgło biegu bezpośredniego

1581 direct gear
 d direkter Gang *m*
 f prise *f* directe
 n directe overbrenging
 p bieg bezpośredni

1582 direct injection
 d direkte Einspritzung *f*
 f injection *f* directe
 n directe inspuiting
 p wtrysk bezpośredni

1583 direction
 d Richtung *f*
 f direction *f*
 n richting
 p kierunek

1584 directional control valve
 d Wegeventil *n*
 f tiroir *m* de distribution
 n grondschuif; verdeelschuif
 p zawór sterujący suwakowy

1585 direction indicator
 d Winker *m*; Fahrtrichtungsanzeiger *m*
 f indicateur *m* de direction
 n richtingaanwijzer
 p kierunkowskaz

1586 direction indicator control light; flasher indicator control light
 d Blinkerkontrolleuchte *f*; Fahrtrichtungsanzeugeleuchte *f*
 f témoin *m* de l'indicateur de direction
 n richtingaanwijzercontrolelicht
 p lampka kontrolna kierunkowskazu

1587 direction indicator switch
 d Blinkleuchtenschalter *m*
 f commande *f* de feux de direction
 n knipperlichtschakelaar
 p wyłącznik kierunkowskazów

1588 direction of flow
 d Strömungsrichtung *f*
 f direction *f* du courant
 n stroominrichting
 p kierunek przepływu prądu

1589 direction of motion
 d Bewegungsrichtung *f*
 f sens *m* de mouvement
 n bewegingsrichting
 p kierunek ruchu

1590 direction of rotation
 d Drehrichtung *f*
 f sens *m* de rotation
 n draairichting
 p kierunek obrotu; kierunek wirowania

1591 direct reading model
 d Modell *n* mit direkter Ablesung
 f modèle *m* à lecture directe
 n model met rechtstreekse aflezing
 p model z odczytem bezpośrednim

1592 direct reading type torque wrench
 d Drehmomentschlüssel *m* mit direkter Ablesung
 f clé *f* dynamométrique à lecture directe
 n momentsleutel met rechtstreekse aflezing
 p klucz momentu obrotowego z odczytem bezpośrednim

1593 disc; disk
 d Scheibe *f*
 f disque *m*
 n schijf
 p tarcza

1594 disc brake
 d Scheibenbremse *f*
 f frein *m* à disque
 n schijfrem
 p hamulec tarczowy

1595 disc clutch
 d Scheibenkupplung *f*
 f embrayage *m* à disque
 n plaatkoppeling
 p sprzęgło tarczowe

1596 discharge
 d Entladung *f*
 f décharge *f*
 n ontladen
 p wyładowywać

1597 discharge curve
 d Entladekurve *f*
 f courbe *f* de décharge
 n ontladingskromme
 p krzywa rozładunku

1598 **discharge lamp**
   d Entladungslampe *f*
   f lampe *f* à discharge
   n ontladingslamp
   p lampa wyładowcza

1599 **discharge loss**
   d Abgasverlust *m*
   f perte *f* d'échappement
   n uitlaatgasverlies
   p strata wylotowa

1600 **disc joint**
   d Scheibengelenk *n*
   f accouplement *m* à disques
   n flenskoppeling
   p tarczowy przegub elastyczny

1601 **discless wheel**
   d scheibenloses Rad *n*
   f roue *f* sans disque
   n schijfloos wiel
   p koło beztarczowe

   * **disc milling cutter** → 4472

1602 **disconnection**
   d Entkupplung *f*
   f désaccouplement *m*
   n afkoppeling
   p rozłączenie

1603 **disc spring**
   d Tellerfeder *f*; Scheibenfeder *f*
   f rondelle *f* Belleville
   n schijfveer
   p sprężyna tarczowa; sprężyna krążkowa

1604 **disc wheel**
   d Scheibenrad *n*
   f roue *f* à disque
   n schijfwiel
   p koło tarczowe

1605 **disengagement**
   d Ausrücken *n*
   f déconnexion *f* mécanique
   n ontkoppelen
   p wyprzęganie; wyłączanie sprzęgła

1606 **dished dolly**
   d Handfaust *f* für gebördelte Kanten
   f tas *m* semelle
   n vingertas
   p klepadło wklęsłe

1607 **dish grinding wheel**
   d Tellerschleifscheibe *f*

   f meule *f* assiette
   n schotelslijpschijf
   p ściernica talerzowa; ściernica tarczowa

1608 **dish shaped spring clutch**
   d Tellerfederkupplung *f*
   f embrayage *m* à ressort à disque
   n schoteldrukveer van koppeling
   p sprzęgło o sprężynie talerzowej

1609 **dish shaped spring type double clutch**
   d Doppeltellerfederkupplung *f*
   f embrayage *m* double à ressort à disque
   n koppeling met dubbele schoteldrukveer
   p sprzęgło podwójne o sprężynie centralnej

   * **disk** → 1593

1610 **dismantling**
   d Demontage *f*
   f démontage *m*
   n demontage
   p demontaż

1611 **displacement**
   d Verlagerung *f*
   f déplacement *m*
   n verplaatsing
   p przemieszczenie

1612 **displacement pump**
   d Verdrängerpumpe *f*
   f pompe *f* aspirante refoulante
   n verdringerpomp
   p pompa wyporowa

1613 **display**
   d Anzeiger *m*
   f afficheur *m*
   n display
   p wyświetlacz

1614 **disposable filter**
   d Wechselfilter *m*; Wegwerffilter *m*
   f filtre *m* jetable
   n wegwerpfilter
   p filtr do jednorazowego użycia

1615 **dissipation of energy**
   d Kraftverbrauch *m* der Bereifung
   f absorption *f* de force des pneus
   n energieverspilling
   p rozproszenie energii

1616 **distance**
   d Distanz *f*
   f distance *f*

n afstand
p dystans

1617 distance bush
d Abstandring *m*
f entretoise *f* d'axe
n afstandsring
p tuleja rozstawcza; tuleja odległościowa

1618 distance gauge
d Spurmessergerät *n*; Spurweite *f*
f gabarit *m* d'écartement
n spoormal
p miernik odległości

1619 distance piece
d Abstandsstück *n*; Abstandshülse *f*
f entretoise *f*
n afstandsbusje
p tulejka rozstawcza

1620 distance plate; clamping plate
d Beilagplatte *f*; Klemmplatte *f*
f plaque *f* de serrage
n klemplaatje; afstandsplaatje
p płytka zaciskowa; płytka przyciskowa

1621 distance post
d Abweisstein *m*
f borne *f* kilométrique
n afstandspaal
p słupek kilometrowy

1622 distance sleeve
d Distanzhülse *f*
f douille *f* d'écartement
n afstandshuls
p tuleja odległościowa

1623 distance washer
d Distanzscheibe *f*
f entretoise *f*
n afstandsring
p podkładka odległościowa

1624 distillation point
d Siedetemperatur *f*
f température *f* de distillation
n kookpunt
p temperatura destylacji

1625 distilled water
d destilliertes Wasser *n*
f eau *f* distillée
n gedistilleerd water
p woda destylowana

1626 distinguishing mark
d Markierung *f*; Marke *f*

f marque *f*
n merkteken
p znak firmowy

1627 distribution clamp
d Klammer *m* für Verteilerkopf
f agrafe *f* de tête
n verdeelkapklem
p klamra głowicy rozdzielczej

1628 distribution duct
d Verteilerschacht *f*
f conduit *m* de répartition
n distributieleiding
p kanał rozdzielczy

1629 distribution of load
d Belastungsverteilung *f*
f distribution *f* du poids
n gewichtsverdeling
p rozłożenie ciężaru

1630 distribution pipe
d Verteilungsrohr *n*
f rampe *f*
n distributieleiding; verdeelpijp
p rura rozdzielcza

1631 distributor
d Verteiler *m*
f distributeur *m* d'allumage
n stroomverdeler
p rozdzielacz

* **distributor body** → 1636

1632 distributor cap; distributor head
d Verteilerkappe *f*
f chapeau *m* d'allumeur
n stroomverdelerkap
p głowica rozdzielacza

1633 distributor cap inner contact
d Verteilerelektrode *f*
f plot *m* intérieur de chapeau
n stroomverdelerkapelektrode
p styk wewnętrzny głowicy rozdzielacza

1634 distributor case
d Verteilergehäuse *n*
f boîtier *m* d'allumeur
n stroomverdelerhuis
p kadłub rozdzielacza

1635 distributor disc
d Verteilerscheibe *f*
f disque *m* de distributeur

n verdelerplaat
p tarcza rozdzielcza

* **distributor head** → 1632

1636 distributor housing; distributor body
d Verteilergehäuse n
f enveloppe f de distributeur d'allumage; corps m du distributeur
n stroomverdelerhuis
p obudowa rozdzielacza

1637 distributor lead
d Verteilerleitung f
f câble m d'allumeur
n verdelerkabel
p przewód rozdzielczy

1638 distributor piece
d Verteilerstück n
f pièce f de distributeur
n verdeelstuk
p element rozdzielczy

1639 distributor port
d Verteilerkanal m
f canal m distributeur
n verdelerkanaal
p kanalik rozdzielczy

1640 distributor rotor arm
d Verteilerfinger m
f doigt m de distributeur
n rotor van stroomverdeler
p palec rozdzielacza

1641 distributor shaft
d Verteilerwelle f
f arbre m de distributeur d'allumage
n stroomverdeleras
p wał rozdzielacza

1642 distributor test bench
d Zündverteilerprüfstand m
f banc m de contrôle de distributeur
n stroomverdelertestbank
p stanowisko badawcze rozdzielacza zapłonu

1643 distributor type injection pump
d Verteilereinspritzpumpe f
f pompe f d'injection à distributeur rotatif
n verdelerinspuitpomp
p rozdzielaczowa pompa wtryskowa

1644 disturb v
d stören
f perturber
n storen
p zakłócać

1645 diversion
d Umleitung f
f détour m
n omleiding
p objazd

1646 dividers
d Spitzzirkel m
f compas m à pointes
n steekpasser
p cyrkiel warsztatowy prosty

1647 division of degrees
d Winkelskala f
f graduation f
n graadverdeling
p podziałka; skala

1648 dog clutch; jaw clutch
d Klauenkupplung f
f accouplement m à griffe
n klauwkoppeling
p sprzęgło zębate kłowe

1649 dog clutch selector shaft
d Schaltkupplungswelle f
f arbre m de commande d'accouplement
n schakelas
p wałek widełek sprzęgła

1650 dog sleeve
d Klauenmuffe f
f douille f à griffes
n klauwmof
p tarcza kłowa

1651 dolly for curvatures
d Handfaust f für Krümmungen
f tas m à courbure
n gewelfde tas
p kowadło ręczne do gięcia

1652 domed dolly
d gewölbter Handtisch m
f table f à main bombée
n bolle handtafel
p klepadło kuliste

1653 domed shrinking blade
d kugelförmiger Spatel m zum Schrumpfen
f spatule m à rétreindre bombée
n bolle spatel om te krimpen
p łopatka kulista do kurczenia

1654 domed single spoon
d kurzes, breites Löffeleisen *n*
f palette *f* simple bombée
n brede gebolde lepel
p szeroka łyżka zgięta

1655 dome head cylinder
d Zylinder *m* mit halbkugelförmigem
   Brennraum
f cylindre *m* à chambre de combustion
   hémisphérique
n cilinder met halfbolvormige
   verbrandingsruimte
p cylinder z komorą spalania o kształcie
   półokrągłym

1656 door
d Tür *f*
f porte *f*
n deur; portier
p drzwi

1657 door ajar sensor
d Fühler *m* für geoffene Tür
f sensor *m* pour porte ouverte
n sensor van geopend portier
p czujnik drzwi otwartych

1658 door bolt
d Türriegel *m*
f targette *f* de porte
n portiergrendel; portierstift
p zasuwa drzwiowa; rygiel u drzwi

1659 door fitting
d Türbeschläg *m*
f accessoire *m* de porte
n deurbeslag
p okucie drzwiowe

1660 door frame
d Türrahmen *m*
f encadrement *m* de la porte
n portierframe
p ościeżnica drzwiowa

1661 door guide
d Türführung *f*
f guidage *m* de porte
n portiergeleider
p prowadnik drzwi

1662 door handle
d Türgriff *m*
f poignée *f* de porte
n portierkruk
p klamka drzwi

1663 door handle push rod
d Türgriffstange *f*
f tige *f* de poignée
n portierkrukdrukstang
p cięgno klamki drzwi

1664 door hinge
d Türscharnier *n*
f charnière *f* de porte
n portierscharnier
p zawiasa drzwi

1665 door hinge bolt
d Türscharnierbolzen *m*
f axe *m* de charnière
n portierscharnierbout
p sworzeń zawiasu drzwiowego

1666 door hinge pillar
d Türscharniersäule *f*
f montant *m* de porte
n portierstijl
p słupek przegubu drzwi

1667 door key illumination
d Türschlossbeleuchtung *f*
f éclairage *m* de la serrure de la portière
n verlichting van portierslot
p oświetlenie zamka drzwiowego

1668 door latch
d Türverriegelung *f*
f verrouillage *m* de porte
n veiligheidsvergrendeling van portier
p blokada drzwi

1669 door lock
d Türschloss *n*
f serrure *f* de porte
n portierslot
p zamek drzwiowy

1670 door lock control lever
d Arretierhebel *m*
f levier *m* de blocage
n vergrendelhefboom
p dźwignia blokady zamka

1671 door lock cylinder
d Türschliesszylinder *m*
f cylindre *m* de fermeture de porte
n portierslotcilinder
p cylinder zamka drzwiowego

1672 door locking switch
d Türverriegelungsschalter *m*
f inverseur *m* condamnation porte

n schakelaar van portiervergrendeling
p wyłącznik blokady drzwi

1673 door lock knob
d Türsperrknopf *m*
f bouton *m* de serrure de sûreté
n portiervergrendelknop
p przycisk blokady zamka

1674 door lock push lever
d Schlossschalterstössel *m*
f poussoir *m* de verrouillage
n blokkeernok
p popychacz wyłącznika zamka

1675 door lock push rod
d Schlosssperrendruckstange *f*
f tige *f* poussoir de serrure de sûreté
n drukstang van vergrendelknop
p cięgno blokady zamka

1676 door lock striker plate
d Schliesskeil *m*
f centreur *m* de porte
n sluitschoot
p prowadnik drzwi

1677 door operation
d Türbetätigung *f*
f manœuvre *f* de porte
n portierbediening
p działanie mechanizmu drzwiowego

1678 door outside handle
d Türaussengriff *m*
f poignée *f* extérieure
n portierkruk aan buitenkant
p klamka zewnętrzna drzwi

1679 door panel
d Türdeckelblech *n*
f panneau *m* de porte
n deurpaneel
p płyta drzwiowa; płat drzwiowy

1680 door stop
d Türanschlag *m*
f arrêt *m* de porte; butée *f* de porte
n portiervanger
p ogranicznik otwarcia drzwi

1681 door striker
d Türdrücker *m*
f poussoir *m*
n deurdrukker
p klamka drzwiowa

1682 door window glass
d Türfensterscheibe *f*

f glace *f* de la porte
n portierraam
p szyba drzwiowa

1683 dosage valve piston
d Dosierkolben *m*
f piston *m* de clapet de dosage
n doseerplunjerklep
p tłoczek zaworu dozującego

1684 dosage valve piston sleeve
d Dosierkolbenhülse *f*
f cylindre *m* de piston de dosage
n doseerplunjerklephuls
p tulejka tłoczka zaworu dozującego

1685 double acting brake
d doppeltwirkende Bremse *f*
f frein *m* à double effet
n dubbelwerkende rem
p hamulec dwustronny

1686 double acting hydraulic cylinder
d doppeltwirkender Hydrozylinder *m*
f cylindre *m* hydraulique à double effet
n dubbelwerkende hydraulische cilinder
p cylinder hydrauliczny obustronnego dzialania

1687 double acting shock absorber
d Doppelwirkungkolbenstossdämpfer *m*
f amortisseur *m* à double effet
n dubbelwerkende schokdemper
p amortyzator dwustronnego działania

1688 double angle cutter
d doppelseitiger Winkelfräser *m*
f fraise *f* angulaire à deux côtes
n frees met dubbele snijkant
p frez kątowy symetryczny

1689 double arm lever
d Doppelhebel *m*
f levier *m* double
n tweearmige hefboom
p dźwignia dwuramienna

1690 double bead construction
d Doppelwulst *m*
f talon *m* double
n dubbele kraalconstructie
p zgrubienie podwójne

1691 double chain drive
d Doppelkettenantrieb *m*
f transmission *f* à double chaîne
n dubbele kettingaandrijving
p podwójny napęd łańcuchowy

1692 double check valve
d Zweiwegventil *m*
f double valve *f* d'arrêt
n tweewegklep
p podwójny zawór zwrotny

1693 double circuit brake system
d Zweikreisbremsanlage *f*
f dispositif *m* de freinage à transmission à circuit double
n tweekringsremsysteem
p dwuzakresowy układ hamulcowy; niezależny układ hamulcowy

1694 double circuit brake valve
d Zweikreisbremsventil *n*
f valve *f* de protection à double circuit
n tweekringsremklep
p dwuobwodowy zawór hamulca

1695 double clip
d Doppellasche *f*
f jumelle *f* double
n dubbele beugel
p wieszak podwójny

1696 double-decker
d Doppeldecker *m*
f autobus *m* à impériale
n dubbeldekker
p piętrobus

1697 double direction ball thrust bearing
d zweiseitig wirkendes axiales Rillenkugellager *n*
f roulement *m* de butée à double rangée de billes
n dubbelwerkend drukkogellager
p łożysko kulkowe wzdłużne obustronnie obciążalne

1698 double door
d Doppeltür *f*
f porte *f* à battants
n dubbele deur
p drzwi podwójne

1699 double ended spanner
d Doppelmaulschlüssel *m*
f clé *f* plate double; clé *f* à écrous double
n tweezijdige steeksleutel
p klucz płaski dwustronny

1700 double faced hammer
d Zweikopfhammer *m*
f marteau *m* à deux têtes
n hamer met twee koppen; tweekopshamer
p młotek dwustronny

1701 double filament light bulb
d Zweifadenlampe *f*
f lampe *f* à filament double
n dubbele-gloeidraadlamp
p żarówka dwuwłóknowa

1702 double flute cutter
d doppelseitiger Fräser *m*
f fraise *f* droite à deux cannelures
n tweezijdige frees
p frez dwustronny

1703 double handed scraper
d Flachschaber *m*
f racloir *m*
n schraapschaaf
p skrobak płaski

1704 double handle
d Doppelhandgriff *m*
f poignée *f* double
n dubbel handvat
p uchwyt podwójny

1705 double joint swing axle
d Doppelgelenkpendelachse *f*
f pont *m* oscillant à deux pivots
n pendelas met twee draaipunten
p dwuprzegubowy most pędny

1706 double lever brake
d Doppelhebelbremse *f*
f frein *m* à levier double
n dubbele handrem
p podwójny hamulec ręczny

1707 double link
d Doppelglied *n*
f maillon *m* double
n dubbele geleiding
p połączenie podwójne

1708 double manifold collar
d Flansch *m* des Doppelkrümmers
f flasque *m* de tuyau double
n flens voor dubbele uitlaatpijp
p kołnierz rury podwójnej

1709 double open ended spanner
d Doppelgabelschlüssel *m*
f clé *f* à fourche à deux becs
n steeksleutel met dubbele bek
p klucz podwójny widlasty

1710 double piston brake cylinder
d doppeltwirkender Bremsnocken *m*
f cylindre *m* de roue à double effet
n remcilinder met twee plunjers
p rozpieracz hamulcowy dwustronny

1711 double piston engine
d Doppelkolbenmotor *m*
f moteur *m* à double piston
n motor met dubbele zuigers
p silnik dwutłokowy

1712 double plate clutch
d Zweischeibenkupplung *f*
f embrayage *m* bidisque
n dubbelplaatkoppeling
p sprzęgło dwutarczowe

1713 double post lift
d Zweisäulenhebebühne *f*
f élévateur *m* à deux colonnes
n tweekolomshefbrug
p podnośnik dwukolumnowy

1714 double reduction rear axle
d Hinterachse *f* mit zwei Untersetzungen
f pont *m* arrière démultiplicateur
n achteras met dubbele reductie
p tylna oś z podwójnym przełożeniem edukcyjnym

1715 double riveted joint
d zweireihige Nietverbindung *f*
f rivure *f* à double rangée
n dubbel klinknagelverband
p szew nitowy dwurzędowy

1716 double roller chain
d Zweifachrollenkette *f*
f chaîne *f* à rouleaux double
n dubbelkettingrol
p podwójny łańcuch rolkowy

1717 double row bearing
d zweireihiges Lager *n*
f roulement *m* à double rangée
n dubbelrijig lager
p łożysko dwurzędowe

1718 double row cylindrical roller thrust bearing
d zweireihiges Axialzylinderrollenlager *n*
f butée *f* à galets cylindriques à double rangée
n tweerijig rollendruklager
p dwurzędowe łożysko walcowe wzdłużne

1719 double row roller bearing
d zweireihiges Wälzlager *n*
f roulement *m* à galets à double rangée
n dubbelrijig wentellager; dubbelrijig rollager
p dwurzędowe łożysko walcowe

1720 double row tapered roller bearing
d zweireihiges Kegelrollenlager *n*

f roulement *m* à galets coniques à double rangée
n dubbelrijig konisch lager
p dwurzędowe łożysko stożkowe

1721 double seat valve
d Doppelsitzventil *n*
f soupape *f* à double siège
n klep met dubbele zitting
p zawór dwugniazdowy; zawór dwusiedzeniowy

1722 double shackle universal joint; spring shackle
d Federlasche *f*
f jumelle *f* de ressort
n veerophangpunt; veerschommel
p wieszak resoru

1723 double spoon
d Hebeleisen *n* doppelt
f palette *f* double
n dubbele lepel
p łyżka podwójna

1724 double start thread
d zweigangiges Gewinde *n*
f filet *m* à deux pas
n dubbele schroefdraad
p gwint dwuzwojny

1725 double thread
d Doppelgewinde *n*
f double filet *m*
n dubbele schroefdraad
p gwint dwuzacięciowy

1726 double tube frame
d Doppelrohrrahmen *m*
f cadre *m* à double tube
n veerophangpunt
p podwójna rama rurowa

1727 double U butt weld
d Doppel-U-Naht *f*
f soudure *f* frontale à U double
n dubbel-U-naad
p spoina czołowa na podwojne U; spoina wustronna na U; spoina kielichowa dwustronna

1728 double V weld
d X-Naht *f*
f soudure *f* à X
n X-las
p spoina czołowa na X

1729 dovetail bit
d Zinkenfräser *m*

f fraise *f* queue d'aronde
n zwaluwstaartfrees
p frez do wczepów

1730 dovetail joint
d Schwalbenschwanz *m*
f assemblage *m* à queue d'aronde
n zwaluwstaartverbinding
p połączenie na wczepy płetwiaste; wczep płetwiasty

1731 dovetail saw
d Zinkensäge *f*
f scie *f* pour couper les queues d'aronde
n toffelzaag
p piła zasuwnica

1732 dowel pin
d Fixierstift *m*; Passstift *m*; Spannstift *m*
f goujon *m* d'assemblage
n paspen; geleidingspen
p kołek ustalający

1733 downdraft carburetter
d Fallstromvergaser *m*
f carburateur *m* universé
n valstroomcarburateur
p gaźnik dolnossący

1734 downtime
d Totzeit *f*
f temps *m* mort
n stilstand
p czas przestoju; okres niesprawności

1735 downward motion
d Abwartsbewegung *f*
f descente *f*
n neerwaartse beweging
p ruch do dołu

1736 drag
d Formunterteil *m*
f partie *f* de dessous
n ondervorm
p spód formy

1737 drag bar
d Zugstange *f*
f barre *f* de remorquage
n trekstang
p cięgło

1738 drag link
d Lenkzwischenstange *f*
f barre *f* intermédiaire de direction
n stuurstang
p drążek kierowniczy

1739 drain cock
d Ablasshahn *m*
f robinet *m* de vidange
n aftapkraan; afvoerkraan
p korek spustowy

1740 drain pipe
d Ablaufrohr *n*
f tuyau *m* de trop-plein
n overlooppijp
p rura spustowa

1741 drain plug
d Ablassschraube *f*
f bouchon *m* de vidange
n aftapplug
p wkręt spustowy

1742 drain plug joint
d Ablassstopfendichtung *f*
f joint *m* de bouchon de vidange
n aftapplugpakking
p uszczelka korka spustowego

1743 draw bench
d Ziehbank *f*
f banc *m* d'étirage
n trekbank; rekbank
p ciągarka

1744 draw bridge
d Zugbrücke *f*
f pont-levis *m*
n ophaalbrug
p most podnoszony

1745 draw filing; smooth finishing
d Feinbearbeitung *f*
f finissage *m* lisse
n glad afwerken
p wygładzanie

1746 drawing pin
d Reisszwecke *f*
f punaise *f*
n punaise
p pluskiewka kreślarska

1747 drawing press
d Ziehpresse *f*
f presse *f* d'emboutissage
n trekpers
p prasa ciągowa

1748 draw knife
d Schnitzmesser *m*; Ziehmesser *m*; Schälmesser *m*
f plane *f* pour écorchage blanc

n trekmes; boommes
p nóż do przeciągania

1749 draw screw
d Tragöse *f*
f piton *m* d'enlèvement
n uitlichtschroefoog
p uchwyt gwintowany

1750 drift hammer
d Treibfäustel *m*
f masse *f*
n drijfhamer
p młotek do wyklepywania blachy

1751 drifting tool
d Splinteintreiber *m*
f chasse *f* pointe
n pendrijver
p wybijak zawleczek

1752 drill
d Bohrer *m*
f foret *m*
n boor
p wiertło

* **drill bit diamond** → **1537**

1753 drill brace; brace
d Bohrkurbel *f*; Bohrwinde *f*
f vilebrequin *m*
n boorbeugel
p korba do wierteł

1754 drill bush
d Klemmhülse *f*
f douille *f* porte-foret
n boorhuls
p tulejka wiertarska

1755 drill chuck
d Bohrerhalter *m*; Bohrfutter *n*
f porte-foret *m*
n boorhouder
p uchwyt wiertarski

1756 drilled ball; center bar
d Zentrierkugel *f*
f bille *f* de centrage
n centreerkogel
p kula środkująca

* **drilled ball dowel** → **1757**

1757 drilled ball pin; drilled ball dowel
d Zentrierkugelstift *m*

f goujon *m* de bille de centrage
n centreerkogelpen
p sworzeń kulki środkującej

1758 drill hole
d Bohrloch *n*
f trou *m* de forage
n boorgat
p otwór wiercony

1759 drilling jig
d Bohrungslehre *f*
f gabarit *m* d'alésage
n borenplaat
p sprawdzian do otworów

1760 drilling machine
d Bohrmaschine *f*
f machine *f* à percer
n boormachine
p wiertarka

1761 drill stand
d Bohrständer *m*
f support *m* de perçage
n boormachinestandaard
p podstawa do wiertarki ręcznej

1762 drill with right hand twist
d rechtsschneidender Spiralbohrer *m*
f foret *m* à hélice à droite
n rechtssnijdende spiraalboor
p wiertło kręte prawotnące

1763 drip feed
d Tropfölschmierung *f*
f graissage *m* à gouttes; lubrication *f* à gouttes
n druppelsmering
p smarowanie kroplowe; olejenie kroplowe

1764 drip moulding
d Regenrinne *f*
f gouttière *f*
n regengoot; watergoot
p rynna dachowa; listwa ściekowa

1765 drive belt
d Antriebsriemen *m*
f courroie *f* d'entraînement
n aandrijfriem; drijfriem
p pas napędowy

1766 drive end
d Antriebsseite *f*
f côte *m* entraînement
n aandrijfzijde
p kierunek napędu

1767 drive gear
d Abtriebszahnrad *n*
f pignon *m* de renvoi d'arbre primaire
n aangedreven tandwiel
p napędzane koło zębate

1768 drive joint housing
d Antriebsgelenkgehäuse *n*
f boîtier *m* de joint de transmission
n huis van aandrijvingskoppelstuk
p obudowa przegubu napędowego

1769 drive line
d Kraftübertragung *f*
f transmission *f*
n aandrijflijn
p przenoszenie energii

1770 driven yoke
d Abtriebskopf *m*
f tête *f* menée
n aangedreven juk
p jarzmo napędzane

1771 drive pinion bearing
d Lagern für Anlassertrieb
f coussinet *m* de lanceur
n rondsellager
p łożysko mechanizmu napędowego

1772 driver's seat
d Führersitz *m*
f siège *m* de chauffeur
n bestuurdersstoel
p siedzenie kierowcy

* **drive selector** → 3982

1773 drive shaft; cardan shaft
d Antriebswelle *f*; Gelenkwelle *f*
f arbre *m* de commande; arbre *m* de roue
n aandrijfas
p wał pędny; wał napędowy

1774 drive shaft bearing
d Achswellenlager *n*
f roulement *m* d'arbre de roue
n aandrijfaslager
p łożysko wału napędowego

1775 drive shaft puller
(tool to draw drive shaft into hub)
d Abzieher *m* für Antriebswelle
f tire *f* cardan
n aandrijfastrekker
p ściągacz do wału napędowego

1776 drive shaft roll pin kit
d Satz *m* der Sicherungsstifte für Antriebswelle
f ensemble *m* pour goupilles de transmission
n set voor borgpennen van aandrijfas
p zestaw do sworzni wału napędzającego

1777 drive shaft stub
d Antriebswellenstumpf *m*
f bout *m* d'arbre d'entraînement
n uiteinde van aandrijfas
p końcówka wału napędowego

1778 drive shaft tube
d Gelenkwellenrohr *n*
f tube *m* d'arbre de transmission
n buisvormige cardanas
p rura wału przegubowego; wał przegubowy rurowy

1779 drive unit
d Antriebsaggregat *n*
f motor-propulseur *m*
n aandrijfaggregaat
p blok pędny

1780 driving ball
d Übertragungskugel *f*
f bille *f* menant
n drijfkogel; transmissiekogel
p kula pędna

1781 driving chain
d Antriebskette *f*
f chaîne *f* d'entraînement
n aandrijfketting
p łańcuch napędowy

1782 driving direction
d Fahrtrichtung *f*
f direction *f* d'avancement
n rijrichting
p kierunek jazdy

1783 driving distance
d Fahrstrecke *f*
f trajet *m*
n traject
p zasięg jazdy; dystans możliwy do przejechania

1784 driving flange
d Antriebsflansch *m*
f flasque *m* menant
n aandrijfflens
p nasadka napędzajaca

1785 driving gear
d Antriebszahnrad *n*
f pignon *m*
n aandrijvend tandwiel
p pędne koło zębate

1786 **driving instruction**
  d Fahrunterricht *m*
  f leçons *fpl* de conduite
  n rij-instructie; rijles
  p nauka jazdy; lekcja jazdy

1787 **driving instructor**
  d Fahrlehrer *m*
  f moniteur *m* d'auto-école
  n rij-instructeur
  p instruktor jazdy

1788 **driving joint**
  d Antriebsgelenk *n*
  f joint *m* de cardan
  n cardankoppeling
  p przegub napędowy

1789 **driving licence**
  d Führerschein *m*; Fahrerlaubnis *f*
  f permis *m* de conduire
  n rijbewijs
  prawo jazdy

1790 **driving link**
  d Antriebsstange *f*
  f bielle *f* de commande
  n aandrijfstang
  p cięgno pędne

1791 **driving link hinge**
  d Gestängescharnier *n*
  f charnière *f* de bielle
  n scharnier van aandrijfstang
  p wieszak cięgna

1792 **driving member**
  d treibendes Teil *n*
  f organe *m* moteur
  n aandrijfeenheid
  p element napędzający

  * **driving mirror** → 2936

1793 **driving pinion**
  d Antriebszahnrad *n*
  f roue *f* dentée de commande
  n drijfrondsel
  p koło koronowe

1794 **driving pinion flange**
  d Ritzelflansch *m*
  f flasque *m* de pignon d'attaque
  n pignonflens
  p kołnierz zębnika

  * **driving plate** → 90

  * **driving pulley** → 1371

1795 **driving resistance**
  d Fahrwiderstand *m*
  f résistance *f* à l'avancement
  n rijweerstand
  p opór jazdy

1796 **driving sprocket**
  d Antriebskettenrad *n*
  f pignon *m* de commande de chaîne
  n aandrijfkettingwiel
  p napędzające koło zębate napędu
      łańcuchowego

1797 **driving technique**
  d Fahrtechnik *f*
  f technique *f* de conduite
  n rijtechniek
  p technika jazdy

1798 **driving yoke**
  d Antriebskopf *m*
  f tête *f* menant
  n aandrijvende gaffelas
  p głowica pędna

1799 **drop arm frame**
  d Tiefbettrahmen *m*
  f châssis *m* surbaissé
  n chassis met geringe bodemspeling
  p rama niskousytuowana

1800 **drop arm shaft**
  d Lenkstockhebelwelle *f*
  f arbre *m* de levier de direction
  n pitman-as
  p wał główny kierownicy

1801 **drop base rim; well base rim; flat base rim**
  d Tiefbettfelge *f*
  f jante *f* à base creuse
  n diepbedvelg
  p obręcz koła o wgłębionym profilu

1802 **drop hammer**
  d Fallhammer *m*
  f mouton *m* à chute libre
  n valhamer
  p młot spadowy

1803 **drop lever**
  d Lenkspurhebel *m*
  f levier *m* de commande de fusée et de roue
  n spoorstangarm
  p dźwignia pośrednicząca mechanizmu
      zwrotnego

1804 **drop point**
  d Tropfpunkt *m*

f point *m* d'égouttement
n druppelpunt
p temperatura kroplenia; punkt kroplenia

1805 drop valve
d hängendes Ventil *n*
f soupape *f* en tête
n kopklep
p zawór górny

1806 drum brake
d Trommelbremse *f*
f frein *m* à tambour
n trommelrem
p hamulec bębnowy

1807 drum brake mechanism
d Trommelbremsmechanismus *n*
f mécanisme *m* de frein à tambour
n trommelremmechanisme
p bębnowy mechanizm hamulcowy

1808 drum heat exchanger
d Trommelwärmetauscher *m*
f échangeur *m* de chaleur à tambour
n trommelwarmtewisselaar
p bębnowy wymiennik ciepła

1809 drum rim
d Trommelkranz *m*
f encadrement *m* de tambour
n buitenloopvlak van de remtrommel
p wieniec bębna

1810 drunken thread
d vielgängiges Gewinde *n*
f filet *m* à pas multiple
n meergangige winding
p gwint falisty

1811 dry cell
d Trockenelement *n*
f pile *f* sèche
n droog element; droge cel
p ogniwo suche

1812 dry clutch
d Trockenkupplung *f*
f embrayage *m* sec
n droge koppeling
p sprzęgło suche

1813 dry cylinder liner
d trockene Zylinderbüchse *f*
f fourrure *f* sèche
n droge cilindervoering; droge cilinderbus
p sucha tuleja cylindra

1814 dry disc joint
d Trockengelenk *n*
f accouplement *m* élastique
n elastische koppeling
p przegub suchy

1815 dry friction
d trockene Reibung *f*
f frottement *m* sec; friction *f* sèche
n droge wrijving
p tarcie suche

1816 drying box
d Abtrocknungsdose *f*
f capacité *f* de denoyage
n droogkast
p komora suszarnicza

1817 drying chamber
d Trockenkammer *f*
f chambre *f* de séchage
n droogkamer
p komora suszarnicza; suszarnia komorowa

1818 drying time
d Trocknungsdauer *f*
f temps *m* de séchage
n droogtijd
p czas suszenia

1819 dry sump lubrication
d Trockensumpfschmierung *f*
f graissage *m* à carter sec
n dry-sump smering
p smarowanie przy suchej komorze korbowej

1820 dual bearing
d Doppellager *n*
f palier *m* double
n tandemlager
p łożysko podwójne

1821 dual bed catalytic converter
d Monolithkatalysator *m*
f catalyseur *m* à monolithe
n monolytische katalysator
p katalizator monolityczny

* dual carriage way → 5230

1822 dual controls
d Doppelbetätigung *f*
f organe *m* à commandes doubles
n dubbel bedieningsorgaan
p zdublowany układ sterowania pojazdu

1823 dual ignition
d Doppelzündung *f*

f allumage *m* double
n ontsteking met twee bougies per cilinder
p zapłon podwójny

1824 dual rear axle
d Tandemhinterachse *f*
f double pont *m* arrière
n tandemachteras
p tylny most z przekładnią dwustopniową

1825 dual seat
d Doppelsitz *m*
f siège *m* à deux sièges
n tweepersoonszitting
p siedzenie dwuosobowe

1826 dual speed wiper
d Scheibenwischer *m* mit zwei
    Geschwindigkeiten
f essuie-glace *m* à deux vitesses
n ruitenwisser met twee snelheden
p wycieraczka z podwójną szybkością

1827 dual stage compressor
d zweistufiger Kompressor *m*
f compresseur *m* à deux etages
n tweetrapscompressor
p sprężarka dwustopniową

* **duct → 1220**

1828 dumb iron
d Federhand *f*
f main *f* de ressort
n veerhand
p przednia podpora resoru

1829 dummy injector
d Zerstäuber *m*
f faux injecteur *m*
n verstuiver
p rozpylacz

1830 dummy pinion
d Einstellritzel *n*
f pignon *m* de réglage
n loos rondsel
p poprawka zębnika

1831 dumper
d Autoschüttlcr *m*
f dumper *m*
n dumper
p dumper

1832 dump truck
d Kippwagen *m*

f camion *m* à benne basculante
n kiepwagen
p wywrotka

1833 duplex brake
d Duplexbremse *f*
f frein *m* duplex
n Duplex-rem
p hamulec o przeciwległych łożyskach szczęk;
    hamulec Duplex

1834 duplex wheel cylinder
d Duplexradzylinder *m*
f cylindre *m* de frein de roue duplex
n tweevoudige wielcilinder
p mechanizm hamulcowy dwucylindrowy

1835 durability
d Haltbarkeit *f*
f durabilité *f*
n levensduur
p trwałość

1836 duralumin
d Duraluminium *n*
f duralumin *m*
n duraluminium
p duraluminium; dural

1837 dust
d Staub *m*
f poussière *f*
n stof
p pył; kurz

1838 dust bag
d Staubsack *m*
f sac *m* de poussière
n stofzak
p worek pyłowy

1839 dust cap
d Staubkappe *f*
f couvercle *m* pare-poussière
n stofkap
p pierścień ochronny

1840 dust catcher
d Staubfänger *m*
f caisse *f* à poussières
n stofvanger
p odpylnik; odpylacz

1841 dust cover
d Staubkappe *f*
f couvercle *m* protecteur de la poussière
n stofdeksel; stofkap
p osłona przeciwpyłowa

1842 dust filter
  d Staubfilter *m*
  f filtre *m* à poussière
  n stoffilter
  p filtr przeciwpyłowy

1843 dusting brush
  d Staubbürste *f*
  f brosse *f* à épousseter
  n stofborstel
  p pędzel do okurzania

1844 dust mask
  d Staubmaske *f*
  f masque *m* à poussière
  n stofmasker
  p maska przeciwpyłowa

1845 dust separator
  d Staubabscheider *m*
  f séparateur *m* de poussière
  n stofafscheider
  p separator pyłu; oddzielacz pyłu

1846 dynamic balancing
  d dynamisches Gleichgewicht *n*
  f équilibre *m* dynamique
  n dynamisch uitbalanceren
  p wyważanie dynamiczne

1847 dynamic brake force distribution
  d dynamische Bremskraftregelung *f*
  f distribution *f* du freinage dynamique
  n dynamische remkrachtverdeling
  p dynamiczne rozłożenie siły hamowania

1848 dynamic characteristic
  d dynamische Charakteristik *f*
  f caractéristique *f* dynamique
  n dynamische karakteristiek
  p charakterystyka dynamiczna

1849 dynamic driving conditions
  d Fahrdynamik *f*
  f conditions *fpl* dynamiques de la conduite
  n dynamische rij-omstandigheden
  p dynamika ruchu

1850 dynamic pressure
  d Staudruck *m*
  f pression *f* dynamique
  n dynamische druk
  p ciśnienie dynamiczne; ciśnienie kinetyczne

1851 dynamic stability
  d dynamische Stabilität *f*
  f stabilité *f* dynamique
  n dynamische stabiliteit
  p stateczność dynamiczna

1852 dynamic viscosity
  d dynamische Viskosität *f*
  f viscosité *f* dynamique
  n dynamische viscositeit
  p lepkość dynamiczna

* dynamo → 2485

1853 dynamo governor
  d Lichtmaschinenregler *m*
  f régulateur *m* de dynamo
  n dynamoregelaar
  p regulator prądnicy

1854 dynamo pulley
  d Lichtmaschinenscheibe *f*
  f poulie *f* de dynamo
  n dynamopoelie; dynamoriemschijf
  p koło pasowe prądnicy

1855 dynamo tester
  d Lichtmaschinenprüfgerät *n*
  f appareil *m* d'essai de dynamo
  n dynamotester
  p wskaźnik pracy prądnicy

E

1856 Earles fork
d Schwingarmgabel *f*
f fourche *f* Earles
n Earles-vork
p widłowe jarzmo zawieszenia wahadłowego

1857 ear muffs
d Ohrenschützer *mpl*
f protèges-oreilles *mpl*
n oorbeschermers
p osłona ucha

1858 earth electrode; ground electrode
d Massaelektrode *f*
f électrode *f* de masse
n massaelektrode
p elektroda masowa

1859 earthing
d Erdung *f*
f mise *f* à la masse
n aarding
p uziemienie

1860 earthing rod
d Erdungsstange *f*
f barre *f* de masse
n aardingspen
p pręt uziemiający

1861 earth moving tyre
d Erdbewegungsreifen *m*
f pneu *m* de génie civil
n band voor grondverzetmachine
p opona do poruszania się po ziemi

1862 earth strap
d Masseband *n*
f bande *f* de connexion à la masse
n massastrip
p taśmowe łącze z masą

1863 earth strip
d Massekabel *n*
f câble *m* de masse
n aardingskabel; aardkabel; aardleiding
p kabel połączeniowy do masy; przewód na masę

1864 earth terminal
d Masseklemme *f*; Erdungsklemme *f*
f cosse *f* de mise à la masse
n aardklem; massaklem
p zacisk uziomowy; zacisk uziemiający

1865 ebonite nut
d Ebonitmutter *f*
f ecrou *m* d'ébonite
n ebonietmoer
p nakrętka ebonitowa

1866 ebullient cooling
d Verdampfungskühlung *f*
f refroidissement *m* par ébullition
n verdampingskoeling
p chłodzenie wyparne

1867 eccentric
d Exzenter *m*
f excentrique *m*
n excentriek
p mimośrodowy

1868 eccentric base rim
d Halbflachbettfelge *f*
f jante *f* à base excentrée
n halfdiepbedvelg
p pół-obręcz z płaską felgą

1869 eccentric bushing
d exzentrische Lagerbüchse *f*
f coussinet *m* excentrique
n excentrische lagerbus
p tuleja łożyskowa niedzielona; tuleja mimośrodowa

1870 eccentric chuck
d exzentrisches Spannfutter *n*
f mandrin *m* excentrique
n excentrische boorhouder
p mimośrodowy uchwyt zaciskowy

1871 eccentric pin
d Exzenterbolzen *m*
f axe *m* excentrique
n excentrische bout
p śruba mimośrodowa

1872 eccentric shaft
d Exzenterwelle *f*
f arbre *m* à excentrique
n excentriekas
p wał mimośrodowy

1873 edge filter
d Spaltfilter *m*
f filtre *m* à lamelles
n spleetfilter
p filtr szczelinowy

1874 edge joint
  d Eckverband *n*
  f assemblage *m* à chanfreins
  n schuine las
  p złącze kątowe

1875 edge of a file
  d Feilenseite *f*
  f bord *m* de lime
  n kant van een vijl
  p krawędź pilnika

1876 edge tool
  d Schneidwerkzeug *n*
  f outil *m* tranchant
  n snijgereedschap
  p narzędzie skrawające; wykrojnik

1877 edging clamp
  d Kantenzwinge *f*
  f serre-joint *m* à coins
  n kantklem
  p ścisk stolarski

1878 edging machine
  d Bördelmaschine *f*
  f machine *f* à border
  n pijpomfelsmachine
  p wyrównywarka

1879 edging saw
  d Besäumsäge *f*
  f scie *f* à rogner
  n randzaag
  p obrzynarka wzdłużna pojedyncza

1880 edging tool
  d Bördeleisen *n*
  f bordoir *m*
  n felsijzer
  p zaginadło; zawijak

1881 effective horsepower
  d Nutzleistung *f*; effektive Leistung *f*
  f puissance *f* effective
  n effectief vermogen
  p moc efektywna; moc użyteczna

1882 effective spring length
  d Federnutzlänge *f*
  f longueur *f* utile de ressort
  n effectieve veerlengte
  p efektywna długość sprężyny

1883 efficiency
  d Wirkungsgrad *m*
  f rendement *m*

  n doelmatigheid; rendement
  p efektywność; skuteczność; wydajność

1884 efficiency test
  d Leistungsfähigkeitsprobe *f*
  f essai *m* d'efficacité
  n doelmatigheidstest
  p próba sprawności

1885 ejection spring
  d Ausstossfeder *f*
  f ressort *m* d'éjection
  n uitstootdrukveer
  p sprężyna wyrzutnika

* elastic insert → 4243

1886 elastic limit
  d Elastizitätsgrenze *f*
  f limite *f* d'élasticité
  n elasticiteitsgrens
  p granica sprężystości

1887 elastic spacer
  d elastische Abstandseinlage *f*
  f entretoise *f* élastique
  n elastisch afstandsstuk; elastisch opvulstuk
  p wkładka elastyczna

1888 elbow
  d Knierohr *n*; Knie *n*
  f tuyau *m* en coude
  n elleboog
  p łącznik kolankowy; rura kolankowa

1889 elbow union
  d gebogener Gummiübergang *m*
  f raccord *m* coudé
  n kniestuk; gebogen stuk
  p połączenie kolankowe

1890 electric circuit
  d Stromkreis *m*
  f circuit *m* électrique
  n stroomkring
  p obwód elektryczny

1891 electric drive
  d elektrischer Antrieb *m*
  f commande *f* électrique
  n elektrische aandrijving
  p napęd elektryczny

1892 electric driven fan
  d Elektroventilator *m*
  f ventilateur *m* à moteur électrique
  n elektrische ventilator
  p wentylator o napędzie elektrycznym

1893 electric hand drilling machine
  d Elektrohandbohrmaschine *f*
  f perceuse *f* électrique à main
  n elektrische handboormachine
  p wiertarka elektryczna ręczna

1894 electric heating
  d elektrische Heizung *f*
  f chauffage *m* électrique
  n elektrische verwarming
  p ogrzewanie elektryczne

1895 electric hoist
  d Elektrohebezeug *m*; Elektrozug *m*
  f palan *m* électrique
  n elektrisch hefwerktuig
  p wciągnik elektryczny

1896 electrician's knife
  d Kabelmesser *m*
  f couteau *m* d'électricien
  n kabelmes
  p nóż do kabli

1897 electrician tool set
  d Elektrikerwerkzeugkasten *m*
  f coffret *m* "électricien"
  n gereedschapskist voor de elektricien
  p zestaw narzędzi elektryka

1898 electric immersion heater
  d Tauchsieder *m*
  f thermoplongeur *m*
  n dompelelement
  p grzałka nurkowa

1899 electric motor
  d Elektromotor *m*
  f moteur *m* électrique
  n elektrische motor
  p silnik elektryczny

1900 electric power
  d elektrische Leistung *f*
  f puissance *f* électrique
  n elektrisch vermogen
  p moc elektryczna

1901 electric pump
  d elektrische Pumpe *f*
  f pompe *f* électrique
  n elektrische pomp
  p pompa elektryczna

1902 electric resistance
  d elektrischer Widerstand *m*
  f résistance *f* électrique

  n elektrische weerstand
  p opór elektryczny

1903 electric retarder
  d elektrischer Verlangsamer *m*
  f ralentisseur *m* électrodynamique
  n wervelstroomvertrager
  p zwalniacz elektromagnetyczny

1904 electric shock
  d elektrischer Ruck *m*
  f choc *m* de tension
  n elektrische schok
  p wstrząs elektryczny

1905 electric valve
  d Elektroventil *n*
  f électrovanne *f*
  n elektrische klep
  p zawór elektromagnetyczny

1906 electric welding
  d elektrische Schweissung *f*
  f soudage *m* électrique
  n elektrisch lassen
  p spawanie elektryczne

1907 electrode
  d Elektrode *f*
  f électrode *f*
  n elektrode
  p elektroda

1908 electrode gap; spark gap
  d Elektrodenabstand *m*
  f écartement *m* d'électrodes
  n elektrodenafstand
  p odstęp elektrod

1909 electrode holder
  d Elektrodenhaltern *n*
  f porte-électrode *m*
  n elektrodehouder
  p uchwyt elektrod

1910 electrolyte
  d Elektrolyt *m*
  f électrolyte *m*
  n elektroliet
  p elektrolit

1911 electrolyte level
  d Elektrolytstand *m*
  f niveau *m* d'électrolyte
  n elektrolietniveau
  p poziom elektrolitu

1912 electromagnetic door locking
  d Elektrozentralverriegelung *f*

f condamnation *f* électromagnétique
n elektromagnetische portiervergrendeling
p elektromagnetyczna blokada drzwi

1913 electromagnetic driven fan
d Elektromagnetlüfter *m*
f ventilateur *m* à embrayage électromagnétique
n door elektromagneet ingeschakelde ventilateur
p wentylator o sprzęgle elektromagnetycznym

1914 electromagnetic hammer
d elektromagnetischer Hammer *m*
f marteau *m* électromagnétique
n elektromagnetische hamer
p młot elektromagnetyczny

1915 electromagnetic injector valve
d elektromagnetisches Einspritzventil *n*
f injecteur *m* électromagnétique
n elektromagnetische injectieklep
p elektromagnetyczny wtryskiwacz roboczy

1916 electromagnetic relay
d elektromagnetisches Relais *n*
f relais *m* électromagnétique
n elektromagnetisch relais
p przekaźnik elektromagnetyczny

1917 electromotive force
d elektromotorische Kraft *f*
f force *f* électromotrice
n elektromotorische kracht
p siła elektromotoryczna

1918 electronical equipment
d elektronische Ausrüstung *f*
f appareillage *m* électronique
n elektronische uitrusting
p wyposażenie elektroniczne

1919 electronically concentrated air suspension
d elektronisches Luftfedersystem *n*
f suspension *f* électronique
n elektronisch geregeld luchtveersysteem
p zawieszenie regulowane elektronicznie

1920 electronic balancing
d elektronisches Auswuchten *n*
f équilibrage *m* électronique
n elektronisch uitbalanceren
p wyważać elektronicznie

1921 electronic comparator
d elektronischer Komparator *m*
f comparateur *m* électronique
n elektronische pulsomvormer
p komparator elektroniczny

1922 electronic controlled key cutting machine
(electronic system suitable to the key cutting
requirements of the automobile and
manufacturing industry)
d elektronisch gesteuerte Schlüsselfräsmaschine
f
f fraiseuse *f* à commande électronique pour la
taille de clés
n elektronisch gestuurde sleutelfreesmachine
p frezarka do kluczy elektronicznie sterowana

1923 electronic control module
d elektronisches Steuergerät *n*
f bloc *m* de commande électrique
n elektronische regeleenheid
p blok sterowania elektronicznego

1924 electronic ignition
d elektronische Zündung *f*
f allumage *m* électronique
n elektronische ontsteking
p zapłon elektroniczny

1925 electronic metering
d elektronische Dosierung *f*
f dosage *m* électronique
n elektronische dosering
p dozowanie elektroniczne

1926 electronic oil
d Elektroniköl *n*
f huile *f* pour système électronique
n olie voor elektronisch systeem
p olej do systemu elektronicznego

1927 electronic pliers
d Elektronikzange *f*
f pince *f* pour l'électronique
n elektronicatang
p szczypce elektroniczne

1928 electronics grease
d Elektronikfett *n*
f graisse *f* électronique
n elektronicavet
p smar do systemu elektronicznego

1929 electroplating
d Galvanisierung *f*
f galvanoplastie *f*
n galvanisering
p galwanizacja

1930 electropneumatic valve
d elektropneumatisches Ventil *n*
f soupape *f* électropneumatique
n elektropneumatische klep
p zawór elektropneumatyczny

1931 electrostatic generator
d Influenzmaschine f
f machine f à induction
n elektrostatische generator
p generator elektrostatyczny

1932 elevated frame
d Hochrahmen m
f châssis m élevé
n verhoogd skelet
p rama górna; rama nadwozia

1933 elevating screw
d Richtschraube f
f vis f de pointage en hauteur
n richtschroef
p śruba korekcyjna

1934 elimination
d Beseitigung f
f élimination f
n eliminering
p usunięcie

1935 Elliot type axle
d Gabelachse f
f essieu m à chape
n vorkas
p oś rozwidlona

1936 elliptic spring
d Ellipsenfeder f
f ressort m à pincette
n elliptische veer
p resor epileptyczny

1937 elongation
d Dehnung f
f allongement m
n verlenging
p wydłużenie

1938 embellisher cap
d Abschlussstopfen m
f bouchon m de fixation
n afsluitstop
p korek zamykający

1939 embossing
d Prägen n
f estampage m
n persen
p wytłaczanie

1940 embossing hammer
d Ausbeulhammer m
f marteau m à emboutir
n uitdeukhamer
p młotek do wypukiwania

1941 emergency brake
d Notbremse f; Notbremsanlage f
f frein m de secours
n noodrem
p hamulec awaryjny

1942 emergency key
d Notschlüssel m
f clé f d'urgence
n noodsleutel
p klucz awaryjny

1943 emergency repair
d Behelfsreparatur f
f réparation f improvisée
n noodreparatie
p naprawa doraźna

1944 emergency tool set
d Reparaturwerkzeugkasten m
f coffret m "dépannage"
n gereedschapskist voor noodreparatie
p zestaw narzędzi na wypadek awarii; zestaw
   naprawczy

1945 emery clotch
d Schmirgelleinwand f
f toile f d'émeri
n schuurlinnen; polijstlinnen
p płótno szmerglowe; płótno ścierne

1946 emery paper
d Schmirgelpapier n
f papier m émerisé
n schuurpapier; polijstpapier
p papier ścierny; papier szmerglowy

1947 emery stone
d Schmirgeleisen n
f meule f en émeri
n polijststeen; amarilsteen
p osełka szmerglowa; pilnik szmerglowy

1948 emery wheel
d Polierscheibe f
f meule-émeri m
n amarilschijf; schuurschijf
p ściernica szmerglowa

1949 emery wheel dresser
d Schleifscheibenreiniger m
f décrasse-meules f
n amarilschijf
p oczyszczalnik ściernicy

1950 emulsion pipe
d Mischrohr n
f tube m d'émulsion

n mengbuis; emulsiebuis
p rurka emulsyjna

1951 emulsion tube
d Mischrohr *n*
f tube *m* d'émulsion
n emulsiebuis; mengbuis
p rurka emulsyjna

1952 enamelled round tube saw with tightener
d Säge *f* mit Stahlrohrbügel, emailliert mit Spanner
f scie *f* à tube rond émaillé à tendeur
n geëmailleerde ronde buiszaag met spanelement
p piła z pałąkiem z rury stalowej

1953 enamelled wire
d Lackdraht *m*
f fil *m* émaillé
n emaildraad
p przewód emaliowany

1954 end block
d Endmass *n*
f cale-étalon *m*
n eindmaat
p płytka wzorcowa

* **end cutter** → 1957

1955 end dump body; end tipper
d Hinterkipper *m*
f benne *f* basculante vers l'arrière
n alleen achterwaarts kiepende wagen
p wywrotka wsteczna

1956 end journal
d Stirnzapfen *m*
f tourillon *m* fini
n eindtap
p czop główny

1957 end milling cutter; end cutter
d Fingerfräser *m*
f fraise *f* en bout
n vingerfrees; kolffrees
p frez palcowy

1958 end of interdiction
d Ende *n* des Verbots
f fin *m* d'interdiction
n einde van het verbod
p koniec zakazu

1959 end outline
d Vorderansicht *f*

f encombrement *m* frontal; encombrement *m* en vue avant
n vooraanzicht
p obrys poprzeczny

1960 end play
d Endspiel *n*
f jeu *m* en bout
n axiale speling
p luz osiowy

1961 end shield
d Lagerschild *n*
f flasque *m*
n lagerschild
p tarcza łożyskowa

1962 end thrust
d Axialdruck *m*
f poussée *f* longitudinale
n axiaaldruk
p nacisk osiowy

* **end tipper** → 1955

* **endurance test** → 2107

1963 engagement
d Einrücken *n*
f connexion *f* mécanique
n koppelen
p wprzęganie; włączanie sprzęgła

1964 engine
d Motor *m*
f moteur *m*
n motor
p silnik

1965 engine block
d Motorblock *m*
f bloc-cylindres *m*
n motorblok
p kadłub silnika

1966 engine bolt
d Motorbolzen *m*
f boulon *m* de fixation du moteur
n motorophangbout
p śruba zawieszenia silnika

1967 engine bracket
d Motorstütze *f*
f support *m* moteur
n motorsteun
p wspornik silnika

1968 **engine brake**
d Motorbremse *f*
f frein-moteur *m*
n motorrem
p hamulec silnikowy

1969 **engine break down; engine failure**
d Motordefekt *m*; Motorpanne *f*
f défaut *m* de moteur; panne *f* de moteur
n motorstoring
p defekt silnika

1970 **engine chamber lighting**
d Motorraumbeleuchtung *f*
f éclairage *m* de compartiment moteur
n motorruimteverlichting
p oświetlenie przegrody silnika

1971 **engine coolant level**
d Motorkühlmittelniveau *n*
f hauteur *f* du liquide de refroidissement du
   moteur
n motorkoelvloeistofniveau
p poziom cieczy chłodzącej silnika

1972 **engine coolant temperature gauge**
d Motorkühlmitteltemperaturmesser *m*
f indicateur *m* de la température de liquide de
   refroidissement du moteur
n motorkoelvloeistoftemperatuurmeter
p wskaźnik temperatury cieczy chłodzącej
   silnika

1973 **engine cranking speed**
d Anlassdrehzahl *f*
f régime *m* démarrage
n starttoerental
p minimalna prędkość obrotowa przy rozruchu

1974 **engine cubic capacity**
d Hubrauminhalt *m* des Motors
f capacité *f* cylindrée du moteur
n cilinderinhoud van de motor
p pojemność skokowa silnika

1975 **engine data**
d Motorkenndaten *npl*
f spécifications *fpl* du moteur
n motorspecificaties
p parametry silnika

1976 **engine diagnosis**
d Motordiagnose *f*
f diagnostic *m* de moteur
n motordiagnose
p diagnostyka silnika

1977 **engine diagnostic plug**
d Motordiagnosestecker *m*

f fiche *f* diagnostic de moteur
n motordiagnosestekker
p gniazdo diagnostyczne silnika

1978 **engine drain plug spanner**
d Schlüssel *m* für Ablassstopfen des Motors
f clé *f* pour bouchon de vidange moteur
n sleutel voor olieaftapstop
p klucz do korka spustowego silnika

1979 **engine dry weight**
d Masse *f* des trockenen Motors
f masse *f* de moteur sec
n motorgewicht zonder olie en koelvloeistof
p masa suchego silnika

1980 **engine dynamometer**
d Motorprüfstand *m*
f banc *m* d'essai moteur
n motortestbank
p hamownia silnika; stanowisko badawcze
   silnika; stanowisko pomiarowe silnika

1981 **engineer's file**
d Werkstattfeile *f*
f lime *f* d'atelier
n bankwerkersvijl
p pilnik ślusarski

* **engine failure** → 1969

1982 **engine flush**
d Motorspülung *f*
f rinçage *m* du moteur
n motorspoeling
p przepłukiwanie silnika

1983 **engine guard plate**
d Motorschutzblech *n*
f tôle *f* de protection de moteur
n carterbeschermingsplaat
p płyta ochronna silnika

1984 **engine inlet port; inlet port**
d Einströmöffnung *f*
f orifice *m* d'admission
n inlaatpoort
p szczelina wlotowa; okno wlotowe

1985 **engine load**
d Höchstbelastung *f*
f charge *f* de pointe
n maximaal vermogen
p obciążenie maksymalne

1986 **engine mounting; engine suspension**
d Motoraufhängung *f*

f   suspension f de moteur
n   motorophanging
p   zawieszenie silnika

1987 engine number
d   Motornummer f
f   numéro m du moteur
n   motornummer
p   numer silnika

1988 engine oil
d   Motoröl n
f   huile f moteur
n   motorolie
p   olej silnikowy

1989 engine oil filter
d   Motorölfilter m
f   filtre m à huile de moteur
n   motoroliefilter
p   filtr oleju silnikowego

1990 engine oil gauge
d   Motorenölniveau n
f   niveau m d'huile moteur
n   motoroliepeil
p   poziom oleju silnikowego

1991 engine oil temperature
d   Motoröltemperatur f
f   température f d'huile moteur
n   motorolietemperatuur
p   temperatura oleju silnikowego

1992 engine outline
d   Motorprofil n
f   contour m de moteur
n   motorprofiel
p   obrys silnika

1993 engine overall height
d   Motorhöhe f
f   hauteur f de moteur
n   totale motorhoogte
p   wysokość silnika kompletna

1994 engine overall width
d   Motorbreite f
f   largeur f de moteur
n   totale motorbreedte
p   szerokość silnika kompletnego

1995 engine plate
d   Motorschild n
f   plaque f de moteur
n   motorplaat
p   płyta silnika

1996 engine power
d   Motorleistung f
f   puissance f du moteur
n   motorvermogen
p   moc silnika

1997 engine reconditioning; engine renovation
d   Motorüberholung f
f   révision f du moteur
n   motorrevisie
p   remont silnika; naprawa silnika

*   **engine renovation → 1997**

1998 engine retarder
d   Motorverlangsamer m
f   ralentisseur m sur moteur
n   motorkrachtvertrager
p   zwalniacz silnika

1999 engine revolution frequency
d   Motorrotationsfrequenz f
f   fréquence f du moteur
n   motorfrequentie
p   częstotliwość obrotów silnika

2000 engine shaft
d   Motorwelle f
f   arbre m du moteur
n   uitgaande as van motor
p   wał silnika

2001 engine side half coupling
d   verstellbare Kupplungshälfte f
f   flasque m sur arbre de commande
n   instelbare koppelingshelft
p   nasadka nastawna

2002 engine simulator
d   Motorensimulator m
f   simulateur m de moteur
n   motorsimulator
p   symulator silnika

2003 engine slap
d   Klopfen n im Motor
f   détonation f dans le moteur
n   detonatie van motor
p   stukanie w silniku

2004 engine speed pick up
d   Motordrehzahlsensor m
f   capteur m de régime du moteur
n   motortoerentalsensor
p   sensor prędkości obrotowej silnika

*   **engine suspension → 1986**

2005 engine suspension sandwich mounting
d Gummikörper *m* der Motoraufhängung
f tampon *m* élastique de suspension moteur
n silentbloc-motorophanging
p poduszka zawieszenia silnika

2006 engine timing gear diagram
d Steuerdiagramm *n*
f diagramme *m* de distribution
n kleppendiagram bij een motor
p wykres rozrządu silnika

2007 engine torque
d Motordrehmoment *n*
f couple *m* moteur
n motordraaimoment
p moment obrotowy silnika

2008 engine tuning
d Motoreinstellung *f*
f mise *f* au point moteur
n motorafstelling
p ustawianie silnika

2009 engine underframe
d Motorhilfsrahmen *m*
f berceau *m* moteur
n subframe voor motor
p rama podsilnikowa

2010 engine unit
d Motoraggregat *n*
f bloc-moteur *m*
n motor
p zespół silnika

2011 engine wear
d Motorverschleiss *m*
f usure *f* du moteur
n motorslijtage
p zużycie silnika

2012 engine working space
d Motorarbeitsraum *m*
f volume *m* total
n werkruimte om motor
p przestrzeń robocza silnika

2013 entering tap; taper tap
d Vorschneider *m*
f taraud *m* ébaucheur
n voorsnijder; conische tap
p gwintownik wstępny

2014 environment protection
d Umweltschutz *m*
f protection *f* de l'environnement
n milieubescherming
p ochrona środowiska

2015 epicyclic third differential; planetary interaxial differential
d Mittelachsplanetenausgleichgetriebe *n*
f différentiel *m* interponts à train épicycloïdal
n planetair differentieel van voor- en achteras
p międzyosiowy mechanizm różnicowy; międzyosiowy mechanizm planetarny

2016 equilibrant force
d Gleichgewichtskraft *f*
f force *f* équilibrée
n evenwichtbiedende kracht
p siła równoważąca

2017 error in reading
d Ablesefehler *m*
f erreur *f* de lecture
n afleesfout
p błąd odczytania

2018 ethyl alcohol
d Ethylalkohol *m*
f alcool *m* éthylique
n ethanol
p etanol; alkohol etylowy

2019 ethyl gasoline; leaded petrol
d Bleibenzin *n*
f carburant *m* avec plomb
n loodhoudende benzine
p benzyna etylizowana; etylina

2020 evaporator
d Verdampfer *m*
f évaporateur *m*
n verdamper
p odparowywacz

2021 evaporator sensor
d Verdampfsensor *m*
f sensor *m* de vaporisation
n verdampersensor
p czujnik wyparowywania

2022 Ewart chain
d Rotarykette *f*
f chaîne *f* à maillons courbés
n ewart-ketting
p łańcuch obrotowy

2023 excess
d Überschuss *m*
f excès *m*; dépassement *m*
n overmaat; overschot
p nadmiar

2024 excess oil space
d Ölausgleichraum *m*

f chambre *f* de compensation
n oliecompensatieruimte
p przestrzeń wyrównawcza

* **excess pressure** → 3636

2025 exchange blades
d Ersatzklingen *fpl*
f lames *fpl* de recharge
n reservemessen
p brzeszczoty zapasowe

2026 exchange engine
d Austauschmotor *m*
f moteur *m* d'échange
n ruilmotor
p silnik wymienny

2027 excitation winding
d Erregerspule *f*
f bobinage *m* d'excitation
n bekrachtigingsspoel
p cewka uzwojenia wzbudzającego

2028 exciter diode
d Erregerdiode *f*
f diode *f* d'excitation
n velddiode
p dioda wzbudzająca

2029 exhaust
d Auspuff *m*
f échappement *m*
n uitlaat
p wydech; wylot

2030 exhaust baffle
d Auspuffblende *f*
f rondelle *f* de l'échappement
n keerschot van uitlaat
p przesłona wydechowa

2031 exhaust brake
d Motorstaudruckbremse *f*
f frein *m* sur échappement
n uitlaatrem
p hamulec silnikowy

2032 exhaust cam
d Auslassnocken *m*
f came *f* d'échappement
n uitlaatnok
p krzywka wylotu

2033 exhaust clip
d Schelle *f*
f collier *m* de tuyau

n uitlaatklem
p jarzmo rury wydechowej

2034 exhaust emission control
d Auspuffgasemissionsregelung *f*
f régulation *f* de gaz d'échappement
n uitlaatgasemissieregeling
p regulacja emisji gazów wylotowych

2035 exhaust gas
d Auspuffgas *n*
f gaz *m* d'échappement
n uitlaatgas
p gaz wydechowy; spaliny

2036 exhaust gas analyser
d Auspuffgasanalysator *m*
f analyseur *m* des gaz d'échappement
n uitlaatgastesterapparaat
p analizator spalin

2037 exhaust gas pressure
d Auspuffgasdruck *m*
f pression *f* des gaz d'échappement
n uitlaatgasdruk
p ciśnienie spalin

2038 exhaust gas purifier
d Abgasentgifter *m*
f purificateur *m* des gaz d'échappement
n uitlaatgasreiniger
p odtruwacz spalin

2039 exhaust hose
d Abgasschlauch *m*
f tube *m* flexible d'échappement
n uitlaatgasafvoerslang
p wąż gazow wydechowych

2040 exhaust line
d Rücklaufleitung *f*
f ligne *f* vidangée
n terugstroomleiding
p przewód powrotny

2041 exhaust manifold joint
d Verbindungsglied *n* der Auspuffrohre
f raccord *m* de tuyaux d'échappement
n verbindingsmof voor uitlaatpijpen
p złącze rur wylotowych

2042 exhaust opacity
d Abgastrübung *f*
f opacité *f* des gaz d'échappement
n ondoorzichtigheid van uitlaatgassen
p zmętnienie gazów wydechowych

2043 exhaust passage
 d Auslasskanal *m*
 f conduit *m* d'échappement
 n uitlaatkanaal
 p kanał wylotowy

2044 exhaust pipe
 d Auspuffrohr *n*; Auspuffleitung *f*
 f tube *m* d'échappement; tuyau *m* d'échappement
 n uitlaatpijp
 p rura wylotowa; rura wydechowa

2045 exhaust pipe flange
 d Auspuffrohrflansch *m*
 f bride *f* de tuyau d'échappement
 n uitlaatpijpflens
 p kołnierz rury wydechowej; kołnierz rury
   wylotowej

2046 exhaust port
 d Auspuffschlitz *m*
 f lumière *f* d'échappement
 n uitlaatpoort
 p szczelina wydechowa

2047 exhaust rubber stretcher
   (tool to refit rubber mountings after replacing
   on exhaust system)
 d Einhängewerkzeug *n*
 f tires *fpl* anneaux d'échappement
 n trekhaak voor uitlaatrubbers
 p hak do wyciągania zamocowanych gum
   systemu wylotowego

2048 exhaust silencer intermediate pipe
 d Schalldämpferverbindungsrohr *n*
 f tuyau *m* de liaison des silencieux
 n uitlaatdemperverbindingspijp
 p rura łącząca tłumiki

2049 exhaust stroke
 d Auspuffhub *m*
 f temps *m* d'échappement
 n uitlaatslag
 p suw wydechu

2050 exhaust system
 d Auspuffanlage *f*
 f système *m* d'échappement
 n uitlaatsysteem
 p system wylotowy; system wydechowy

2051 exhaust valve
 d Auspuffventil *n*
 f soupape *f* d'échappement
 n uitlaatklep
 p zawór wydechowy

2052 exhaust valve closing period
 d Schliessungsdauer *m* des Auspuffventils
 f fermeture *f* de soupape d'échappement
 n sluitingsduur van uitlaatklep
 p zamknięcie zaworu wydechowego

2053 exhaust valve head diameter
 d Durchmesser *m* des Auslassventiltellers
 f diamètre *m* de tête de soupape d'échappement
 n doorsnede van de uitlaatklepschotel
 p średnica grzybka zaworu wylotowego

2054 exhaust valve opening period delay
 d Verzögerung *f* bei der Öffnung des
   Auspuffventils
 f retard *m* d'ouverture de la soupape
   d'échappement
 n vertraging bij het openen van de uitlaatklep
 p opóźnienie otwarcia zaworu wydechowego

2055 exhaust valve tappet clearance
 d Ventilspiel *n* zwischen Auslassventil und
   Hebel
 f jeu *m* de soupape d'échappement
 n speling van de uitlaatklep
 p luz zaworu wylotowego

2056 exit
 d Autobahnausfahrt *f*
 f fin *m* d'autoroute
 n afrit
 p zjazd z autostrady

2057 expander cam
 d Spannocken *m*
 f came *f* d'expansion
 n uitzetnok
 p krzywka rozpierająca

2058 expander lever
 d Spannvorrichtungshebel *m*
 f levier *m* d'expandeur
 n remlever
 p dźwignia rozpieracza

2059 expander two lobe cam
 d Bremsdaumen *m*
 f came *f* d'expandeur
 n remnok
 p krzywka rozpieracza

2060 expanding bit
 d Bohrer *m* mit verstellbarem Durchmesser;
   Stellbohrer *m*
 f foret *m* à épaisseur réglable
 n boor met verstelbare diameter
 p wiertło z nastawną średnicą

2061 **expanding machine reamer**
 d Maschinenreibahle *f*
 f alésoir *m* mécanique
 n verstelbare machineruimer
 p rozwiertak maszynowy nastawny

2062 **expanding mandril**
 d Spanndorn *m*; Spreizdorn *m*
 f mandrin *m* à expansion
 n verstelbare boorhouder
 p trzpień rozprężny

2063 **expanding reamer with long knives**
 d einstellbare Langmesserreibahle *f*
 f alésoir *m* expansible à couteaux longs
 n verstelbare ruimer met lange messen
 p rozwiertak rozprężny z długimi nożami

2064 **expansion hole**
 d Ausgleichloch *n*
 f orifice *m* de dilatation
 n expansiegat
 p otwór wyrównawczy

 * **expansion stroke** → 3852

2065 **expansion tank pipe**
 d Schlauch *m* des Ausgleichbehälters
 f tuyau *m* de réservoir d'expansion
 n expansievatpijp
 p przewód zbiornika rozprężnego

2066 **expansion valve**
 d Expansionsventil *n*
 f détendeur *m*
 n reduceerklep
 p reduktor ciśnienia

2067 **expert**
 d Sachverständiger *m*
 f expert *m*
 n expert
 p ekspert

2068 **explosive rivet**
 d Sprengniet *m*
 f rivet *m* explosé
 n plofnagel
 p nit wybuchowy

2069 **extension**
 d Verlängerung *f*
 f rallonge *f*
 n verlengstuk
 p przedłużenie

2070 **extension coil spring**
 d Schraubenzugfeder *f*
 f ressort *m* hélicoïdal de traction
 n schroeftrekveer
 p rozciągana sprężyna śrubowa

2071 **extension cord**
 d Verlängerungskabel *n*
 f corde *f* d'allongement
 n verlengsnoer
 p przedłużacz

2072 **extension ladder**
 d Schiebeleiter *m*
 f échelle *f* coulissante
 n schuifladder
 p drabina rozsuwana

2073 **extension rule**
 d verschiebbares Lineal *n*
 f jauge *f* à coulisse
 n uitschuifbare duimstok
 p linijka przesuwalna

2074 **external lighting**
 d Aussenbeleuchtung *f*
 f éclairage *m* extérieur
 n buitenverlichting
 p oświetlenie zewnętrzne

2075 **external marking**
 d aussere Markierung *f*
 f identification *f* extérieure
 n markering aan de buitenzijde
 p oznaczenie zewnętrzne

2076 **external thread**
 d Aussengewinde *n*
 f filet *m* extérieur
 n buitenschroefdraad
 p gwint zewnętrzny

2077 **extractor**
 d Abziehvorrichtung *f*; Abzieher *m*
 f extracteur *m*
 n trekgereedschap
 p ściągacz

2078 **extreme pressure lubricant; hypoid oil**
 d Höchstdruckschmiermittel *n*
 f lubricant *m* de pression extrême
 n hogedruksmeermiddel
 p olej hipoidalny do smarowania mechanizmów
    przenoszących duże naciski

2079 **eye bolt**
 d Augbolzen *m*
 f vis *f* à œil
 n oogbout
 p śruba złączna o łbie oczkowym

2080 eyelet pliers; eyelet punch
 d Ösenzange *f*
 f pince *f* à poser les œillets
 n ponstang
 p szczypce oczkowe

 * **eyelet punch** → **2080**

F

2081 fabric joint
d Hardyscheibe *f*
f joint *m* Hardy
n Hardy-koppeling; Hardy-schijf
p tarcza Hardy'ego

2082 face
d Stirnflache *f*
f about *m*
n kops vlak
p powierzchnia czołowa

2083 face cutter
d Stirnfräser *m*
f fraise *f* de face
n kopfrees
p frez czołowy

2084 face of tooth
d Zahnkopf *m*
f tête *f* de dent
n tandkop
p głowa zęba

2085 face shield
d Schweisserschutzschild *m*
f masque *m* de soudeur; écran *m* de soudeur
n lasmasker; laskap
p tarcza spawacza

* **facia → 1460**

2086 factor of safety
d Sicherheitsfaktor *m*
f coefficient *m* de sécurité
n veiligheidsfactor
p współczynnik bezpieczeństwa

2087 falling rock pieces
d Steinschlag *m*
f chutes *m* de pierres
n vallend gesteente
p spadające odłamki skalne

2088 fan
d Lüfter *m*
f ventilateur *m*
n ventilator
p wentylator

2089 fan belt
d Ventilatorriemen *m*
f courroie *f* du ventilateur
n ventilatorriem
p pasek wentylatora

2090 fan belt idler
d Lüfterriemenspannrolle *f*
f poulie-tendeur *m* de courroie
n spanrol van ventilatorriem
p rolka naciągowa pasów wentylatora

2091 fan blade
d Ventilatorflügel *m*
f ailette *f* de ventilateur
n ventilatorblad
p łopatka wentylatora

2092 fan centrifugal clutch
d Ventilatorfliehkraftkupplung *f*
f embrayage *m* centrifuge de ventilateur
n centrifugaalkoppeling van ventilator
p sprzęgło odśrodkowe wentylatora

2093 fan clutch
d Lüfterkupplung *f*
f embrayage *m* de ventilateur
n ventilatorkoppeling
p sprzęgło wentylatora

2094 fan cowl
d Lüfterhaube *f*
f capotage *m* de ventilateur
n ventilatortunnel
p tunel wentylatora

2095 fan disc washer
d Fächerscheibe *f*
f rondelle *f* crantée
n vlakke tandveerring
p podkładka podatna płatkowa; podkładka ząbkowana

2096 fan drive shaft
d Lüfterwelle *f*
f arbre *m* de ventilateur
n ventilatoras
p wał wentylatora

2097 fan driving pulley
d Ventilatorriemenscheibe *f*
f poulie *f* de ventilateur
n ventilatorriemschijf
p koło pasowe wentylatora

2098 fan fixed shaft
d Lüfterachse *f*
f axe *m* de ventilateur
n vaste as van ventilator
p oś wentylatora

2099 fan fixing hub
  d Ventilatornabe *f*
  f moyeu *m* de ventilateur
  n ventilatornaaf
  p piasta wentylatora

2100 fan housing
  d Lüftergehäuse *n*
  f carter *m* de ventilateur
  n ventilatorhuis
  p obudowa wentylatora

2101 farm tractor
  d Ackerschlepper *m*
  f tracteur *m* agricole
  n landbouwtrekker
  p ciągnik rolniczy

2102 fast acting valve
  d Schnellschaltventil *n*
  f soupape *f* à fermeture rapide
  n snelsluitklep
  p zawór rozdzielczy szybkiego działania

2103 fastener
  d Federklammer *m*
  f arrêtoir *m*
  n spanband
  p opaska resoru

2104 fastener removal tool
  d Werkzeug *n* zum Entfernen von
       Befestigungsclips
  f spatule *m* d'extraction des clips
  n gereedschap voor demontage van
       bevestigingsklemmen
  p narzędzie do usuwania zacisków montujących

2105 fastener thread
  d Befestigungsgewinde *n*
  f filet *m* de fixation
  n bevestigingsschroefdraad
  p gwint złączny

2106 fastening bolt
  d Schlossschraube *f*
  f boulon *m* de fixation
  n bevestigingsbout
  p wkręt umocowujący

2107 fatigue test; endurance test
  d Prüfung *f* der Ermüdung
  f essai *m* de fatigue
  n duurproef
  p próba trwałości

2108 feather
  d Spaltkeil *m*

  f coin *m* à fendre
  n splijtwig
  p klin rozszczepiający

2109 feather joint
  d Federverbindung *f*
  f assemblage *m* à languette
  n losse veer- en groefverbinding
  p połączenie na pióro i wpust

2110 feed duct
  d Zuführschlauch *m*
  f conduit *m* d'alimentation
  n toevoerslang; toevoerleiding
  p przewód zasilający

2111 feeding gallery; spill gallery
  d Angleichraum *m*
  f chambre *f* d'alimentation
  n toevoergalerij
  p przestrzeń wyrównawcza; komora
       wyrównawcza

2112 feeding spring
  d Druckfeder *f*
  f ressort *m* de refoulement
  n pompdrukveer
  p sprężyna tłocząca

2113 feeler gauge
  d Spaltlehre *f*; Fühllehre *f*
  f cale *f* d'épaisseur; jauge *f*; palpeur *m*
  n voelermaatje
  p szczelinomierz

2114 felt
  d Filtz *m*
  f feutre *m*
  n vilt
  p filc

2115 felt gasket
  d Filzdichtung *f*
  f joint *m* de feutre
  n vilten pakking
  p uszczelnienie filcowe

2116 felt pad
  d Filzplatte *f*
  f manchon *m* feutre filtrant
  n viltplaat
  p płat filcu

2117 felt washer
  d Filtzunterlagscheibe *f*
  f rondelle *f* de feutre
  n viltring; vilten onderlegring
  p podkładka filcowa

2118 **female connection thread**
d Anschlussgewinde *n*
f taraudage *m* de raccordement
n inwendige schroefdraad van verbinding
p gwint przełączenia

2119 **female inner joint; slotted inner joint**
d Führungshalbgelenk *n*
f noix *f* à mortaise
n geleidingsgewricht
p ślizgacz prowadzący

2120 **fencing pliers**
d Hammerzange *f*
f pince *f* marteau
n hamertang
p szczypce płaskie do przewodów

2121 **fender**
d Kotflügel *m*
f garde-boue *m*
n spatscherm; spatbord
p błotnik

2122 **ferrite**
d Ferrit *m*
f métal *m* ferrique
n ferriet
p żelazo

2123 **F head engine**
d Motor *m* mit übereinander angeordneten Ventilen
f moteur *m* à soupapes en F
n motor met een hangende en een staande klep per cilinder
p silnik z zaworami rozmieszczonymi jeden nad drugim

2124 **fibre**
d Faser *f*
f fibre *f*
n vezel
p włókno

2125 **fibre insulation**
d Fiberisolation *f*
f isolation *f* de filament
n vezelisolatie
p izolacja włóknista

2126 **fibre joint**
d Fiberdichtung *f*
f joint *m* fibre
n fiberring
p uszczelnienie z włókna

2127 **fiddle drill**
d Bogenbohrer *m*
f drille *m* à archet et corde
n drilboor; fidelboor
p furkadło smyczkowe

2128 **field coil**
d Erregungswicklung *f*
f bobine *f* d'excitation
n veldwikkeling
p uzwojenie wzbudzania

2129 **field magnet**
d Feldmagnet *m*
f induit *m* de magnéto
n veldmagneet
p magneśnica

2130 **fifth wheel coupling**
d Kupplung *f*
f sellette *f* d'attelage
n opleggerkoppeling
p sprzęg siodłowy

2131 **fifth wheel kingpin**
d Kupplungszapfen *m*
f pivot *m* d'attelage
n koppelpen van opleggerkoppeling
p czop sprzęgłowy

2132 **filament of bulb**
d Faden *m* der Glühlampe
f filement *m* de lampe
n gloeidraad van gloeilamp
p drut żarówki

2133 **file**
d Feile *f*
f lime *f*
n vijl
p pilnik

2134 **file brush**
d Feilenbürste *f*
f brosse *f* à limes
n vijlborstel; vijlenborstel
p szczotka do pilników

2135 **file cutting**
d Feilbehauung *f*
f taillage *m* de lime
n kappen van de vijl
p nacinanie zębów pilnika

2136 **file face**
d Feilenfläche *f*
f face *f* de lime
n vlak van een vijl
p powierzchnia pilnika

2137 file rasp
d Feilraspel *f*
f lime *f* râpe; lime *f* mordante
n raspvijl; raspenvijl
p tarnik

2138 filing
d Feilen *n*
f limage *m*
n vijlen
p pilnikowanie; obróbka pilnikiem

2139 filing machine
d Feilmaschine *f*
f limatrice *f*; limeuse *f*
n vijlmachine
p pilnikarka

2140 filler
d Füllung *f*
f matière *f* active
n vulstof; vulmateriaal
p wypełniacz

2141 filler opening
d Einfülloch *n*
f trou *m* de remplissage
n vulopening
p otwór do napełniania

2142 filler sealing plate
d Verschlussdichtung *f*
f fermeture *f* d'orifice de remplissage
n vulopeningafdichting
p uszczelnienie otworu do napełniania

2143 fillet of weld
d Schweissnaht *f*
f cordon *m* de soudure
n lasrups
p spoina

2144 fillet radius
d Fassrundung *f*
f rayon *m* de l'arrondi
n afrondingsstraal
p promień zaokrąglenia

2145 filling bottle
d Füllflasche *f*
f bombe *f* à air comprimé
n vulfles
p butla do napełniania

2146 filling knife
d Malermesser *m*
f couteau *m* de peintre
n stopmes
p szpachelka do kitowania

2147 fillister head screw
d Linsenflachkopfschraube *f*
f vis *f* à tête fendue cylindrique bombée
n cilinderkopbout; lenskopschroef
p wkręt o soczewkowym łbie walcowym

*   **filter body → 2153**

2148 filter bracket
d Filterträger *m*
f support *m* de filtre
n filtersteun
p wspornik filtru

2149 filter cartridge
d Filterpatrone *f*
f cartouche *f* de filtre
n filterpatroon; filterelement
p wkładka filtra

2150 filter cover base
d Filtergehäusefuss *m*
f pied *m* de couvercle de filtre
n aansluitflens
p podstawa pokrywy filtru

2151 filter element
d Filtereinsatz *m*
f élément *m* filtrant
n filterelement
p wkład filtru

2152 filter head
d Filterkopf *m*
f tête *f* de filtre
n filterkop
p głowica filtru

2153 filter housing; filter body
d Filtergehäuse *n*
f corps *m* de filtre; boîtier *m* de filtre
n filterhuis
p obudowa filtru

2154 filtering disc
d Lamelle *f*
f lamelle *f* filtrante
n lamel van filter
p płytka filtracyjna

2155 filtering element bolt
d Stehbolzen *m*
f goujon *m* de cartouche filtre
n afstandsbout
p sworzeń wkładu filtracyjnego

2156 filter securing strap
d Halteband *n* für Filter

f sangle *f* maintien filtre
n filterklemband
p opaska przytrzymująca filtr

2157 filter shaft
d Filtereinsatzachse *f*
f broche *f* de filtre
n filterspil
p trzpień filtru

2158 final check
d Schlusskontrolle *f*
f contrôle *m* final
n eindcontrole
p kontrola końcowa

2159 final squeezing
d Nachpressen *n*
f pressage *m* résiduel
n napersen
p prasowanie wykańczające

2160 fine boring
d Feinbohren *n*
f perçage *m* de précision
n precisieboren
p wytaczanie dokładne

2161 fine machining
d maschinelle Feinbearbeitung *f*
f finissage *m* mécanique
n machinaal fijnen
p dokładna obróbka maszynowa

2162 fine thread
d Feingewinde *n*
f filet *m* mince
n fijne schroefdraad
p gwint drobnozwojny

2163 finishing hammer
d Planierhammer *m*
f marteau *m* à planer
n vlakhamer
p młotek wykańczający

2164 finishing reamer
d Fertigreibahle *f*
f alésoir *m* finisseur
n fijnruimer
p rozwiertak wykańczak

2165 finishing ring
d Ziehring *m*
f jonc *m* de finition
n trekring; treksteen
p pierścień wykańczający

2166 finishing trowel; laying-on trowel
d Glättekelle *f*
f plâtroir *m* à poignée ouverte
n plekspaan; pleisterspaan
p kielnia tynkarska

2167 finned sleeve
d Rippenhülse *f*
f douille *f* à ailettes
n van ribben voorziene huls
p tuleja użebrowana

2168 fire engine; fire fighting vehicle
d Feuerwehrfahrzeug *n*
f pompe *f* à incendie automobile
n brandweerauto
p samochód pożarniczy; pojazd pożarniczy

2169 fire extinguisher
d Löschgerät *n*
f extincteur *m*
n brandblusser
p gaśnica

* **fire fighting vehicle** → **2168**

2170 firing order; firing sentence
d Zündfolge *f*
f ordre *m* d'allumage
n ontstekingsvolgorde
p kolejność zapłonu

* **firing pressure** → **1171**

* **firing sentence** → **2170**

2171 firmer chisel
d Stemmeisen *n*
f fermoir *m*
n fermoorbeitel; haakbeitel
p dłuto ciesielskie płaskie

2172 firm joint
d feste Verbindung *f*
f assemblage *m* résistant
n hechte verbinding
p mocne połączenie

2173 first aid kit
d Verbandkasten *m*
f nécessaire *m* de premier secours
n verbandkist; verbandtrommel
p apteczka podręczna

2174 first and second speed fork rod
d Gabelwelle *f* für 1. und 2. Gang
f axe *m* de fourchette de première et deuxième vitesse

n schakelvorkas voor eerste en tweede versnelling
p wałek widełek biegów pierwszego i drugiego

2175 first and second speed synchronizer
d Synchronvorrichtung *f* für 1. und 2. Gang
f synchroniseur *m* de première et deuxième vitesse
n synchromesh voor eerste en tweede versnelling
p synchronizator biegów pierwszego i drugiego

2176 first and second speed synchronizing sleeve
d Synchronmuffe *f* für 1. und 2. Gang
f douille *f* de synchroniseur de première et deuxième vitesse moyeu
n schakelmof voor eerste en tweede versnelling
p tuleja synchronizatora biegów pierwszego i drugiego

2177 first speed driving gear
d treibendes Zahnrad *n* 1. Ganges
f pignon *m* menant de première vitesse
n aandrijftandwiel voor eerste versnelling
p napędzające koło biegu pierwszego

2178 fishtail bit; blade bit
d Fischschwanzbohrer *m*
f trépan *m* à queue de poisson
n visstaartbeitel
p świder płaski; świder "rybi ogon"

* **fissure** → 1351

2179 fitter's hammer
d Installateurshammer *m*
f marteau *m* de monteur
n bankwerkershamer; installateurshamer
p młotek ślusarski

2180 fitting
d Anpassung *f*
f ajustement *m*
n aanpassing
p dopasowanie

2181 fitting dimension
d Einbaumass *n*
f cote *f* de montage
n inbouwmaat
p wymiary wbudowania

2182 fitting grease
d Montagepaste *f*
f mixture *f* de montage
n montagepasta
p smar montażowy

2183 fitting material
d Installationsmaterial *n*
f matériau *m* d'installation
n installatiemateriaal
p materiał instalacyjny

2184 fitting tool
(tool to fit drive shaft spiders)
d Werkzeug *n* für Antriebsachse
f outil *m* de montage
n aandrijfasgereedschap
p przyrząd do ustalania krzyżaków wału napędowego

2185 fixed caliper housing
d Bremssattel *m*
f corps *m* d'étrier de frein
n remzadel
p obudowa zaciskacza

* **fixed conical disc** → 2887

2186 fixed contact
d unbeweglicher Kontakt *m*
f contact *m* fixe
n vast contact
p styk nieruchomy; styk stały

2187 fixed pivot
d feststehender Zapfen *m*
f pivot *m* fixe
n bevestigingsnok
p ustalony czop

2188 fixing device
d Befestigungselement *n*
f élément *m* de fixation
n bevestigingselement
p element mocujący

2189 fixing moment
d Anzugsdrehmoment *n*
f couple *m* de serrage
n aanhaalmoment
p moment dokręcania

2190 flame propagation
d Ausbreitung *f* der Zündflamme
f propagation *f* de flammes
n voortschrijding van vlamfront
p rozchodzenie się płomienia

2191 flame restraining ring
d Flammeneinschnürring *m*
f limiteur *m* de flamme
n plaat voor concentratie van de vlam
p pierścieniowy ogranicznik płomieni

2192 flange
d Flansch m
f flasque m
n flens; kraag
p kołnierz

2193 flange bearing
d Flanschlager n
f palier m à flasque
n flenslager; kraaglager
p łożysko kołnierzowe

* flange boss → 2201

2194 flange clutch; flange coupling
d Flanschkupplung f
f accouplement m à plateaux
n flenskoppeling
p sprzęgło kołnierzowe

* flange coupling → 2194

2195 flanged flare nut wrench
d offener Ringschlüssel m mit Anschlagkante
f clé f polygonale à tuyauter avec toile
n open ringsleutel met kraag
p klucz oczkowy z kołnierzem

2196 flanged hub
d Flanschnabe f
f moyeu m à flasque
n flensnaaf
p piasta kołnierza

2197 flanged shaft
d Flanschwelle f
f arbre m à brides
n flensas
p wał kołnierzowy

2198 flanged tubular rivet
d Rohrniet m
f rivet m creux à tête plate
n felsnagel met doorlopend gat
p nit drążony

2199 flange joint
d Flanschverbindung f
f accouplement m à bride; joint m à brides
n flensverbinding
p złącze kołnierzowe

2200 flange screw
d Schraube f für Flansch
f vis f de bride
n flensschroef
p śruba do kołnierza

2201 flange sleeve; flange boss
d Flanschmuffe f
f manchon m de poussée; manchon m bride
n flensmof; flensbus
p tuleja kołnierzowa

2202 flange support
d Flanschträger m
f support m de bride
n flensdrager
p podpora kołnierza

2203 flange yoke
d Kreuzgelenkgabel f
f fourche f de joint à croisillon
n juk van cardankoppeling
p rozwidlone złącze kołnierzowe

2204 flap
d Felgenband n
f bande f de fond de jonte
n velglint
p ochraniacz dętki

2205 flare nut wrench
d Ringschlüssel m für Überwurfmuttern und Rohrleitungen
f clé f polygonale pour raccords de tuyauterie
n flensmoersleutel; wartelmoersleutel
p klucz oczkowy do nakrętek kołpakowych i przewodów rurowych

2206 flaring tool set
d Bördelgerät n
f appareil m à collets
n set ruimergereedschap
p przyrząd do wywijania obrzeży; przyrząd do zawinięcia obrzeża

2207 flasher control light; flashing indicator
d Fahrtrichtungsblinkleuchtenanzeige f
f répétiteur m feux de direction
n verklikkerlamp voor knipperlichten
p wskaźnik działania kierunkowskazów

* flasher indicator control light → 1586

* flashing indicator → 2207

2208 flashing point
d Flammpunkt m
f point m d'éclair
n vlampunt
p temperatura płomienia

2209 flat bar
d Flacheisen n

f méplat *m*
n vlakke stang
p płaskownik

* **flat base rim** → 1801

2210 flat belt drive
d Flachriementrieb *m*
f transmission *f* par courroie plate
n vlakke-riemoverbrenging
p napęd pasowy taśmowy

2211 flat countersunk head screw
d Senkkopfschraube *f*
f vis *f* à tête fendue fraisée
n schroef met platverzonken kop
p wkręt o płaskim łbie stożkowym

2212 flat dolly
d Handstock *m* gerade
f table *f* à main plate
n platte handtafel
p klepadło płaskie

2213 flat dresser
d Klempnerhammer *m*
f batte *f* plate
n platte plethamer
p miękki młotek blacharski

2214 flat faced follower
d Flachstössel *m*
f poussoir *m* à plateau
n klepstoter met vlakke kop
p popychacz płaski

2215 flat file
d Flachfeile *f*
f lime *f* plate
n blokvijl; platte vijl
p pilnik płaski

2216 flat fuse
d flache Schmelzsicherung *f*
f fusible *m* plat
n lamelzekering
p płytkowy bezpiecznik topikowy

2217 flat guideway
d Flachführung *f*
f guidage *m* plan
n vlakke geleiding
p prowadnica płaska

2218 flat key
d Nasenflachkeil *m*

f clavette *f* inclinée plate
n platte kopspie
p klin płaski

2219 flat lever
d Flachhebel *m*
f levier *m* plat
n platte hefboom
p dźwignia płaska

2220 flatness
d Flachheit *f*
f planéité *f*
n vlakheid
p płaskość

2221 flat spot
d Flachstelle *f*
f zone *f* plate
n platte kant
p miejsce płaskie

2222 flatting colour
d Mattfarbe *f*
f couleur *m* mate
n matverf
p farba matowa

2223 flat washer
d Flachscheibe *f*
f rondelle *f* plate
n platte ring
p podkładka płaska

2224 flat wedge dolly
d Handfaust *f* Keilform
f tas *m* épinçoir
n vlak tas
p klepadło płaskie w kształcie klina

2225 flex handle magnetic holder
d biegsamer Magnetsucher *m*
f doigt *m* magnétique avec tige à déformation
n magnetische houder met buigzame handgreep
p giętki uchwyt magnetyczny

2226 flexibility
d elastische Nachgiebigkeit *f*
f souplesse *f*; flexibilité *f*
n flexibiliteit
p giętość; elastyczność

2227 flexible coupling
d elastisches Gelenk *n*
f joint *m* élastique
n flexibele koppeling
p przegub elastyczny

2228 flexible duct
d elastische Leitung *f*
f conduit *m* souple
n flexibele leiding
p przewód giętki

2229 flexible facing
d Verkleidung *f* von Schaumgummi
f revêtement *m* en mousse
n schuimrubberen bekleding
p pokrycie z gumy piankowej; pokrycie z pianogumy

2230 flexible mounting
d elastisches Zwischenglied *n*
f entretoise *f* élastique
n flexibel tussenstuk
p wkład elastyczny

2231 flexible mounting plate
d Stützplatte *f*
f plaque *f* support
n ophangingsplaat
p płytka wieszaka

2232 flexible screwdriver
d flexibler Schraubendreher *m*
f tournevis *m* flexible
n flexibele schroevendraaier
p giętki śrubokręt

2233 flexible shaft
d biegsame Welle *f*
f arbre *m* flexible
n flexibele as
p wał giętki

2234 flexible sleeve
d elastische Hülse *f*
f bague *f* élastique
n elastische huls
p tuleja elastyczna

2235 flexible sleeve insert
d Hülseneinlage *f*
f douille *f* de bague élastique
n buigzame manchetinzet
p wkładka tulei elastycznej

2236 flexible spark plug fitter
(tool to thread in spark plugs in awkward locations)
d Gelenkschlüssel *m* zum Einschrauben von Zündkerzen
f pose-bougies *f* flexible
n flexibel hulpstuk voor aanbrengen van bougies
p łącznik elastyczny do mocowania świec

2237 flexible support
d elastischer Halter *m*
f support *m* élastique
n elastisch ophangstuk
p wieszak sprężysty

2238 flexible tube end
d Panzerendstück *n*
f embout *m* de gaine
n elastisch buseind
p elastyczna końcówka pancerza

2239 flexional strength
d Biegungsfestigkeit *f*
f résistance *f* à flexion
n weerstand tegen doorbuiging
p odporność na zginanie

2240 flexural spring
d Biegefeder *f*
f ressort *m* travaillant en flexion
n buigzame veer
p sprężyna zginana

2241 float
d Schwimmer *m*
f flotteur *m*
n vlotter
p pływak

2242 float chamber
d Schwimmerkammer *f*
f chambre *f* de flotteur
n vlotterkamer
p komora pływakowa

2243 float feed
d Schwimmerzuführung *f*
f alimentation *f* par flotteur
n vlotterregelaar van toevoer
p zasilanie pływakowe

2244 floating piston pin
d schwimmender Kolbenbolzen *m*
f axe *m* de piston flottant
n zwevende zuigerpen
p pływający sworzeń tłokowy

2245 floating shoe
d Schwimmbacke *f*
f mâchoire *f* flottante
n zwevende remschoen
p szczęka pływająca

2246 float level
d Schwimmerstand *m*
f niveau *m* du flotteur

n vlotterniveau
p poziom pływaka

2247 float lever
d Schwimmerhebel *m*
f levier *m* de flotteur
n vlotterarm
p dźwignia pływaka

2248 float needle valve
d Schwimmernadelventil *m*
f ensemble *m* pointeau
n vlotternaaldventiel
p zawór iglicowy pływaka

2249 float pivot pin
d Schwimmerachse *f*
f axe *m* de flotteur
n scharnieras van vlotter
p oś pływaka

2250 floor carpet
d Mittelbodenbelag *m*
f tapis *m* milieu
n bodembedekking; vloerbedekking; middelste vloerbedekking
p dywanik środkowy; wykładzina podłogi

2251 floor cross member
d Querstrebe *f* im Fussboden
f traverse *f* de plancher
n dwarsligger in de vloer
p poprzeczka podłogi

2252 floor frame
d Bodenrahmen *m*
f cadre *m* de plancher
n vloerdraagconstructie
p podstawa nadwozia

2253 floor level
d Ladehöhe *f*
f hauteur *f* du plan de chargement
n laadvloerhoogte
p wznios powierzchni ładowania

2254 floor moulding
d Bodenformen *n*
f moulage *m* au sol
n bodemvormen
p formowanie w gruncie; formowanie w podłożu

2255 floor panel
d Fussbodenplatte *f*
f panneau *m* de plancher
n bodemplaat; vloerplaat
p płyta podłogowa

2256 floor space
d Bodenfläche *f*
f surface *f* de sol
n vloeroppervlak
p powierzchnia podłogi

2257 flow
d Strömung *f*
f écoulement *m*
n stroming
p przepływ

2258 flow bench
d Vergaserprüfstand *m*
f banc *m* d'écoulement de carburateur
n testbank voor carburateurs
p stanowisko badawcze gaźnika

2259 flow chart
d Ablaufplan *m*
f diagramme *m* de fonctionnement
n stroomschema
p schemat przebiegu

2260 flowing pressure
d Förderdruck *m*
f pression *f* d'écoulement
n spuitdruk
p ciśnienie tłoczenia

2261 flow meter
d Strömungsmengenmesser *m*
f rhéomètre *m*
n doorstroomhoeveelheidsmeter
p przepływomierz

2262 fluid chamber; transfer chamber
d Flüssigkeitskammer *f*
f chambre *f* à liquide
n vloeistofkamer
p przestrzeń cieczy

* **fluid clutch** → 5422

2263 fluid friction
d flüssige Reibung *f*
f frottement *m* fluide
n wrijving tussen vloeistofdelen onderling
p tarcie płynne

* **fluid jet** → 5390

2264 fluid reservoir
d Flüssigkeitsbehälter *m*
f réservoir *m* de liquide
n vloeistofreservoir
p zbiornik cieczy

2265 fluorescent tube inspection lamp
d Arbeitslampe *f* mit Leuchtröhre
f baladeuse *f* à tube fluorescent
n looplamp met TL-buis
p lampa przenośna ze świetlówką

2266 flushing oil
d Spülöl *n*
f huile *f* de rinçage; huile *f* de lavage
n spoelolie
p olej płuczący; olej zmywający

2267 flush rivet
d Senkniet *m*
f rivet *m* noyé
n verzonken klinknagel
p nit kryty; nit płaski; nit z łbem wpuszczanym

* fly nut → 5471

2268 flyweight
d Fliehgewicht *n*
f masse *f* centrifuge
n centrifugaalgewichtje
p ciężarek wirujący

2269 flyweight carrier
d Fliehgewichtbolzen *m*
f axe *m* de masse
n tap voor centrifugaalgewichten
p czop ciężarka

2270 flywheel
d Schwungrad *m*
f volant *m*
n vliegwiel
p koło zamachowe

2271 flywheel end
d Schwungradseite *f*
f côté *m* volant
n eind van vliegwiel
p strona koła zamachowego

2272 flywheel housing
d Schwungradgehäuse *n*
f carter *m* de volant
n vliegwielhuis
p obudowa koła zamachowego

2273 flywheel lock plate
d Riegelflansch *m*
f flasque *m* de fixation de volant
n borgplaat voor vliegwielbouten
p tarcza osadcza koła zamachowego

2274 flywheel magneto
d Schwungmagnetzünder *m*
f volant *m* magnétique

n vliegwiel met magneten
p iskrownik w kole zamachowym

2275 flywheel stay
d Schwungradsperre *f*
f bloque *m* volant moteur
n vliegwielblokkering
p blokada koła zamachowego

2276 foaming
d Verschäumung *f*
f formation *f* de mousse; moussage *m*
n schuimvorming
p spienianie

2277 foaming inhibitor
d Antischaumzusatz *m*
f additif *m* antimousse
n antischuimdope
p dodatek przeciwpieniący

2278 foam pad
d Schaumelement *n*
f élément *m* mousse
n schuimelement
p element piankowy

2279 foam plastic cement
d Schaumstoffkleber *m*
f colle *f* pour plastique mousse
n lijm voor schuimplastic
p klej do tworzywa piankowego

2280 foam rubber
d Schaumgummi *m*
f mousse *f* de caoutchouc
n schuimrubber
p guma piankowa; pianoguma; guma porowata; guma gąbczasta

2281 foam suppressor
d Schaumdämpfungszusatz *m*
f additif *m* antimousse
n antischuimdope
p środek przeciwpieniący

2282 fog lamp
d Nebellampe *f*; Nebelscheinwerfer *m*
f projecteur *m* pour brouillard
n mistlamp
p reflektor przeciwmgłowy

2283 fog lamp mounting bracket
d Nebelscheinwerferhalter *m*
f support *m* du projecteur pour brouillard
n mistlampsteun
p wspornik reflektora przeciwmgłowego

2284 fog lamp switch
  d Nebelscheinwerferschalter *m*
  f interrupteur *m* des phares antibrouillard
  n mistlampschakelaar
  p wyłącznik reflektora przeciwmgłowego

  * **foldable roof → 3402**

2285 folding fitter's table
  d zusammenklappbarer Monteurtisch *m*
  f table *f* de monteur
  n opvouwbare fitterswerkbank
  p składany stół monterski

2286 folding measure; foot rule
  d Gliedermassstab *m*
  f mètre *m* pliant
  n duimstok
  p przymiar kreskowy; miarka składana

2287 folding rear seat
  d umklappbare Rücksitzbank *f*
  f banquette *f* arrière rabattable
  n neerklapbare achterbank
  p składane siedzenie tylne

2288 folding seat
  d Klapsitz *m*
  f strapontin *m*; siège *m* rabattable
  n opklapbare zitting; klapstoeltje
  p siedzenie składane

  * **foot rule → 2286**

2289 forced
  d Zwangs...
  f forcé
  n geforceerd
  p wymuszony

2290 forced air cooling
  d Gebläseluftkühlung *f*
  f refroidissement *m* par circulation d'air forcée
  n geforceerde luchtkoeling
  p wymuszone chłodzenie powietrza

2291 forced circulation
  d Zwangsumlauf *m*
  f circulation *f* forcée
  n geforceerde circulatie
  p obieg wymuszony

2292 forced fitting
  d Gewaltmontage *f*
  f montage *m* forcé
  n vaste montage
  p montaż siłowy

2293 forced lubrication
  d Druckschmierung *f*
  f graissage *m* sous pression
  n druksmering
  p smarowanie ciśnieniowe

2294 forging
  d Schmieden *n*
  f forgeage *m*
  n smeden
  p kucie

2295 forging machine
  d Schmiedepresse *f*
  f presse *f* à forger
  n smeedmachine; smeedpers
  p prasa kuźnicza; maszyna kuźnicza

2296 fork
  d Gabel *f*
  f fourche *f*
  n gaffel; vork
  p widełki

2297 fork bolt
  d Gabelbolzen *m*
  f boulon *m* de fourchette
  n gaffelpen
  p śruba ze sworzniem rozwidlonym

2298 forked end
  d Gabelende *n*
  f chape *f*
  n gevorkt verbindingstuk
  p łącznik rozwidlony

2299 forked rod
  d Gabelpleuel *n*
  f bielle *f* à fourche
  n vorkvormige stang
  p korbowód widlasty

  * **forked stub axle → 2300**

2300 forked swivel axle; forked stub axle
  d Gabelachsschenkel *m*
  f fusée *f* à chape
  n vorkvormige fusee
  p zwrotnica rozwidlona

2301 fork head
  d Gabelkopf *m*
  f tête *f* de fourche
  n gaffel; gevorkt uiteinde van as
  p głowica widlasta

2302 fork joint
  d Gabelgelenk *n*

f chape *f*
n vorkverbinding
p przegub widełkowy

2303 fork lever
d Gabelhebel *m*
f levier *m* à fourche
n vorkhefboom
p dźwignia rozwidlona

2304 fork separator
(tool used for steering and suspension ball joints)
d Kugelgelenkabzieher *m*
f extracteur *m* de rotule
n stuurkogeltrekker
p ściągacz do przegubów kulowych

2305 fork set
d Federhalter *m*
f jeu *m* de coupelles
n veerhouder
p uchwyt do sprężyn

2306 fork springs cover plate
d Deckel *m* der Sperreinrichtung
f couvercle *m* de ressorts de cliquets
n deksel van arreteerinrichting
p pokrywa sprężyn zatrzasków

2307 form grinding; profile grinding
d Profilschleifen *n*
f rectification *f* de profil
n profielslijpen
p szlifowanie ksztaltowe; szlifowanie profilowe

2308 forming
d Profilieren *n*
f profilation *f*
n profileren
p profilowanie

2309 forward inertia
d dynamische Achslastverlagerung *f*
f changement *m* dynamique de la charge sur l'essieu
n dynamische aslastverplaatsing
p przemieszczenie obciążenia osi

2310 foundry
d Giesserei *f*
f fonderie *f*
n gieterij
p odlewnia

2311 foundry defect
d Gussfehler *m*
f défaut *m* de fonderie

n gietfout
p wada odlewnicza

2312 four jaw chuck
d Vierbackenbohrfutter *n*
f mandrin *m* à quatre mors
n boorhouder met vier bekken
p uchwyt czteroszczękowy

2313 four stroke engine
d Viertaktmotor *m*
f moteur *m* à quatre-temps
n vierslagmotor; viertaktmotor
p silnik czterosuwowy

2314 frame
d Rahmen *m*
f cadre *m*
n frame
p rama

2315 frame bottom member
d unterer Längsträger *m*
f longeron *m* inférieur
n onderste ligger van chassis
p podłużnica dolna

2316 frame center rest
d Rahmenbrille *f*
f lunette *f* de châssis
n steun op midden van frame
p oczko osadcze ramy

2317 frame central member
d Zentralträger *m*
f longeron *m* central
n centrale draagbalk
p podłużnica centralna

2318 frame fork
d Rahmengabel *f*
f fourche *f* de châssis
n vorkstuk van frame
p rozwidlenie ramy

2319 frame for milled files
d Karosseriefeilenhalter *m*
f monture *f* pour limes fraisées
n frame voor metaalrasp; houder voor carrosserievijl
p uchwyt do pilników frezowanych

2320 frame panel
d Rahmenplatte *f*
f panneau *m* de châssis
n raamplaat
p płyta ramy

* **frame saw** → 4586

2321 frame side member
 d Rahmenlängsträger *m*
 f longeron *m* de cadre
 n langsdraagbalk; gevormde langsdrager
 p podłużnica ramy

2322 frame upper member
 d oberer Längsträger *m*
 f longeron *m* supérieur
 n bovenste ligger van chassis
 p podłużnica górna

2323 free ball
 d Freikugel *f*
 f bille *f* libre
 n vrijkogel
 p kula swobodna

2324 free height; free length
 d freie Länge *f*
 f hauteur *f* libre
 n vrije lengte
 p długość swobodna

* **free length** → 2324

2325 free piston
 d Freikolben *m*
 f piston *m* libre
 n vrije zuiger
 p tłok swobodny

2326 free spring
 d Gleitfeder *f*
 f ressort *m* libre
 n schuifspie
 p resor ze ślizgaczem

2327 freewheel
 d Freilauf *m*
 f roue *f* libre
 n vrijwiel
 p wolne koło

2328 freewheel hub
 d Freilaufnabe *f*
 f moyeu *m* à roue libre
 n vrijloopnaaf
 p piasta z wolnym kołem

2329 freezing point
 d Stockpunkt *m*; Gefrierpunkt *m*
 f point *m* de congélation
 n stollingstemperatuur
 p temperatura krzepnięcia

2330 freight transport
 d Warentransport *m*
 f transport *m* des marchandises
 n vrachtvervoer
 p przewóz towarów

2331 fret saw
 d Dekupiersäge *f*; Schweifsäge *f*
 f scie *f* à découper
 n figuurzaag; fretzaag; decoupeerzaag
 p piła wyrzynarka

2332 friction
 d Reibung *f*
 f frottement *m*
 n wrijving
 p tarcie

2333 frictional loss
 d Reibungsverlust *m*
 f perte *f* par frottement
 n wrijvingsverlies
 p strata wskutek tarcia

2334 friction clutch
 d Reibungskupplung *f*
 f embrayage *m* à friction
 n wrijvingskoppeling
 p sprzęgło cierne

2335 friction disc; thrust disc
 d Reibscheibe *f*
 f disque *m* de friction
 n frictieschijf; koppelingsschijf
 p tarcza dociskowa

2336 friction force
 d Reibungskraft *f*
 f force *f* de friction
 n wrijvingskracht
 p siła tarcia

2337 friction gearing
 d Reibradtrieb *m*
 f transmission *f* à friction
 n frictieoverbrenging
 p napęd cierny

* **friction pad** → 674

2338 friction plate
 d Reibscheibe *f*
 f plaque *f* de friction
 n frictieplaat; wrijvingsplaat
 p płytka cierna

2339 friction power
 d mechanische Verlustleistung *f*

f puissance *f* de frottement
n vermogenverlies
p moc strat mechanicznych

2340 friction press
d Reibspindelpresse *f*
f presse *f* à friction
n frictiepersstuk
p prasa śrubowa cierna

2341 friction shock absorber
d Reibungsstossdämpfer *m*
f amortisseur *m* à friction
n wrijvingsschokdemper
p amortyzator cierny

2342 friction sleeve
d Reiblager *n*
f manchon *m* de frottement
n glijlager; wrijvingslager
p tuleja cierna

2343 friction surface
d Reibungsfläche *f*
f surface *f* de frottement
n wrijvingsoppervlakte
p powierzchnia tarcia

2344 friction wheel
d Reibkörper *m*
f roue *f* de friction
n wrijvingswiel
p koło cierne

2345 front axle
d Vorderachse *f*
f essieu *m* avant
n vooras
p oś przednia

2346 front axle beam
d Vorderachskörper *m*
f corps *m* de l'essieu avant
n voorste aslichaam; voorste ashuis; stijve
    vooras
p belka osi przedniej

*   **front axle Elliot type head** → 2348

2347 front axle shaft
d Vorderachswelle *f*
f arbre *m* d'essieu avant
n vooras met voorwielaandrijving
p przednia półośka

2348 front axle yoke; front axle Elliot type head
d Gabelkopf *m* der Vorderachse

f chape *f* d'essieu avant
n gaffel van de vooras
p rozwidlony łeb osi przedniej

2349 front brake actuator
d Vorderbremsenhydrozylinder *m*
f récepteur *m* de frein avant
n werkcilinder van voorrem
p siłownik przedniego hamulca

2350 front brake lining
d vorderer Bremsbelag *m*
f garniture *f* de frein d'avant
n voorste remvoering
p okładzina przednich szczęk hamulca

2351 front bumper
d Vorderstossfanger *m*
f pare-chocs *m* avant
n voorbumper
p zderzak przedni

2352 front clutch actuator
d Vorderkupplungshydrozylinder *m*
f récepteur *m* d'embrayage avant
n voorste-koppelingwerkcilinder
p siłownik przedniego sprzęgła

2353 front cover
d vorderer Deckel *m*
f couvercle *m* avant
n voorste deksel
p pokrywa przednia

2354 front cross member
d vordere Querstrebe *f*
f traverse *f* avant
n voorste dwarsbalk
p poprzeczka przednia

2355 front damage
d Frontschaden *m*
f dommage *m* avant
n voorschade
p szkoda przednia

2356 front door
d vordere Tür *f*
f porte *f* avant
n voorportier
p drzwi przednie

2357 front door panel
d vorderes Türdeckblech *n*
f panneau *m* de porte avant
n voorste plaatwerkdeel; voorste plaatstalen
    paneel
p przednia blacha pokrywy drzwiowej

2358 front drive
d Frontantrieb *m*
f traction *f* avant
n voorwielaandrijving
p napęd przedni

2359 front engine support
d vorderes Motorlager *n*; vorderes Motorträger *m*
f support *m* avant de moteur
n voorste motorsteun
p przednie łoże silnika

2360 front exhaust pipe
d vorderes Auspuffrohr *n*
f tuyau *m* d'échappement avant
n voorste uitlaatpijp
p rura wylotowa przednia

2361 front fork spring
d Vordergabelfeder *f*
f ressort *m* de fourchette avant
n veer van voorvork
p sprężyna przednich widełek

2362 front hose
d Vorderleitung *f*
f tuyau *m* avant
n slang aan voorzijde
p wąż przedni

2363 front mudguard; front wing
d Vorderkotflügel *m*
f aile *f* avant; garde-boue *m* avant
n voorspatbord; voorspatscherm
p błotnik przedni

2364 front multidisc clutch
d vordere Lamellenkupplung *f*
f embrayage *m* avant à disques juxtaposés
n voorste lamellenkoppeling
p sprzęgło wielotarczowe przednie

2365 front overhang
d vordere Überhanglänge *f*
f porte *f* à faux avant
n vooroverhang
p zwis przedni

2366 front panel
d vordere Verkleidung *f*
f panneau *m* avant
n voorste paneel
p osłona przednia

2367 front pillar
d Vordersäule *f*

f pied *m* avant
n voorstijl; voorste carrosseriestijl
p słupek przedni

2368 front pillar shield
d Fenstersäulenschild *n*
f revêtement *m* de pied avant
n bekleding van de raamstijl
p nakładka słupka przedniego

2369 front reaction rod
d vordere Reaktionstange *f*
f barre *f* de réaction avant
n voorste reactiearm; voorste geleidearm; voorste trekstang
p drążek reakcyjny przedni

2370 front roof rail
d vordere Dachluppe *f*
f becquet *m* avant de pavillon
n voorste daklangsdrager
p przednia szyna dachowa

2371 front shock absorber
d vorderer Stossdämpfer *m*
f amortisseur *m* avant
n voorste schokdemper
p zderzak przedni

2372 front silencer
d vorderer Schalldämpfer *m*
f pot *m* de détente; silencieux *m* avant
n voordemper; voorste uitlaatdemper
p tłumik przedni

2373 front spring assembly
d Vorderfedergruppe *f*
f ensemble *m* ressort d'avant
n voorste bladveerpakket
p zespół przedniego resoru

2374 front spring bracket; leaf spring bracket
d vorderer Federbock *m*
f butée *f* avant de ressort
n voorste veerdrager; voorste veerhand
p wspornik resoru przedniego

2375 front stabilizer bar
d Vorderachsschubstange *f*
f bielle *f* stabilisatrice d'avant
n voorste stabilisatorstang
p drążek popychający przedniej osi

2376 front support bracket
d vordere Tragpratze *f*
f support *m* avant
n voorste draagsteun
p przednia łapa zawieszenia silnika

2377 front suspension
  d Vorderradaufhängung f
  f suspension f avant
  n voorste wielophanging
  p zawieszenie kół przednich

2378 front wheel
  d Vorderrad n
  f roue f avant
  n voorwiel
  p koło przednie

2379 front wheel brake
  d Vorderradbremse f
  f frein m de roue avant
  n voorwielrem
  p hamulec koła przedniego

2380 front wheel driving shaft
  d Vorderradantriebswelle f
  f arbre m de commande de roue avant
  n voorste wielaandrijfas
  p półoś przedniego koła

2381 front wheel hub
  d Vorderradnabe f
  f moyeu m de roue avant
  n voorste wielnaaf
  p piasta koła przedniego

2382 front wheel valance
  d vordere Seitentrennwand f
  f coffrage m de roue avant
  n voorwielkuip
  p boczna przegroda przednia

  * front wing → 2363

2383 frow
  d Spaltkeil m
  f coin m à fendre le bois
  n splijtwig
  p klin rozdzielnik

2384 fuel
  d Brennstoff m
  f combustible m
  n brandstof
  p paliwo

2385 fuel accumulator
  d Kraftstoffspeicher m; Benzinspeicher m
  f accumulateur m de pression de carburant
  n brandstofaccumulator
  p zasobnik paliwa

  * fuel cock → 2425

2386 fuel consumption quota
  d Kraftstoffverbrauchsnorm f
  f norme f de consommation
  n brandstofverbruiknorm
  p norma zużycia paliwa

2387 fuel density
  d Kraftstoffdichte f
  f densité f de combustible
  n brandstofdichtheid
  p gęstość paliwa

2388 fuel discharge port
  d Förderstutzen m
  f orifice m de sortie
  n brandstofdoorvoerleiding
  p króciec tłoczny pompy

2389 fuel distillation range
  d Kraftstoffaktionzusammensetzung f
  f gamme f de distillation de carburant
  n destillatietraject van de brandstof
  p montaż destylacyjny paliwa

2390 fuel distributor
  d Brennstoffverteiler m
  f doseur-distributeur m de carburant
  n brandstofdoseerinrichting
  p dozownik-rozdzielacz paliwa

2391 fuel durability
  d Haltbarkeit f des Kraftstoffs
  f durabilité f de combustible
  n brandstofduurzaamheid
  p trwałość paliwa

2392 fuel economy
  d Brennstoffeinsparung f
  f économie f de combustible
  n brandstofbesparing
  p oszczędność paliwa

  * fuel engine → 2956

2393 fuel feed chamber
  d Speicher m; Druckraum m
  f chambre f de réserve
  n drukkamer; drukruimte
  p przestrzeń tłoczna

2394 fuel feeding injection unit
  d Einspritzung f mit Pumpedüsen
  f alimentation f à injecteurs-pompes
  n brandstofinspuitsysteem
  p zasilanie pompowtryskiwaczami

2395 fuel feed pump; supply pump
  d Förderpumpe f; Kraftstoffpumpe f

f pompe *f* d'alimentation
n brandstofopvoerpomp
p pompa zasilająca paliwa

2396 fuel filter
   d Kraftstoffilter *m*
   f filtre *m* à carburant
   n brandstoffilter
   p filtr paliwa

2397 fuel gauge; petrol gauge
   d Kraftstoffvorratsanzeiger *m*; Kraftstoffuhr *f*
   f indicateur *m* de niveau d'essence; jauge *f* de carburant
   n brandstofniveaumeter; brandstofvoorraadmeter
   p wskaźnik poziomu zapasu paliwa

2398 fuel heater
   d Kraftstoffheizung *f*
   f appareil *m* de réchauffage du carburant
   n brandstofvoorverhitter
   p podgrzewacz paliwa

2399 fuel inflow; supply of fuel
   d Brennstoffzufluss *m*
   f arrivée *f* de combustible
   n brandstoftoevoerstroming
   p dopływ paliwa

2400 fuel injection
   d Kraftstoffeinspritzung *f*
   f injection *f* de combustible
   n brandstofinspuiting
   p wtrysk paliwa

2401 fuel injection engine
   d Einspritzmotor *m*
   f moteur *m* à injection
   n motor met brandstofinspuiting
   p silnik wtryskowy

2402 fuel injection nozzle
   d Einspritzdüse *f*
   f injecteur *m*
   n brandstofverstuiver
   p dysza wtryskiwacza paliwa

2403 fuel inlet neck
   d Saugstutzen *m*
   f tubulure *f* d'amenée de combustible
   n brandstoftoevoerbuis
   p króciec ssawny

2404 fuel inlet passage
   d Druckkanal *m*
   f canal *m* d'arrivée

n brandstofdrukkanaal
p kanalik dopływowy

2405 fuel leak off passage
   d Leckölbohrung *f*
   f canal *m* des fuites de combustible
   n brandstofoverstroomkanaal; lekbrandstofkanaal
   p kanalik nadmiarowy paliwa

2406 fuel level
   d Brennstoffstandhöhe *f*; Kraftstoffstandhöhe *f*
   f niveau *m* de combustible; niveau *m* d'essence
   n brandstofniveau
   p poziom paliwa

2407 fuel line
   d Sperrlinie *f*
   f ligne *f* continue
   n markeringslijn
   p linia ciągła

2408 fuel mixture
   d Kraftstoffgemisch *n*
   f carburant *m* mélangé
   n brandstofmengsel
   p mieszanka paliwowa

2409 fuel nozzle assembly
   d Kraftstoffzerstäuber *m*
   f distributeur *m* de combustible
   n brandstofverstuiver
   p rozpraszacz paliwa

2410 fuel pipe
   d Kraftstoffleitung *f*
   f tuyau *m* de carburant
   n brandstofleiding
   p przewód paliwa

2411 fuel pressure pipe
   d Brennstoffdruckleitung *f*
   f conduit *m* de refoulement du combustible
   n brandstofdrukpijp
   p przewód wysokociśnieniowy paliwa

2412 fuel pressure valve
   d Kraftstoffdruckregler *m*
   f régulateur *m* de pression d'essence
   n brandstofpersklep
   p regulator ciśnienia paliwa

2413 fuel pump
   d Kraftstoffpumpe *f*
   f pompe *f* à combustible
   n brandstofpomp
   p pompa paliwa

2414 fuel pump body
d Kraftstoffpumpengehäuse *n*
f corps *m* de pompe à combustible
n brandstofpomphuis; brandstofpomplichaam
p kadłub pompy paliwa

2415 fuel pump bracket
d Kraftstoffpumpenbock *m*
f support *m* de pompe à combustible
n ophanging van de brandstofpomp
p wspornik pompy paliwa

2416 fuel pump cam
d Kraftstoffpumpennocken *m*
f came *f* de pompe à carburant
n brandstofopvoerpompnok
p krzywka pompy paliwowej

2417 fuel pump contact
d Kraftstoffpumpenkontakt *m*
f contact *m* de pompe à combustible
n brandstofpompcontact
p styk pompy paliwa

2418 fuel pump push rod
d Stössel *m* der Kraftstoffpumpe
f poussoir *m* de pompe à combustible
n stootstang van de brandstofpomp
p popychacz pompy paliwa

2419 fuel pump tappet
d Kraftstoffpumpenstössel *m*
f poussoir *m* de pompe d'alimentation
n klepstoter van brandstofopvoerpomp
p popychacz pompy paliwowej

2420 fuel reserve control light
d Kraftstoffmangelkontrollampe *f*
f témoin *m* de minimum d'essence
n brandstofreservevoorraadlicht
p lampka kontrolna zapasu paliwa

2421 fuel return pipe
d Rücklaufleitung *f*
f tube *m* de retour
n brandstofretourleiding
p przewód nadmiarowy paliwa

2422 fuel system
d Brennstoffsystem *n*; Kraftstoffsystem *n*
f système *m* d'alimentation
n brandstoftoevoersysteem
p instalacja paliwowa

2423 fuel tank
d Kraftstoffbehälter *m*; Benzintank *m*
f réservoir *m* à carburant

n brandstoftank
p zbiornik paliwa

2424 fuel tank cap
d Verschlussdeckel *m* des Kraftstoffbehälters
f chapeau *m* de réservoir à carburant
n brandstoftankvuldop
p pokrywa zbiornika paliwa

2425 fuel tap; fuel cock
d Kraftstoffhahn *m*
f robinet *m* de carburant
n brandstofkraantje
p kurek dopływu paliwa

2426 fuel vapour pressure
d Kraftstoffdampfdruck *m*
f pression *f* de vapeur de carburant
n dampdruk van de brandstof
p prężność par paliwa

2427 fulcrum
d Stützpunkt *m*
f point *m* d'oscillation
n steunpunt
p punkt podparcia

2428 full flow filter
d Hauptstromfilter *m*
f filtre *m* à huile sur circuit principal
n hoofdstroomfilter
p filtr pełnego przepływu; filtr szeregowy

2429 full gloss
d Hochglanz *m*
f poli *m* fin
n hoogglans
p wysoki połysk

2430 full load contact
d Vollastkontakt *m*
f contact *m* pleine charge
n vollastcontact
p styk całkowitego obciążenia

2431 full load needle
d Vollastnadel *f*
f aiguille *f* de pleine charge
n vollastnaald
p igła pełnego obciążenia

2432 full scale model
d naturgrosses Modell *n*
f modèle *m* en vraie grandeur
n proefmodel
p model naturalnej wielkości

2433 functioning
d Funktionierung *f*

f   fonctionnement *m*
n   werking
p   funkcjonowanie

2434 furniture van; removal van
d   Möbelkofferwagen *m*
f   fourgon *m* de déménagement; déménageuse *f*
n   verhuiswagen; verhuisauto
p   samochód meblowy

2435 fuse
d   Sicherung *f*
f   fusible *m*
n   zekering
p   bezpiecznik

2436 fuse box
d   Sicherungskasten *m*
f   boîte *f* à fusible
n   zekeringskast
p   skrzynka bezpiecznikowa

2437 fuse cylinder
d   Sicherungszylinder *m*
f   cylindre *m* de fusible
n   zekeringscilinder
p   cylinder bezpiecznika

2438 fuse schema
d   Sicherungsschema *n*
f   tableau *m* des fusibles
n   zekeringtabel; schema met zekeringen
p   tablica bezpiecznikowa

2439 fuse tongs
d   Sicherungsringzangen *fpl*
f   pince *f* pour fusibles
n   zekeringentang
p   kleszcze bezpiecznikowe

G

2440 galvanometer stroboscope
d analoge Einstellampe *f*
f lampe *f* stroboscopique à galvanomètre
n analoge afstellamp
p stroboskop wskaźnikowy
galwanometryczny

2441 gang drilling machine
d Mehrspindelbohrmaschine *f*;
Vielspindelbohrmaschine *f*
f perceuse *f* à broches multiples
n boormachine met meerdere spillen
p wiertarka wielowrzecionowa

2442 gang milling
d Gruppenfräsen *n*
f fraisage *m* combiné
n combinatiefrezen
p frezowanie zespołem frezów

* **gangway** → 962

2443 garage
d Garage *f*
f garage *m*; remise *f*
n garage
p garaż

2444 garage costs
d Garagekosten *mpl*
f frais *mpl* de garage
n garagekosten
p koszta garażu

2445 garageman
d Garagebezitzer *m*
f garagiste *m*
n garagehouder
p właściciel garażu

2446 garage wall panel
d Garagewerkzeugtafel *m*
f tableau *m* "mécanique garage"
n gereedschapsbord in werkplaats
p tablica na narzędza w garażu

2447 gas cylinder
d Gaszylinder *m*
f bouteille *f* à gaz
n gascilinder
p butla do gazów

* **gasket** → 3646

2448 gas oil
d Gasöl *n*
f huile *f* à gaz
n gasolie
p olej gazowy

2449 gasoline; petrol
d Benzin *n*; Gasolin *n*
f essence *f*
n benzine
p benzyna

2450 gasoline engine; petrol engine
d Benzinmotor *m*
f moteur *m* à essence
n benzinemotor
p silnik benzynowy

2451 gasoline filter; petrol filter
d Benzinfilter *m*
f filtre *m* à essence
n benzinefilter
p filtr benzyny

2452 gasoline pump
d Benzinpumpe *f*
f pompe *f* à essence
n benzinepomp
p pompa benzyny

2453 gas thread
d Gasgewinde *n*
f filet *m* des tuyaux à gaz
n gasdraad; gasschroefdraad
p gwint rurowy drobnozwojny

2454 gas turbine
d Gasturbine *f*
f turbine *f* à gaz
n gasturbine
p silnik turbospalinowy

2456 gate
d Sägerahmen *m*
f cadre *m*
n frame; raam
p rama

2455 gate
d Absperrschieber *m*
f registre *m*
n afsluiter
p zasuwa zamykająca

2457 gauge set
(tool to check and adjust carburettor butterfly opening)
d Messstift *m*
f jeu *m* de piges
n meetstift
p kołek pomiarowy

2458 gauze tube
d perforierte Hülse *f*
f tube *m* perforé
n geperforeerde binnenhuls; gaasbuis
p tuleja perforowana

2459 gavelock; iron crowbar
d eisernes Brecheisen *n*
f verdillon *m* en fer
n ijzeren koevoet
p łom żelazny

* **gear lubricant → 2476**

2460 gearbox; transmission gear
d Wechselgetriebe *n*; Schaltgetriebe *n*
f boîte *f* de vitesses
n versnellingsbak
p skrzynka przekładniowa; skrzynka biegów

2461 gearbox cover
d Wechselgetriebedeckel *m*
f couvercle *m* de boîte de vitesses
n versnellingsbakdeksel
p pokrywa skrzynki biegów

2462 gearbox flange
d Getriebeflansch *m*
f bride *f* de fixation de boîte de vitesses
n versnellingsbakflens
p kołnierz skrzynki biegów

* **gearbox fluid → 2476**

2463 gearbox isolating switch
d Wechselgetriebeschalter *m*
f interrupteur *m* de boîte de vitesses
n wisselbakschakelaar
p wyłącznik skrzynki biegów

2464 gearbox oil cooler
d Getriebeölkühler *m*
f refroidisseur *m* de huile pour boîtes de vitesses
n versnellingsbakoliekoeler
p chłodnica oleju przekładniowego

2465 gearbox tunnel
d Wechselgetriebetunnel *m*
f tunnel *m* de boîte de vitesses

n cardantunnel
p tunel skrzynki biegów

2466 gear change ball
d Schaltkugel *f*
f rotule *f*
n kogel van versnellingshefboom
p kulka zmiany biegów

* **gear change lever → 2480**

2467 gear change lever bracket
d Schalthebellagerbock *m*
f support *m* de levier de changement de vitesses
n klembeugel voor schakelhandel
p wspornik dźwigni zmiany biegów

2468 gear change lever location
d Anordnung *f* des Gangschalthebels
f disposition *f* de levier de changement de vitesses
n plaats van de versnellingshandel
p usytuowanie dźwigni zmiany biegów

2469 gear cluster
d Zahnrädersatz *m*
f couple *m* d'engrenages
n tandwielgroep
p zespół kół zębatych

2470 gear coupling
d Bogenzahnkupplung *f*
f accouplement *m* à denture
n tandkoppeling
p sprzęgło zębate

2471 gear drive
d Zahnradantrieb *m*
f commande *f* par engrenages
n distributietandwielpaar; tandwielpaar
p napęd zębaty

2472 geared axle shaft
d Zahnkranzachswelle *f*
f arbre *m* à dentures coniques
n tandkranssteekas
p półoś z wieńcem zębatym

2473 geared flywheel
d Zahnkranz *m* der Schwungscheibe
f couronne *f* dentée de volant
n vliegwiel met starterkrans
p koło zamachowe zębate

2474 gear hydraulic motor
d Zahnradmotor *m*
f moteur *m* à engrenage

n tandradmotor
p silnik zębaty

2475 gear milling cutter
d Modulfräser *m*
f fraise *f* à module
n moduulfrees
p frez modulowy krążkowy

2476 gear oil; gear lubricant; gearbox fluid
d Getriebeöl *n*
f huile *f* pour engrenages
n versnellingsbakolie; tandwielolie
p olej przekładniowy

2477 gear puller
d Zahnradzieher *m*
f arrache-pignon *m*
n tandwieltrekker
p ściągacz do kół zębatych

2478 gear pump
d Zahnradpumpe *f*
f pompe *f* à engrenage
n tandradpomp
p pompa zębata

2479 gear shift fork
d Schaltstangengabel *f*
f fourchette *f* de tirette
n schakelvork
p widełki zmiany biegów

2480 gear shift lever; gear change lever
d Getriebeschalthebel *m*
f levier *m* de changement de vitesse
n versnellingshefboom
p dźwignia zmiany biegów

2481 gear synchronization
d Synchronisierung *f* der Gänge
f synchronisation *f* des vitesses
n versnellingssynchronisatie
p synchronizacja biegów

2482 general map
d Übersichtskarte *f*
f plan *m* d'ensemble
n overzichtskaart
p mapa przeglądowa

2483 general purpose bowl
d Universalkugel *f*
f boule *f* universelle
n universele bol
p kulka uniwersalna

2484 generation
d Entwicklungsphase *f*

f génération *f*
n ontwikkelingsfase
p wytwarzanie

2485 generator; dynamo
d Dynamo *m*; Lichtmaschine *f*
f dynamo *f*; générateur *m*
n dynamo; generator
p generator; prądnica

2486 generator cap
d Lichtmaschinenhaube *f*
f capot *m* de dynamo
n dynamokap
p kołpak prądnicy

2487 generator frame
d Lichtmaschinensockel *m*
f support *m* de génératrice
n dynamobevestiging
p cokół prądnicy

2488 generator test bench
d Lichtmaschinenprüfgerät *n*
f banc *m* d'essai du générateur
n dynamotestbank
p urządzenie kontrolne do prądnicy

2489 geoptic
(tool to control completely and precisely the
geometry of passengers cars and trucks)
d Geoptik *n*
f géoptic *m*
n geoptiek
p aparat do kontroli geometrii samochodu

2490 gibhead
d Keilnase *f*
f talon *m* de clavette
n wigvormig uitsteeksel
p nosek klina

2491 girder truck
d Sperrgutlastwagen *m*
f porteur *m* de pièces longues
n langslader
p samochód dłużycowy

2492 gland
d Stopfbuchse *f*
f presse *f* à étoupe
n pakkingbus
p dławnica

2493 glare *v*; dazzle *v*
d blenden
f éblouir

n verblinden
p oślepiać

2494 glass
d Glas *n*
f verre *m*
n glas; glasruit
p szkło; szyba

2495 glass bowl
d Filterbecher *m*
f boîtier *m* de filtre
n brandstofkolf
p osadnik

2496 glass bulb
d Glühlampenglaskolbe *f*
f ampoule *f*
n glasbol
p szklana bańka żarówki

2497 glass cutter
d Glasschneider *m*
f coupe-verre *m*
n glassnijder
p krajak; diament szklarski

2498 glass fragment
d Glassplitter *m*
f fragment *m* de glace
n glassplinter
p odłamek szkła

2499 glass parting element
d Glasscheiben *fpl* für Regale
f glaces *fpl* pour rayons
n glazen scheidingselement
p szyba do regału

2500 glazed frost
d Glatteis *n*
f verglas *m*
n ijzel
p gołoledź

* **glove box** → 2501

2501 glove compartment; glove box
d Handschuhkasten *m*; Handschuhfach *n*
f plage *f* à gants
n handschoenenkastje
p schowek podręczny

2502 glow
d Vorglühen *n*
f préchauffage *m*
n voorgloeien; voorverwarming
p żarzenie wstępne

2503 glycerine
d Glycerin *n*
f glycérine *f*
n glycerine
p gliceryna

2504 gooseneck
d Schwanenhals *m*
f col-de-cygne *m*
n zwanenhals
p gęsia szyjka

* **gouge bit** → 4434

2505 governor; regulator
d Regler *m*
f gouverneur *m*; régulateur *m*
n regelaar
p regulator

2506 governor base
d Reglergrundplatte *f*
f socle *m* de régulateur
n regulateurplaat
p płyta regulatora

2507 governor collar
d Reglermuffe *f*
f manchon *m* de régulateur
n regelmof
p tuleja regulatora

2508 governor cone
d Reglerkegel *m*
f cône *m* de régulateur
n regelkegel
p stożek regulatora

2509 governor engagement control
d Schaltkorrektor *m*
f correcteur *m* de passage des vitesses
n regulateurkoppelmechanisme
p regulator przekładni zmiany biegów

2510 governor housing
d Reglergehäuse *n*
f carter *m* de régulateur
n regulateurhuis
p obudowa regulatora

2511 governor lever
d Reglerhebel *m*
f levier *m* de régulateur
n hefboom van regulateur
p dźwignia regulatora

2512 governor spring
d Reglerfeder *f*

f ressort *m* de régulateur
n regulateurveer
p sprężyna regulatora

2513 gradation
d Abstufung *f*
f gradation *f*
n gradatie
p stopniowanie

2514 gradientmeter; clinometer
d Steigungsmesser *m*
f gradomètre *m*
n hellingshoekmeter
p klinometr

* **grading instrument** → 2515

2515 gradiograph; grading instrument
d Gefällmesser *m*
f clinomètre *m*
n hellingshoekmeter
p pochyłomierz

2516 gradometer
d Steigungsmesser *n*
f indicateur *m* de pente
n stijgingsmeter
p pochyłomierz

2517 gradual
d stufenweise
f gradué
n trapsgewijs
p stopniowy

2518 graphite
d Graphit *m*
f graphite *m*
n grafiet
p grafit

2519 graphite bearing
d Kohlelager *n*
f palier *m* en graphite
n grafietglijlager
p łożysko grafitowe

2520 graphite ring
d Graphitring *m*
f anneau *m* en graphite
n grafietring
p pierścień grafitowy

2521 graphite tube
d Graphitzerstäuber *m*
f tube *m* de graphite

n grafiettube
p rozpylacz grafitowy

* **gratel** → **3983**

* **grating of gear** → **1075**

2522 gravel
d Schotter *m*
f gravier *m*
n grind
p żwir

2523 graving tool
d Gravurstift *m*
f crayon *m* graveur
n graveerpen
p ołówek do grawerowania

2524 grease
d Schmierfett *n*
f graisse *f*
n smeervet
p smar

2525 grease bag
d Lederstulpe *f*
f manchon *m* en cuir souple
n vetzak
p pierścień samouszczelniający skórzany

2526 grease chamber
d Schmierfettspeicher *m*
f réservoir *m* de graisse
n vetkamer
p zasobnik smaru

2527 grease connection
d Schmieranschluss *m*
f raccord *m* de graissage
n doorsmeeraansluiting
p połączenie smarowane

2528 grease cup
d Schmierbüchse *f*; Schmiertopf *m*
f chapeau-graisseur *m*
n smeerpot
p smarownica; smarowniczka

2529 grease fittings
(connectors to fit all popular grease nipples)
d Sortiment *n* mit Schmiermundstücken
f composition *f* embouts de graissage
n set mondstukken
p łącznik rozgałęźny przewodów
smarowniczych

2530 **grease gun**
  d Fettpresse *f*; Schmierpresse *f*
  f compresseur *m* à graisse; pompe *f* à graisse
  n smeerspuit; vetspuit
  p smarownica tłokowa; smarownica ciśnieniowa

2531 **grease nipple**
  d Schmierkopf *m*; Abschmiernippel *m*
  f graisseur *m* à graisse
  n smeernippel; vetnippel
  p smarowniczka

2532 **grease remover**
  d Fettentfernungsmittel *n*
  f dégraisseur *m*
  n ontvetter
  p odtłuszczacz

  * **greasing pad** → 3167

2533 **greasing pit**
  d Abschmiergrube *f*
  f fosse *f* de grainage
  n smeerkuil
  p rów do przeprowadzania robót smarowych

2534 **greasing station**
  d Abschmierdienst *m*
  f station *f* de graissage
  n smeerstation
  p stanowisko do smarowania

2535 **grib head taper key**
  d Nasenkeil *m*
  f clavette *f* inclinée à talon
  n kopspie
  p klin wzdłużny noskowy

  * **grid** → 2536

2536 **grille; grid**
  d Grill *m*; Rost *m*
  f grille *f*
  n grille; rooster
  p krata; okratowanie

2537 **grilled tube radiator**
  d Rippenrohrkühler *m*
  f radiateur *m* à tuyaux à ailettes
  n radiateur met koelribben
  p chłodnica z rurkami użebrowanymi

2538 **grind** *v*
  d schleifen
  f rectifier; meuler
  n slijpen
  p szlifować

2539 **grinding machine**
  d Schleifmaschine *f*
  f machine *f* à rectifier
  n slijpmachine
  p szlifierka

2540 **grinding of the engine**
  d Motoreinlaufen *n*
  f rodage *m* du moteur
  n inrijproces van motor
  p docieranie silnika

2541 **grinding set**
  d Schleifkörpersatz *m*
  f assortiment *m* de meules
  n slijpset
  p zestaw ściernic

2542 **grinding stone**
  d Schleifstein *m*
  f meule *f*
  n hoonsteen; slijpsteen
  p kamień szlifierski

2543 **grinding wheel dresser; abrasive wheel dresser**
  d Schleifscheibenabrichter *m*
  f décrasse-meules *f*
  n slijpsteenscherper
  p obciągacz ściernicowy

2544 **gripping jaw**
  d Spannbacke *f*
  f griffe *f* de serrage
  n spanklauw
  p szczęka mocująca

2545 **gripping kit**
  d Zugarmsatz *m*
  f ensemble *m* crampon
  n trekarmset
  p komplet do wyciągania

  * **grit sprayer** → 4280

2546 **groove**
  d Schlitz *m*
  f rainure *f*; cannelure *f*; encoche *f*
  n groef
  p rowek; żłobek

2547 **grooved straight pin**
  d Kerbstift *m*
  f goupille *f* cannelée
  n kerbstift
  p kołek z korbami

* **groove milling tool** → 4556

2548 groove plane
d Nuthobel *m*
f bouvet *m* femelle
n moerploeg; groefschaaf
p strug wpustnik; strug kątnik

2549 groove seal
d Spaltdichtung *f*
f joint *m* gorgé
n groefafdichting
p uszczelnienie szczelinowe

2550 grooving hammer
d Pinnhammer *m*
f marteau *m* à suage
n hamer om groeven te maken
p młotek do rowków

2551 grooving saw
d Nutsäge *f*
f scie *f* à rainurer
n groefzaag
p piła tarczowa

* **grooving tool** → 3683

2552 gross weight
d zulässiges Gesamtgewicht *n*
f poids *m* total en plein charge
n maximaal toelaatbaar totaalgewicht
p maksymalny dopuszczalny ciężar całkowity

2553 ground clearance
d Bodenfreiheit *f*
f dégagement *m* entre le véhicule et le sol
n bodemvrijheid; bodemspeling
p prześwit

2554 ground clearance compensator
d Höhenkorrekturgerät *n*; Niveauregler *m*
f régulateur *m* de hauteur; correcteur *m* de hauteur
n niveauregelklep van hydropneumatisch veersysteem; hoogteregelaar
p regulator wzniosu nadwozia

* **ground electrode** → 1858

2555 ground plan
d Grundriss *m*
f plan *m* horizontal
n bovenaanzicht
p rzut poziomy

2556 grub screw
d Stiftschraube *f*
f vis *f* de pointage
n stelschroef; tapbout
p wkręt dociskowy

2557 guarded railway crossing
d Bahnübergang *m* mit Schranke
f passage *m* à niveau sans barrage
n bewaakte spoorwegovergang
p przejazd kolejowy z zaporami

* **gudgeon pin** → 3771

2558 gudgeon pin slide
d Gleitpilz *m* des Kolbenbolzens
f glissière *f* d'axe de piston
n zuigerpenborging
p ślizgacz sworznia tłokowego

2559 guide
d Führung *f*
f guidage *m*
n geleider; geleiding
p prowadnica

2560 guide lever
d Regelhebel *m*
f levier *m* de guide
n regelhendel
p dźwignia regulująca

2561 guide lever bracket
d Führungshebelstütze *f*
f support *m* de levier-guide
n lagering van de hulpstuurstang
p wspornik drążka poprzecznego

2562 guide pin
d Führungbolzen *m*
f boulon *m* de guidage
n geleidingspen; geleidingsbout
p kołek prowadzący

2563 guide plate
d Führungsplatte *f*
f plaque *f* de guidage
n geleidingsplaat
p płyta wiodąca; płyta prowadząca

2564 guide rod
d Führungsstange *f*
f barre *f* de conductrice
n geleidstang
p drążek prowadzący

2565 guide roller
d Umlenkrolle *f*
f poulie *f* de renvoi

n geleiderol
p krążek prowadzący; rolka prowadząca

2566 guillotine shears
 d Rahmenschere *f*
 f cisaille *f* à guillotine
 n guillotineschaar
 p nożyce gilotynowe

H

2567 hacksaw
  d Bügelsäge *f*
  f scie *f* alternative
  n metaalzaag; ijzerzaag
  p piła do metali

2568 hair compasses
  d Bogen *m* und Stellschraube *f*
  f compas *m* à pointes à charnière
  n steekpasser met scharnier en stelschroef
  p odmierzacz

2569 half elliptic spring
  d Halbfeder *f*
  f semi-elliptique *m*
  n halfelliptische veer
  p resor półepileptyczny

2570 half round chisel
  d halbrunder Beitel *m*
  f ciseau *m* demi-rond
  n halfronde beitel
  p dłuto półokrągłe

2571 half round file
  d Halbrundfeile *f*
  f lime *f* demi-ronde
  n halfronde vijl
  p pilnik półokrągły

2572 half round sharpening stone
  d Halbrundschleiffeile *f*
  f pierre *f* à affuter demi-ronde
  n halfronde aanzetsteen
  p pilnik półokrągły ścierny

2573 half shaft
  d Achshälfte *f*
  f demi-essieu *m*
  n steekas
  p półoś napędowa

2574 half smooth file
  d Halbschlichtfeile *f*
  f lime *f* demi-douce
  n halfzoetvijl
  p pilnik półgładzik

2575 half truck vehicle
  d Halbkettenfahrzeug n
  f half-track *m*
  n halfrupstrekker
  p pojazd półgąsienicowy

2576 halogen bulb; iodine bulb; quartz iodine bulb
  d Halogenlampe *f*; Jodlampe *f*
  f lampe *f* halogène; lampe *f* à iode
  n halogeenlamp
  p żarówka halogenowa; żarówka jodowa

2577 hammer
  d Hammer *m*
  f marteau *m*
  n hamer
  p młotek

2578 hammer grip
  d Hammerhandgriff *m*
  f poignée *f* de marteau
  n hamerhandgreep
  p uchwyt młotka

2579 hammer head bolt
  d Hammerkopfschraube *f*
  f vis *f* à tête rectangulaire
  n hamerkopbout; verbindingsbout met T-kop
  p śruba złączna o łbie młoteczkowym

2580 hammer pane
  d Hammerfinne *f*
  f panne *f* du marteau
  n hamerpen
  p nosek młotka

2581 hand brake
  d Handbremse *f*
  f frein *m* à main
  n handrem
  p hamulec ręczny

2582 hand brake bracket
  d Halter *m* der Handbremse
  f support *m* de frein à main
  n handremsteun
  p wspornik hamulca ręcznego

2583 hand brake lever
  d Handbremshebel *m*
  f levier *m* de frein à main
  n handremhendel
  p dźwignia hamulca ręcznego

2584 hand chisel
  d Handbeitel *m*
  f ciseau *m* à main
  n handbeitel
  p dłuto ręczne

2585 hand crimper
d Handwürgezange *f*
f pince *f* à soyer
n kantzettang
p ręczne szczypce do obciskania

2586 hand file
d Handfeile *f*
f lime *f* à main
n handvijl; blokvijl
p pilnik ręczny

2587 hand hammer
d Handhammer *m*
f marteau *m* à main
n handhamer
p młot ręczny

* **hand lamp** → 3837

2588 handle
d Handgriff *m*
f poignée *f*
n handgreep
p rękojeść; uchwyt

2589 hand level
d Handwasserwage *f*
f niveau *m* portatif
n handwaterpas
p poziomnica ręczna

2590 hand milling machine
d Fräsmaschine *f* mit Handbedienung
f fraiseuse *f* à main
n handfrees
p frezarka ręczna

2591 hand rasp
d Handraspel *f*
f râpe *f* à main
n handrasp
p tarnik ręczny

2592 hand reamer
d Handreibahle *f*
f alésoir *m* à main
n handruimer
p rozwiertak ręczny

2593 hand rest
d Werkzeughalter *m*
f support *m* d'outils à main
n handgereedschapshouder
p uchwyt narzędziowy

2594 handsaw
d Handsäge *f*

f scie *f* à main
n handzaag
p piła ręczna jednochwytowa

2595 hand sounder
d Handlot *n*
f sonde *f* à main
n handlood
p sonda ręczna

2596 hand stamp machine
(machine for cylinder and car keys)
d Handprägevorrichtung *f*
f estampilleuse *f*
n handpersmachine
p urządzenie do ręcznego wytłaczania

2597 hand tap
d Handgewindebohrer *m*
f taraud *m* à main
n handtap
p gwintownik ręczny

2598 hand tools
d Handwerkzeuge *npl*
f outillage *m* à main
n handgereedschap
p narzędzia ręczne

2599 handwheel
d Handrad *n*
f volant *m*
n handwiel
p koło ręczne

2600 harden *v*
d härten
f tremper
n harden
p hartować

2601 hardener
d Härter *m*
f agent *m* de durcissement
n verharder
p utwardzacz; środek utwardzający

2602 hardening paste
d Härterpaste *f*
f pâte *f* pour durcissement
n harderpasta
p pasta utwardzająca

2603 hardening temperature
d Härtetemperatur *f*
f température *f* de durcissement
n hardingstemperatuur
p temperatura twardnienia

2604 hard faced valve
d Panzerventil *n*
f soupape *f* blindée
n gepantserde klep
p zawór z przylgniami natapianymi stopami
twardymi

2605 hardness
d Härte *f*
f dureté *f*
n hardheid
p twardość

2606 hardness of a spring
d Federhärte *f*
f dureté *f* de ressort
n hardheid van een veer
p twardość sprężyny

2607 hardness test
d Härteprüfung *f*
f test *m* de dureté
n hardheidsproef
p próba twardości

* **hard plug** → 1139

2608 hard solder
d Hartlot *n*
f soudure *f* pour braser
n hardsolderen
p lutowanie twarde

2609 hard top
d Hardtop *m*
f décapotable *m*
n auto met afneembaar dak
p hard top auto; samochód ze zdejmowalnym
dachem

2610 harmonic balancer
d harmonischer Stabilisator *m*
f stabilisateur *m* harmonique
n torsietrillingsdemper
p stabilizator harmoniczny

2611 hatchet iron
d Hammerlötkolben *m*
f fer *m* à souder à marteau
n hamersoldeerbout
p lutownica kątowa

2612 hatchet stake
d Umschlageisen *n*
f tranche *f* de ferblantier
n plaatwerkerbeitel
p zaginadło blacharskie proste

* **haul** *v* → 5128

2613 hazardous materials
d Gefahrengüter *npl*
f matériaux *mpl* dangereux
n gevaarlijke stoffen
p materiały niebezpieczne

2614 head cloth
d Dachüberspannung *f*; Dachauskleidung *f*
f garniture *f* du toit; revêtement *m* du toit
n dakhemel; dakhemelbekleding
p poszycie dachu

2615 headlamp
d Hauptscheinwerfer *m*
f projecteur *m* principal
n koplamp
p reflektor główny

2616 headlamp aligner
d Einstellgerätn für Scheinwerfer
f contrôleur *m* d'éclairage
n koplampafstelapparaat
p aparat ustawiający światła główne

2617 headlamp bezel; headlamp rim
d Deckelring *m* des Scheinwerfers
f bride *f* de projecteur
n rand van de koplamp
p obramowanie żarówki

2618 headlamp body
d Scheinwerfergehäuse *n*
f carcasse *f* de projecteur
n koplichthuis
p obudowa reflektora

2619 headlamp control light; headlamp indicator
d Fernlichtanzeige *f*
f témoin *m* de feux de route; témoin *m* de
phares; témoin *m* de projecteurs
n controlelicht voor groot licht
p lampa kontrolna reflektorów

2620 headlamp holder
d Scheinwerferfassung *f*
f douille *f* de projecteur
n gloeilampfitting
p oprawka żarówki

* **headlamp indicator** → 2619

* **headlamp rim** → 2617

2621 headlamp socket
d Scheinwerferanschluss *m*

f culot *m* de phare
n fitting voor koplamp
p gniazdo reflektora

2622 headless screw; cup point set screw
d Gewindestift *m*
f vis *f* sans tête fendue
n stelschroef
p wkręt bez łba

2623 head on collision
d Frontalzusammenstoss *m*
f choc *m* frontal; impact *m* frontal
n frontale botsing
p zderzenie czołowe

2624 headrest
d Kopstütze *f*; Kopflehne *f*
f appui-tête *m*
n hoofdsteun
p podgłówek

2625 head wind; wind ahead
d Gegenwind *m*
f vent *m* de bout
n tegenwind
p wiatr przeciwny

2626 heat absorption capacity
d Wärmeaufnahmefähigkeit *f*
f capacité *f* thermique
n warmteopnemingcapaciteit;
   warmteabsorptiecapaciteit
p zdolność pochłaniania ciepła

2627 heat capacity
d Wärmekapazität *f*
f capacité *f* calorifique
n warmtecapaciteit
p pojemność cieplna

2628 heat conductivity; thermal conductivity
d Wärmeleitungsvermögen *n*
f conductibilité *m* de la chaleur
n warmtegeleidingsvermogen;
   warmtegeleidbaarheid
p przewodność cieplna

2629 heat dissipation
d Wärmeabführung *f*
f dissipation *f* de chaleur
n warmteafgifte; warmteafvoer
p odprowadzanie ciepła

2630 heated rear window
d heizbare Heckscheibe *f*
f vitre *f* arrière à filament chaud

n verwarmde achterruit
p szyba tylna ogrzewana

2631 heater blower
d Heizluftgehäuse *n*
f soufflante *f* d'air réchauffé
n aanjager voor warme lucht
p dmuchawa ogrzewanego powietrza

2632 heater starter switch
d Glühanlassschalter *m*
f commutateur *m* de préchauffage-démarrage
n voorgloeistartschakelaar
p przełącznik uruchomienia żarzenia

2633 heater switch
d Wärmeschalter *m*
f contacteur *m* de chauffage
n verwarmingschakelaar
p wyłącznik ogrzewania

2634 heat exchanger
d Wärmetauscher *m*
f échangeur *m* de chaleur
n warmtewisselaar
p wymiennik ciepła

2635 heat gun
d Heizpistole *f*
f pistolet *m* thermique
n heteluchtpistool
p dmuchawa gorącego powietrza

2636 heating
d Heizung *f*
f chauffage *m*
n verwarming
p ogrzewanie

2637 heating element
d Heizelement *n*
f élément *m* chauffant
n verwarmingselement
p element grzejny

2638 heating fan
d Heizgebläse *n*
f soufflante *f*
n kachelaanjager
p dmuchawa

2639 heating system
d Heizsystem *n*
f circuit *m* de chauffage
n verwarmingssysteem
p instalacja ogrzewcza

2640 heat insulation
d Wärmeisolierung *f*

   f calorifugeage *m*
   n warmte-isolatie
   p izolacja cieplna; izolacja termiczna

2641 heat resistance; thermal stability
   d Wärmebeständigkeit *f*
   f stabilité *f* à la chaleur; résistance *f* à la chaleur
   n hittebestendigheid
   p wytrzymałość cieplna

2642 heat transmission
   d Wärmeübertragung *f*
   f échange *m* calorifique
   n warmteoverdracht
   p wymiana ciepła; przenoszenie ciepła

2643 heat treatment
   d Wärmebehandlung *f*
   f traitement *m* thermique
   n warmtebehandeling
   p obróbka cieplna

2644 heavy oil
   d Schweröl *n*
   f huile *f* lourde
   n zware olie
   p olej ciężki

2645 heel dolly
   d Handfaust *f* Absatzform
   f tas *m* talon
   n hakvormtas
   p klepadło do krawędzi

2646 height measuring lever
   d Niveauregelventilhebel *m*
   f levier *m* de correction d'assiette
   n niveauregelingshefboom van luchtveersysteem
   p dźwignia stabilizacyjna; dźwignia poziomująca

2647 helical gear
   d Schrägstirnrad *n*
   f roue *f* dentée cylindrique à denture hélicoïdale
   n schuin tandwiel
   p koło zębate walcowe skośne

2648 helical roller
   d Federrolle *f*
   f rouleau *m* en hélice
   n schroefrol; spiraalvormige rol
   p wałek sprężysty; krążek sprężysty

2649 helical spring
   d Spiralfeder *f*
   f ressort *m* en spirale

   n spiraalveer
   p sprężyna śrubowa

2650 helper leaf; overload leaf
   d Zusatzfederblatt *n*
   f lame *f* compensatrice; lame *f* auxiliaire
   n hulpveerblad
   p pióro resoru

2651 hemispherical ball pin
   d Halbkugelzapfen *m*
   f queue *f* de rotule sphérique
   n halvekogeltap
   p sworzeń półkulowy

2652 hemispherical chamber in piston
   d halbkugelige Kammer *f* des Kolbens
   f chambre *f* hémisphérique du piston
   n halfbolvormige uitsparing in zuiger
   p półkulista przestrzeń tłoka

2653 heptane
   d Heptan *n*
   f heptane *m*
   n heptaan
   p heptan

2654 herringbone gear
   d Pfeilradgetriebe *n*
   f engrenage *m* à chevrons
   n tandwiel met pijlvertanding; tandwiel met V-vormige tanden
   p koło zębate daszkowe; koło zębate strzałkowe

2655 hexagonal head
   d Sechskantkopf *m*
   f tête *f* à six pans
   n zeskantige kop
   p łeb sześciokątny

2656 hexagonal nut
   d Sechskantmutter *f*
   f écrou *m* à six pans
   n zeskantige moer
   p nakrętka sześciokątna

2657 hexagon castellated nut
   d Kronenmutter *f*
   f écrou *m* crénelé; écrou *m* à créneaux dégagés
   n zeskante kroonmoer
   p nakrętka koronowa sześciokątna

2658 hexagon fastening bolt
   d Sechskantkopfverbindungsschraube *f*
   f vis *f* à tête hexagonale
   n verbindingsbout met zeskantige kop
   p śruba złączna o łbie sześciokątnym

2659 hexagon set screw
  d Sechskantkopfdruckschraube *f*
  f vis *f* de pression à tête hexagonale
  n bout met zeskantige kop
  p śruba dociskowa o łbie sześciokątnym

2660 hexagon socket head screw
  d Innensechskantzylinderkopfschraube *f*
  f vis *f* à tête cylindrique à six pans creux
  n cilinderkopbout met binnenzeskant
  p śruba złączna o łbie walcowym i gnieździe sześciokątnym

2661 high frequency generator
  d Hochfrequenzgenerator *m*
  f générateur *m* de haute fréquence
  n generator met hoge frequentie
  p generator wysokiej częstotliwości

2662 high gear; top gear
  d hoche Übersetzung *f*
  f multiplication *f* forte; prise *f* directe
  n hoge overbrenging
  p przekładnia biegu szybkiego

2663 high grade steel
  d Stahl *m* von hoher Qualität
  f acier *m* fin
  n hoogwaardig staal
  p stal wysokiej jakości

2664 high pressure lubricator
  d Hochdruckschmierapparat *n*
  f équipement *m* de graissage forcé
  n hogedruksmeernippel; hogedrukoliespuit
  p smarownica wysokociśnieniowa

2665 high pressure oil pump
  d Hochdrucköl pumpe *f*
  f pompe *f* à huile à haute pression
  n hogedrukoliepomp
  p wysokociśnieniowa pompa oleju

2666 high pressure pipe
  d Hochdruckleitung *f*
  f tube *m* à haute pression
  n hogedrukpijp
  p przewód wysokociśnieniowy

2667 high pressure pump
  d Hochdruckpumpe *f*
  f pompe *f* à haute pression
  n hogedrukpomp
  p pompa wysokociśnieniowa

2668 high pressure tyre
  d Hochdruckreifen *m*

  f pneumatique *m* à haute pression
  n hogedrukautoband
  p opona wysokociśnieniowa

2669 high revolutions
  d hoche Touren *fpl*
  f hautes fréquences *fpl*
  n hoge frequenties
  p wysokie obroty

2670 high speed downshift valve
  d Hochgeschwindigkeitsrückschaltventil *n*
  f soupape *f* de rétrogradation à grande vitesse
  n snelwerkende terugschakelklep
  p zawór zwrotnego przełączenia dużej prędkości

2671 high speed reduction gear
  d Vorgelege *f* der kleinen Untersetzung
  f réducteur *m* de régime rapide
  n reductietandwielpaar met lage overbrengingsverhouding
  p przekładnia małego przełożenia

2672 high tension cable
  d Zündkerzenleitung *f*
  f câble *m* de haute pression
  n hoogspanningskabel
  p kabel wysokociśnieniowy

2673 high tension distributor
  d Hochspannungsverteiler *m*
  f distributeur *m* de haute tension
  n hoogspanningsverdeler
  p rozdzielacz wysokiego napięcia

  * **high tension winding** → 4349

2674 high voltage
  d Hochspannung *f*
  f haute tension *f*
  n hoogspanning
  p wysokie napięcie

2675 hinge
  d Gelenk *n*
  f charnière *f*
  n scharnier
  p zawiasa

2676 hinged fork spring compressor
       (instrument used when changing a strut in situ or on the workbench)
  d Federbeinspanner *m* mit verstellbaren Klauen
  f compresseur *m* de ressort à griffes orientables
  n veerspanner met draaibare houders
  p przyrząd do demontażu sprężyn z uchwytami obrotowymi

2677 **hinged wrench**
 d Zündkerzensteckschlüssel *m* mit Gelenk
 f clé *f* articulée
 n sleutel met kniegewricht
 p klucz z przegubem

2678 **hinge framing chisel**
 d Fitschenbeitel *m*
 f ciseau *m* pour charnières
 n scharnierbeitel
 p dłuto gniazdowe do zawias wpuszczanych

2679 **hinge pin; pivot pin**
 d Scharnierstift *m*; Scharnierbolzen *m*
 f pivot *m* de charnière
 n scharnierpen
 p sworzeń zawiasy

2680 **hinge pin tool**
 d Scharnierstiftdübel *m*
 f outil *m* pour les axes de charnières
 n scharnierpendrevel;
    scharnierpengereedschap
 p przyrząd do sworzni zawiasów

2681 **hired car; rental car**
 d Mietwagen *m*
 f voiture *f* de louage
 n huurauto
 p samochód wynajęty

2682 **hiss**
 d Geräusch *n*
 f bruit *m*
 n geruis
 p szum

 * **hob** → 2683

2683 **hobbing cutter; hob**
 d Abwalzfräser *m*
 f fraise *f* vis-mère
 n genererende frees
 p frez ślimakowy

2684 **holdfast**
 d Spannkluppe *f*
 f crampon *m*
 n klamp
 p imadło drewniane; trzymadełko; urządzenie
    przytrzymujące

2685 **holding fixture**
 d Haltevorrichtung *f*
 f dispositif *m* de fixation; bride *f* de montage
 n montageklem
 p docisk

2686 **hole cutting shear**
 d Lochschere *f*
 f cisaille *f* coupe-trou
 n schaar om gaten te snijden
 p dziurkarka

2687 **hole nozzle**
 d Lochdüse *f*
 f injecteur *m* à trou
 n eengatsverstuiver
 p rozpylacz jednootworkowy

2688 **hollowing plane**
 d Hohlkehlhobel *m*
 f rabot *m* à gorge
 n holschaaf
 p strug żłobik

2689 **hollow jack**
 d Hohlschraubenwinde *f*
 f vérin *m* creux
 n holle vijzel
 p podnośnik

 * **hollow punch** → 476

2690 **hollow screw**
 d hohle Schraube *f*
 f vis *f* raccord creuse
 n banjoschroef
 p śruba wydrążona

2691 **hollow shaft**
 d Hohlwelle *f*
 f arbre *m* creux
 n holle as
 p wał pusty; wał przewiercony; wał drążony

2692 **hone** *v*
 d honen
 f honer
 n honen
 p gładzić; honować

2693 **honeycomb radiator; cellular radiator**
 d Lamellenkühler *m*; Wabenkühler *m*;
    Zellenkühler *m*
 f radiateur *m* nid d'abeilles
 n radiateur met honingraatvormig koelblok
 p chłodnica ulowa

2694 **honeycomb radiator core**
 d Lamellenkühlnetz *n*
 f radiateur *m* à ailettes
 n lamellenkoelblok
 p rdzeń komorowy

2695 **honing**
 d Schleifen *n*

f rectification *f*; meulage *m*
n aanzetten; wetten
p szlifowanie; gładzenie

2696 honing machine
d Honmaschine *f*
f machine *f* à roder
n uitslijpmachine; hoonmachine
p szlifierka-wygładzarka; honownica

2697 hook spanner
d Hakenschlüssel *m*
f clé *f* à crochet; clé *f* à téton
n haaksleutel; pennensleutel
p klucz do nakrętek okrągłych z wcięciami

2698 hook type electrode
d Hakenelektrode *f*
f électrode *f* à crochet
n hoekelektrode
p elektroda hakowa

2699 hoop iron
d Bandeisen *n*
f fer *m* feuillard
n bandenlichter; bandijzer
p stal obręczowa; bednarka

2700 hopper semitrailer
d Bunker-Auflieger *m*
f semi-remorque *f* à déchargement par gravité
n oplegger met onderlosinrichting
p naczepa-bunker zsypowy

2701 horizontal carburetter
d Horizontalvergaser *m*
f carburateur *m* horizontal
n horizontale carburateur
p gaźnik poziomy

2702 horizontal cylinder
d liegender Zylinder *m*
f cylindre *m* horizontal
n liggende cilinder
p cylinder poziomy

* horizontal miller → 2703

**2703 horizontal milling machine; horizontal
     miller**
d Wagerechtfräsmaschine *f*
f fraiseuse *f* horizontale
n horizontale freesmachine
p frezarka pozioma

2704 horizontal outline
d Draufsicht *f*
f contour *m* en vue de dessus

n bovenaanzicht
p obrys poziomy

2705 horn
d Signalhorn *m*; Signalhupe *f*
f avertisseur *m* sonore
n hoorn; claxon
p klakson; sygnał dźwiękowy

2706 horn button
d Horndruckknopf *m*
f bouton *m* d'avertisseur
n claxonknop
p przycisk sygnału

2707 horn button spring
d Horndruckknopffeder *f*
f ressort *m* de bouton de cornet
n claxonknopveer
p sprężyna przycisku sygnału

2708 horn support
d Träger *m* für Signalhorn
f support *m* d'avertisseur
n claxonbevestigingsbeugel
p wspornik klaksonu

2709 horsepower formula
d Leistungsformel *f*
f formule *f* de puissance
n vermogenformule
p wymiar mocy; formuła mocy

2710 horseshoe
d Hufeisen *n*
f fer *m* à cheval
n hoefijzer
p podkowa

2711 hose
d Schlauch *m*
f boyau *m*; tubulure *f*
n slang
p wąż; przewód giętki

2712 hose clamp
d Schlauchanschlussschelle *f*
f collier *m* à tubulure
n slangklem
p opaska zaciskowa węża

2713 hose nipple
d Schlauchstützen *m*
f embout *m*
n slangnippel
p króciec rury

2714 hot air inlet
d Heisslufteintritt *m*

f　arrivée *f* d'air chaud
n　heteluchtinlaat
p　otwór wlotowy ciepłego powietrza

2715　hot air pipe
d　Warmluftansaugstutzen *m*
f　tubulure *f* d'air chaud
n　inlaatbuis voor warme lucht
p　króciec ciepłego powietrza

2716　hot bearing
d　heissgelaufenes Lager *n*
f　palier *m* échauffé
n　heetgelopen lager
p　łożysko nagrzane

2717　hot bulb
d　Glühkopf *m*
f　boule *f* incandescente
n　gloeilamp
p　gruszka żarowa

2718　hot bulb engine
d　Glühkopfmotor *m*
f　moteur *m* à boule chaude
n　gloeikopmotor
p　silnik żarowy

2719　hot cutting chisel
d　Warmschrotmeissel *m*
f　tranche *f* à chaud
n　warmbeitel
p　przecinak na gorąco

2720　hot plug; soft plug
d　heisse Zündkerze *f*
f　bougie *f* chaude
n　warme bougie
p　świeca "gorąca"

2721　hot riveted
d　warmgenietet
f　rivé à chaud
n　warmgeniet
p　nitowany na gorąco

2722　hot rolled section
d　Warmwalzprofil *n*
f　profilé *m* laminé à chaud
n　warmgewalst profiel
p　profil walcowany na gorąco

2723　hot spot chamber
d　Vorwärmkammer *f*
f　chambre *f* de réchauffage
n　voorverwarmingskamer
p　komora podgrzewcza

2724　hour glass screw
d　Kugelschnecke *f*
f　vis *f* globique
n　hol wormwiel
p　ślimak globoidalny

2725　house trailer
d　Wohnanhänger *m*
f　remorque *f* home mobile; roulotte *f*
n　verblijfsaanhanger
p　przyczepa mieszkalna

2726　housing
d　Gehäuse *n*
f　boîte *f*; corps *m*
n　huis; ommanteling
p　obudowa; oprawa; obsada; osłona

2727　H type engine
d　H-Motor *m*
f　moteur *m* à cylindres en H
n　H-type motor
p　silnik o układzie cylindrów H

2728　hub
d　Nabe *f*
f　moyeu *m*
n　naaf
p　piasta

*　**hub bolt → 5446**

2729　hub cap; hub cover; wheel gap
d　Zierdeckel *m*; Radkappe *f*; Raddeckel *m*
f　enjoliveur *m*; cache-roue *m*
n　naafdop; wieldop
p　kołpak koła; kołpak piasty

*　**hub cover → 2729**

2730　hub puller
d　Nabenzieher *m*
f　arrache-moyeu *m*
n　naaftrekker
p　ściągacz piasty koła

2731　hydraulic accumulator
d　Hydrospeicher *m*
f　accumulateur *m* hydraulique
n　hydraulische accumulator
p　akumulator hydrauliczny

2732　hydraulic annular piston
d　Hydraulikringkolben *m*
f　piston *m* hydraulique annulaire
n　hydraulische ringzuiger
p　hydrauliczny tłok pierścieniowy

2733 hydraulic brake
d Flüssigkeitsbremse *f*
f frein *m* hydraulique
n hydraulische rem
p hamulec hydrauliczny

2734 hydraulic circuit
d hydraulisches Schaltbild *n*
f circuit *m* hydraulique
n hydraulisch circuit
p obwód hydrauliczny

2735 hydraulic clutch operation
d hydraulischer Kupplungsausrücker *m*
f commande *f* hydraulique de débrayage
n hydraulisch ontkoppelmechanisme;
   hydraulische bediening van de koppeling
p wyprzęgnik hydrauliczny

2736 hydraulic control
d hydraulische Steuerung *f*
f commande *f* hydraulique
n hydraulische aandrijving
p sterowanie hydrauliczne

2737 hydraulic cylinder
d hydraulischer Zylinder *m*
f cylindre *m* hydraulique
n hydraulische cilinder
p cylinder hydrauliczny

2738 hydraulic damper piston
d Dämpferkolben *m*
f piston *m* amortisseur
n plunjer van hydraulische demper
p tłoczek amortyzatora hydraulicznego

2739 hydraulic drive
d hydraulischer Antrieb *m*
f commande *f* hydraulique
n hydraulische aandrijving
p napęd hydrauliczny

* **hydraulic fluid** → 2743

2740 hydraulic lash adjuster
d hydraulischer Ventilstössel *m*
f poussoir *m* hydraulique
n hydraulische klepstoter
p hydrauliczny popychacz zaworu

2741 hydraulic lift
d hydraulischer Wagenheber *m*
f élévateur *m* hydraulique
n hydraulische krik
p podnośnik hydrauliczny

2742 hydraulic motor
d Hydromotor *m*

f moteur *m* hydraulique
n hydromotor
p silnik hydrostatyczny

2743 hydraulic oil; hydraulic fluid
d Hydrauliköl *n*
f fluide *f* hydraulique
n hydraulische olie
p olej hydrauliczny

2744 hydraulic pipe
d Flüssigkeitsleitung *f*
f ligne *f* hydraulique
n hydraulische leiding
p przewód hydrauliczny

2745 hydraulic piston
d Hydraulikkolben *m*
f piston *m* hydraulique
n hydraulische plunjer
p tłoczek wspieracza hydraulicznego

2746 hydraulic power drive
d hydraulischer Antrieb *m*
f commande *f* hydraulique
n hydraulische aandrijving
p napęd hydrauliczny

2747 hydraulic regulator
d hydraulischer Regler *m*
f régulateur *m* hydraulique
n hydraulische regulateur
p regulator hydrauliczny

2748 hydraulic reservoir
d Hydraulikbehälter *m*
f réservoir *m* hydraulique
n vloeistoftank
p zbiornik z cieczą roboczą

2749 hydraulic resistance
d hydraulischer Widerstand *m*
f résistance *f* hydraulique
n stromingsweerstand
p opór hydrauliczny

2750 hydraulic shock absorber
d Flüssigkeitsdämpfer *m*
f amortisseur *m* hydraulique
n hydraulische schokdemper
p amortyzator hydrauliczny

2751 hydraulic stabilizer
d hydraulischer Stabilisator *m*
f stabilisateur *m* hydraulique
n hydraulische stabilisator
p stabilizator hydrauliczny

2752 hydraulic switch
d hydraulischer Schalter *m*
f interrupteur *m* hydraulique
n hydraulische schakelaar
p wyłącznik hydrauliczny

2753 hydraulic system
d hydraulische Betätigung *f*
f système *m* hydraulique
n hydraulisch systeem
p układ hydrauliczny

2754 hydraulic torque converter
d Flüssigkeitsgetriebe *n*
f transmission *f* hydraulique
n hydraulische transmissie
p przekładnia hydrauliczna

2755 hydraulic transformer
d hydraulischer Druckübersetzer *m*
f transformateur *m* hydraulique
n hydraulische drukomzetter
p hydrauliczny przełącznik ciśnienia

2756 hydraulic unit
d hydraulische Regeleinheit *f*
f groupe *m* régulateur hydraulique
n hydraulische regeleenheid
p zespół hydrauliczny

2757 hydrodynamic bearing
d hydrodynamisches Lager *n*
f palier *m* hydrodynamique
n hydrodynamisch lager
p łożysko hydrodynamiczne

2758 hydrodynamic coupling
d hydrodynamische Leistungsübertragung *f*
f accouplement *m* hydrodynamique
n hydrodynamische koppeling
p sprzężenie hydrokinetyczne

2759 hydrodynamic retarder
d hydrodynamische Dauerbremse *f*
f ralentisseur *m* hydrodynamique
n hydropneumatische vertrager
p zwalniacz hydrodynamiczny

2760 hydrodynamic torque converter
d hydrodynamisches Getriebe *n*
f transmission *f* hydrodynamique
n hydrodynamische koppelomvormer
p przekładnia hydrodynamiczna

2761 hydroelastic suspension
d hydroelastische Radaufhängung *f*
f suspension *f* hydroélastique

n ophanging bij hydro-elastische vering;
  ophanging met hydro-elastisch veersysteem
p zawieszenie hydroelastyczne

2762 hydromechanical transaxle
d hydromechanische Antriebsachse *f*
f ensemble *m* transmission automatique et
  différentiel
n hydromechanische achteraseenheid
p hydromechaniczny most pędny

2763 hydromechanical transmission
d hydromechanisches Getriebe *n*
f transmission *f* automatique classique
n hydromechanische transmissie
p hydromechaniczna skrzynka biegów

2764 hydromechanic brake
d hydromechanische Bremse *f*
f frein *m* hydromécanique
n hydromechanische rem
p hamulec hydromechaniczny

* **hydrometer** → 1515

2765 hydrophilic
d hydrophil; wasseranziehend
f hydrophile
n wateraantrekkend
p hydrofilowy

2766 hydropneumatic chamber
d hydropneumatische Kammer *f*
f sphère *f* hydropneumatique
n hydropneumatische kamer
p komora hydropneumatyczna

2767 hydropneumatic pump
d hydropneumatische Pumpe *f*
f pompe *f* hydropneumatique
n luchthydralische pomp
p pompa hydropneumatyczna

2768 hydropneumatic spring
d hydropneumatische Feder *f*
f ressort *m* hydropneumatique
n hydropneumatische veer
p resor hydropneumatyczny

2769 hydropneumatic suspension
d hydropneumatische Aufhängung *f*
f suspension *f* hydropneumatique
n hydropneumatische ophanging
p zawieszenie hydropneumatyczne

2770 hydrostatic
d hydrostatisch

f hydrostatique
n hydrostatisch
p hydrostatyczny

2771 hydrostatic guideway
d hydrostatische Führung *f*
f glissière *f* hydrostatique
n hydrostatische geleiding
p prowadnica hydrostatyczna

2772 hydrosteering gear
d Hydrolenkung *f*
f direction *f* hydromécanique
n hydrostuurreductie
p hydromechaniczna przekładnia kierownicza

2773 hygrometer
d Luftfeuchtigkeitsmesser *m*
f hygromètre *m*
n hygrometer
p higrometr; wilgotnościomierz

2774 hypoid final drive
d Hypoidachsgetriebe *n*
f couple *m* à taille hypoïde
n hypoïde aandrijving
p hipoidalna przekładnia główna

2775 hypoid gear wheel
d Hypoidkegelrad *n*
f roue *f* conique hypoïde
n hypoïd tandwiel
p koło zębate hypoidalne

*   **hypoid oil → 2078**

I

2776 ice scraper
d Eiskratzer *m*
f grattoir *m*
n ijskrabber
p skrobak do lodu

2777 identification number
d Kennummer *f*
f numéro *m* d'identification
n identificatienummer
p numer seryjny; numer fabryczny

2778 identification plate
d Typenschild *n*
f plaque *f* constructeur
n typeplaatje; identificatieplaat
p tabliczka identyfikacyjna

2779 idle gear
d Leergang *m*
f point *m* mort
n neutrale stand
p bieg jałowy

2780 idle metering jet; slow running jet
d Leerlaufdüse *f*
f gicleur *m* de ralenti
n stationaire sproeier
p dysza paliwa biegu jałowego

2781 idle mixture
d Leerlaufgemisch *n*
f mélange *m* de ralenti
n mengsel bij stationair toerental
p mieszanka jałowa

2782 idle mixture cut off valve
d Leerlaufgemischabschaltventil *n*
f soupape *f* d'arrêt de mélange pour le ralenti
n sluitklep voor stationair mengsel
p zawór mieszanki biegu jałowego

2783 idle position
d Ruhestellung *f*
f position *f* de repos
n rusttoestand
p położenie spoczynkowe

2784 idle running characteristic
d Leerlaufcharakteristik *f*

f caractéristique *f* à vide
n nullastkarakteristiek
p charakterystyka biegu jałowego

2785 idler valve
d Leerlaufventil *n*
f étouffoir *m* de ralenti
n stationaire afsluiter
p zawór ruchu jałowego

2786 idling adjustment
d Leerlaufeinstellung *f*
f réglage *m* du ralenti
n afstelling stationair draaien
p regulacja biegu jałowego

2787 idling and maximum speed centrifugal governor
d Leerlauf- und Enddrehzahlregler *m*
f régulateur *m* de ralenti et de la vitesse maximum
n centrifugaalregulateur voor stationair en maximum toerental
p dwuzakresowy regulator odśrodkowy

2788 idling control unit
d Leerlaufdose *f*
f capsule *f* ralenti accéléré
n regelaar voor stationair draaien
p zespół kontroli biegu jałowego

2789 idling speed
d Leerlaufgeschwindigkeit *f*
f vitesse *f* de ralenti
n stationair toerental
p prędkość biegu jałowego

2790 idling speed control range
d Leerlaufregelung *f*
f régime *m* de régulation de ralenti
n traject van de stationaire afstelling
p zakres regulacji biegu jałowego

2791 idling spring
d Leerlauffeder *f*
f ressort *m* de ralenti
n veer voor stationair toerental
p sprężyna biegu jałowego

2792 ignitability
d Zündwilligkeit *f*
f inflammabilité *f*
n ontstekingsgewilligheid
p zapłonność

2793 ignite *v*
d entzünden

  f allumer
  n ontsteken
  p zapalać

2794 igniter
  d Zünder *m*
  f allumeur *m*
  n ontstekingsmodule
  p zapalarka; zapłonnik

2795 ignition
  d Zündung *f*
  f allumage *m*
  n ontsteking
  p zapłon

2796 ignition advance
  d Frühzündung *f*
  f avance *f* à l'allumage
  n voorontsteking
  p wyprzedzenie zapłonu

2797 ignition advance lever
  d Frühzündungshebel *m*
  f levier *m* de réglage d'avance à l'allumage
  n ontstekingsvervroegingshefboom
  p dźwignia wyprzedzenia zapłonu

2798 ignition aid
  d Starthilfe *f*; Starterleichterung *f*
  f aide *f* au démarrage
  n starthulp
  p urządzenie ułatwiające rozruch

2799 ignition analyser
  d Zündungseinsteller *m*
  f vérificateur *m* d'allumage
  n ontstekingstestapparaat
  p przyrząd do ustawiania zapłonu

2800 ignition assembly
  d Zündanlage *f*
  f équipement *m* d'allumage
  n ontstekingssysteem
  p urządzenie zapłonowe

2801 ignition coil
  d Zündspule *f*
  f bobine *f* d'allumage
  n ontstekingsspoel
  p cewka zapłonowa

2802 ignition coil cable terminal
  d Zündspulenkabelanschluss *m*
  f prise *f* de câble haute tension
  n ontstekingsspoelkabelaansluiting
  p gniazdo przewodu cewki zapłonowej

2803 ignition coil head
  d Zündspulenkopf *m*
  f tête *f* de bobine d'allumage
  n bobine-isolatiekop
  p głowica cewki zapłonowej

2804 ignition cycle
  d Zündzyklus *m*
  f changement *m* de la charge
  n verbrandingscyclus
  p cykl spalania

2805 ignition device
  d Zündeinrichtung *f*; Zündapparat *n*
  f dispositif *m* d'allumage
  n ontstekingsinrichting
  p urządzenie zapłonowe

2806 ignition distributor
  d Zündverteiler *m*
  f allumeur *m*
  n stroomverdeler
  p rozdzielacz zapłonu

2807 ignition distributor mounting flange
  d Zündverteilerflansch *m*
  f support *m* d'allumeur; support *m* de distributeur d'allumage
  n grondplaat voor stroomverdeler
  p gniazdo rozdzielacza zapłonu

2808 ignition failure
  d Zündaussetzer *m*
  f raté *m* d'allumage
  n overslaan van motor door onregelmatige ontstekingsvonken
  p zapłon opuszczony

2809 ignition impulse
  d Zündimpuls *m*
  f impulsion *f* d'allumage
  n ontstekingsimpuls
  p impuls wyzwalający

  * **ignition key** → **2816**

2810 ignition lag
  d Zündverzug *m*
  f délai *m* d'allumage; délai *m* d'inflammation
  n ontstekingsvertraging
  p opóźnienie zapłonu; zwłoka zapłonu

2811 ignition lock
  d Zündschloss *n*
  f serrure *f* de contact d'allumage
  n contactslot
  p wyłącznik zapłonu

2812 **ignition moment**
d Zündungsmoment n
f couple m d'allumage
n ontstekingsmoment
p moment zapłonowy

2813 **ignition setting**
d Zündeinstellung f
f calage m de l'allumage
n ontstekingsafstelling
p ustawianie zapłonu

2814 **ignition starter switch**
d Anlasszündschalter m
f commutateur m allumage démarreur
n ontstekings- en startschakelaar
p wyłącznik zapłon-rozrusznik

2815 **ignition switch**
d Zündschalter m
f interrupteur m d'allumage
n contactschakelaar
p wyłącznik zapłonu

2816 **ignition switch key; ignition key**
d Zündschlussel m; Schaltschlussel m
f clé f de contact
n contactsleutel
p klucz wyłącznika zapłonu

2817 **ignition system**
d Zündsystem n
f système m d'allumage
n ontstekingssysteem
p układ zapłonowy

2818 **ignition temperature**
d Entzündungstemperatur f
f température f d'inflammabilité
n ontstekingstemperatuur
p temperatura zapłonu

2819 **ignition wire**
d Zündkabel m; Zündleitung f
f câble m à haute tension
n hoogspanningskabel
p przewód wysokiego napięcia

2820 **I head engine**
d Motor m mit hängenden Ventilen
f moteur m à soupapes commandées par le haut
n kopklepmotor
p silnik o górnym usytuowaniu cylindrów

2821 **illumination; lighting**
d Beleuchtung f

f éclairage m
n verlichting
p oświetlenie

2822 **immersion pump**
d Fasspumpe f
f pompe f moyée
n pomp voor onderwater
p pompa z nożnym napędem

2823 **impact test**
d Stossversuch m
f essai m au choc
n botsproef; slagproef
p próba udarowa

2824 **impact wrench**
d Schlagschraubeschlüssel m
f clé f à chocs
n slagmoersleutel
p klucz udarowy

2825 **impregnant**
d Imprägnierungsmittel n
f imprégnant m
n impregneermiddel
p impregnat; środek impregnujący

2826 **impulse generator**
d Impulsgenerator m
f générateur m d'impulsions
n impulsgenerator
p generator impulsów

2827 **impurity**
d Verunreinigung f
f colmatage m; pollution f
n verontreiniging
p zanieczyszczenie

2828 **inboard brake disc**
d innenliegende Bremsscheibe f
f frein m à disque à l'intérieur
n "binnen" gemonteerde remschijf
p tarcza hamulcowa

2829 **inclination of steering knuckle**
d Neigung f des Achsschenkels
f inclinaison f des pivots de fusée
n fuseepenhelling
p pochylenie przegubu zwrotnicy

2830 **inclined engine**
d geneigter Motor m
f moteur m incliné
n dwarsingebouwde motor
p silnik skośny

2831 inclined knurl
 d Kordel *f*
 f moletage *m*
 n kartel
 p radełko skośne

2832 incombustible
 d unverbrennbar
 f incombustible
 n onbrandbaar
 p niepalny

2833 incomplete combustion
 d Teilverbrennung *f*
 f combustion *f* incomplète
 n onvolledige verbranding
 p spalanie niezupełne

2834 increased
 d erhöht
 f élevé
 n verhoogd
 p podwyższony

2835 indent *v*; dent *v*
 d einbeulen
 f enfoncer
 n indeuken
 p wgnieść

2836 independent front wheel suspension
 d Einzelvorderradaufhängung *f*
 f suspension *f* de roues avant indépendantes
 n onafhankelijke voorwielophanging
 p niezależne zawieszenie kół przednich

2837 index
 d Zeiger *m*
 f indicateur *m*
 n wijzer
 p wskazówka

2838 indicated horsepower
 d indizierte Leistung *f*
 f puissance *f* indiquée en chevaux
 n indicateurvermogen
 p moc indykowana

2839 indicated pressure
 d indizierter Druck *m*
 f pression *f* indiquée
 n geïndiceerde druk
 p ciśnienie indykowane

2840 indicator circuit armature
 d Anker *m* für Blinklichtstromkreis
 f armature *f* de circuit de clignotants

 n anker van richtingwijzercircuit
 p zwora obwodu kierunkowskazów

2841 indicator diagram
 d Indikatordiagramm *n*
 f diagramme *m* d'indicateur
 n indicateurdiagram
 p wykres indykatorowy

2842 indicator housing
 d Blinkergehäuse *n*
 f boîtier *m* d'indicateur
 n indicateurhuis
 p obudowa wskaźnika

2843 indicator needle
 d Messwertzeiger *m*
 f aiguille *f* d'indicateur
 n wijzernaald
 p igła wskaźnika

2844 indicator plate
 d Anzeigeblech *n*
 f plaque *f* indicatrice de commande
 n aanduidingsplaat
 p tarcza wskaźnika

2845 indirect illumination
 d indirekte Beleuchtung *f*
 f illumination *f* indirecte
 n indirecte verlichting
 p oświetlenie pośrednie

2846 induced current
 d induzierter Strom *m*
 f courant *m* induit
 n geïnduceerde stroom
 p prąd indukowany

2847 induction hardening
 d Induktionshärten *n*
 f trempe *f* par induction
 n inductieharden
 p hartowanie indukcyjne

2848 induction pipe mounting flange
 d Sitz *m* der Saugrohrstützflansch
 f bossage *m* de bride de conduit d'admission
 n montageflens voor inlaatspruitstuk
 p nadlew kołnierza rury dolotowej

2849 industrial area
 d Industriegebiet *n*
 f zone *f* industrielle
 n industriegebied
 p obszar przemysłowy; teren przemysłowy

2850 inertia
 d Trägheit *f*

f inertie *f*
n traagheid; inertie
p bezwładność

2851 inflation pressure
d Reifenluftdruck *m*
f pression *f* de gonflage
n bandspanning
p ciśnienie ogumienia

2852 inflow filter
d Zuflussfilter *m*
f filtre *m* à arrivée
n toestroomfilter
p filtr dopływowy

2853 information screen
d Informationschirm *m*
f information *f* électronique visuelle
n informatiescherm
p ekran informacyjny

2854 information table
d Informationsschild *n*
f panneau *m*
n informatiebord
p tablica informacyjna

2855 infrared drying
d Infrarottrocknung *f*
f séchage *m* infrarouge
n infrarooddroging
p suszenie promiennikowe

2856 initial pressure
d Anfangdruck *m*
f pression *f* initiale
n aanvangsdruk
p ciśnienie początkowe

2857 initial voltage
d Anfangsspannung *f*
f tension *f* initiale
n beginspanning
p napięcie początkowe

2858 injection
d Einspritzung *f*
f injection *f*
n injectie; inspuiting
p wtrysk

2859 injection advance
d Voreinspritzung *f*
f avance *f* à l'injection
n inspuitvervroeging
p wyprzedzenie wtrysku

2860 injection advance timing device
d Spritzversteller *m*
f dispositif *m* d'avance à l'injection
n inrichting om inspuitmoment aan te passen
p przestawiacz wyprzedzenia wtryskiwania

2861 injection nozzle
d Einspritzdüse *f*
f injecteur *m*
n verstuiver
p wtryskiwacz

2862 injection point
d Einspritzpunkt *m*
f point *m* d'injection
n inspuitmoment; inspuittijdstip
p moment wtrysku; punkt wtrysku

2863 injection pump
d Einspritzpumpe *f*
f pompe *f* d'injection
n injectiepomp; inspuitpomp
p pompa wtryskowa

2864 injection pump governor
d Einspritzpumpenregler *m*
f régulateur *m* pour pompe à injection
n brandstofinspuitpompregelaar
p regulator pompy wtryskowej

2865 injection pump wrench
d Schlüssel *m* für Einspritzpumpe
f clé *f* pour les tuyaux d'injection
n sleutel voor inspuitpomp
p klucz do pompy wtryskowej

2866 injection sequence
d Einspritzfolge *f*
f ordre *m* d'injection
n inspuitvolgorde
p kolejność wtryskiwania

2867 injection system; injection unit
d Einspritzsystem *n*
f système *m* d'injection
n inspuitsysteem
p zespół wtryskowy

2868 injection tester
d Einspritzprüfgerät *n*
f appareil *m* de contrôle de l'injection
n inspuittester
p próbnik wtrysku

2869 injection timing collar
d Spritzverstellermuffe *f*
f manchon *m* de commande d'avance à

l'injection
n aandrijfmof van inspuitmomentverstelinrichting
p tuleja wyprzedzania wtryskiwania

2870 injection timing housing
d Spritzverstellergehäuse n
f carter m de commande d'avance à l'injection
n huis van inspuitmomentverstelinrichting
p obudowa regulatora wtrysku

2871 injection timing hub
d Spritzverstellernabe f
f moyeu m de commande d'avance à l'injection
n naaf van inspuitmomentverstelinrichting
p piasta regulatora wtrysku

2872 injection timing lever
d Spritzverstellhebel m
f levier m de commande d'avance à l'injection
n regelhefboom van brandstofinspuiting
p dźwignia przestawiacza wyprzedzenia wtryskiwania

* injection unit → 2867

2873 injection yoke
d Einspritzdüsenbügel m
f étrier m de fixation de l'injecteur
n draagjuk voor inspuitventiel
p jarzmo wtryskiwacza

2874 injector test pump
(tool to check injector leak back and injection pressure)
d Austarierpumpe f für Einspritzdüsen
f pompe f à tarer les injecteurs
n verstuivertester
p tester pomp wtryskowych

2875 inlet cam
d Einlassnocken m
f came f d'admission
n inlaatnok
p krzywka wlotu

2876 inlet closing
d Einlassende f
f fermeture f d'admission
n einde inlaten
p koniec dolotu

2877 inlet manifold
d Ansaugkrümmer m; Ansaugsammelrohr n
f tubulure f d'admission
n inlaatspruitstuk
p otwór wlotowy przewodu dolotowego

* inlet port → 1984

2878 inlet rocker
d Einlasskiphebel m
f culbuteur m admission
n inlaattuimelaar
p dźwignia zaworu ssącego

2879 inlet valve
d Einlassventil n
f soupape f d'aspiration
n inlaatklep
p zawór dolotowy; zawór wlotowy

2880 inlet valve closing period
d Schliessungsdauer f des Einlassventils
f fermeture f de soupape d'admission
n sluitingsduur van inlaatklep
p zamknięcie zaworu dolotowego

2881 inlet valve head diameter
d Durchmesser m des Ansaugventiltellers
f diamètre m de tête de soupape d'admission
n doorsnede van de inlaatklepschotel
p średnica grzybka zaworu dolotowego

2882 inlet valve opening period
d Öffnungsdauer f des Einlassventils
f ouverture f de soupape d'admission
n openingsduur van inlaatklep
p otwarcie zaworu dolotowego

2883 inlet valve tappet clearance
d Ventilspiel n zwischen Ansaugventil und Hebel
f jeu m de soupape d'admission
n speling van de inlaatklep
p luz zaworu dolotowego

2884 inlet velocity
d Ansauggeschwindigkeit f
f vitesse f d'aspiration
n aanzuigsnelheid
p prędkość wlotowa

2885 inline engine
d Reihenmotor m
f moteur m à ligne
n lijnmotor
p silnik rzędowy

2886 inner articulated shaft
d innere Gelenkwelle f
f arbre m articulé intérieur
n gelede binnenas
p wałek przegubowy wewnętrzny

2887 inner conical disc; fixed conical disc
 d innere Kegelscheibe *f*; befestigte Kegelscheibe *f*
 f flasque *m* intérieur; flasque *m* fixe
 n inwendige conische schijf
 p wewnętrzna tarcza stożkowa; wewnętrzna tarcza ustalona

* **inner dead center** → 5106

2888 inner door panel
 d Türverkleidung *f*
 f panneau *m* intérieur de porte
 n portierbekleding
 p płat wewnętrzny drzwi

2889 inner race
 d Innenring *m*
 f bague *f* intérieure
 n binnenring
 p pierścień wewnętrzny

2890 inner scavenger rotor
 d Innenrotor *m*
 f pignon *m* intérieur
 n binnenrotor
 p wirnik wewnętrzny

2891 inner valve spring
 d innere Ventilfeder *f*
 f ressort *m* intérieur de soupape
 n binnenste klepveer
 p wewnętrzna sprężyna zaworu

2892 input gear
 d Antriebszahnradpaar *n*
 f couple *m* d'engrenage primaire
 n primaire tandwielen
 p przekładnia wejściowa

2893 input link
 d Antriebsglied *n*
 f élément *m* menant
 n aandrijvend element
 p człon czynny; człon napędzający

2894 inscription
 d Beschriftung *f*
 f inscription *f*; désignation *f*
 n opschrift
 p napis

2895 insect remover
 d Insektenentferner *m*
 f produit *m* pour enlever les insectes
 n insektenverwijderingsmiddel
 p środek usuwający owady

2896 inserted valve seat
 d Ventilsitzring *m*
 f siège *m* de soupape rapporté
 n klepzittingring; klepzetelring
 p pierścień gniazda zaworu

2897 inside board height
 d Pritscherhöhe *f*
 f hauteur *f* des ridelles
 n laadbakhoogte
 p wysokość skrzyni ładunkowej

2898 inside diameter of minimum turning circle
 d Kleinsterinnenwendekreisdurchmesser *m*
 f diamètre *m* minimal intérieur de braquage
 n draaicirkel van binnenachterwiel
 p najmniejsza wewnętrzna średnica zawracania

2899 inside diameter of thread
 d Kerndurchmesser *m*
 f diamètre *m* du corps cylindrique
 n kerndiameter
 p średnica wewnętrzna gwintu

2900 inside door handle
 d Innertürgriff *m*
 f poignée *f* intérieure de porte
 n binnendeurkruk
 p klamka drzwi wewnętrznych

2901 inside door handle pivot pin
 d Innentürgriffbolzen *m*
 f axe *m* de poignée intérieure
 n as van de binnenkruk
 p sworzeń klamki wewnętrznej

2902 inside micrometer caliper
 d Innenmikrometer *m*
 f micromètre *m* d'intérieurs
 n binnenmicrometer
 p mikrometr do pomiarów wewnętrznych

2903 insignia
 d Markenzeichen *n*
 f insigne *m*; emblème *f*
 n embleem
 p emblemat

2904 inspection; overhaul
 d technische Überprüfung *f*
 f révision *f* technique
 n controlebeurt
 p przegląd techniczny

2905 inspection pit
 d Inspektionsgrube *f*
 f fosse *f* d'inspection

n werkkuil
p kanał rewizyjny

2906 install *v*
d einsetzen; installieren
f installer
n monteren; inbouwen
p instalować

2907 installation drawing
d Einbauzeichnung *f*
f dessin *m* d'encombrement
n bouwtekening
p rysunek montażowy

* **instruction book** → 4405

* **instrument** → 1532

2908 instrument panel support
d Instrumententafelstütze *f*
f support *m* de tableau de bord
n dashboardsteun
p wspornik tablicy rozdzielczej

2909 insulated bolt
d isolierter Bolzen *m*
f boulon *m* isolé
n geïsoleerde bout
p sworzeń izolowany

2910 insulating
d isolierend
f isolant
n isolerend
p izolujący

2911 insulating material
d Isoliermaterial *n*
f matière *f* isolante
n isolatiemateriaal
p materiał izolacyjny

2912 insulating sleeve
d Isolierhülse *f*
f tube *m* isolant; bague *f* isolante
n isolatiemof
p tulejka izolacyjna

2913 insulating washer
d Isolierscheibe *f*
f rondelle *f* isolante
n isoleerplaatje
p podkładka izolacyjna

2914 insulation disc
d Isolierscheibe *f*

f disque *m* d'isolement
n isolatieschijf
p krążek izolacyjny

2915 insulation gasket
d Isolierdichtung *f*
f joint *m* isolant
n isolerende pakking
p uszczelnienie izolujące

2916 insulation tape
d Isolierband *n*
f ruban *m* isolant
n isolatieband
p taśma izolująca

* **insulation tester** → 3301

2917 insulator
d Isolator *m*
f isolateur *m*
n isolator
p izolator

2918 insurance
d Versicherung *f*
f assurance *f*
n verzekering
p ubezpieczenie

2919 intake edge
d Einlasskante *f*
f arête *f* d'admission
n kant van inlaat
p krawędź wlotowa

2920 intake passage
d Ansaugkanal *m*
f conduit *m* d'admission; canal *m* d'admission
n inlaatkanaal
p kanał dolotowy

* **intake pressure** → 4870

2921 intake pressure sensor
d Saugdruckfilter *m*
f détecteur *m* de pression d'admission
n aanzuigdruksensor
p czujnik ciśnienia ssania

2922 intake pulsation
d Einlasspulsation *f*
f pulsation *f* à l'admission
n inlaatpulsering
p bicie wlotowe

* **intake stroke** → 4872

2923 intake tube; silencer tube; snorkel tube
d Ansaugstutzen *m*
f tuyau *m* d'aspiration; tube *m* de silencieux
d'aspiration
n aanzuigstuk
p króciec ssawny

2924 integral body
d selbsttragender Aufbau *m*
f carrosserie *f* autoportante
n zelfdragende carrosserie
p nadwozie samoniosące

2925 integral construction
d Verbundkonstruktion *f*
f construction *f* intégrale
n compoundconstructie
p konstrukcja zespolona

2926 integrated circuit
d integrierte Schaltung *f*
f circuit *m* intégré
n geïntegreerde schakeling
p układ scalony; obwód scalony

2927 integrator
d Ausgleichhebel *m*
f intégrateur *m*
n compensatiehefboom
p integrator

2928 intensifier
d Multiplikator *m*
f multiplicateur *m*
n multiplicator
p multiplikator

2929 intensity
d Intensität *f*
f intensité *f*
n intensiteit
p intensywność

2930 intercell connector
d Zellenverbinder *m*; Polbrücke *f*; Polschiene *f*
f barrette *f* de connexion
n verbindingsstrip tussen accucellen
p łącznik międzyogniowy

2931 interchangeable
d auswechselbar
f interchangeable
n onderling uitwisselbaar
p zamienny; wymienny

2932 intercooler
d Zwischenkühler *m*

f refroidisseur *m* intermédiaire
n tussenkoeler
p chłodnica międzystopniowa

2933 interference suppressor
d Entstörer *m*
f antiparasitef *m*
n ontstoorder
p eliminator zakłóceń

2934 interior equipment
d Innenausstattung *f*
f équipement *m* intérieur
n interieuruitrusting
p wyposażenie wnętrza

2935 interior light switch
d Innenlichtschalter *m*
f commutateur *m* d'éclairage intérieur
n binnenlichtschakelaar
p włącznik oświetlenia wnętrza

2936 interior mirror; driving mirror
d innerer Rückblickspiegel *m*; Innenspiegel *m*
f rétroviseur *m* intérieur
n binnenspiegel
p lusterko wsteczne wewnętrzne

2937 interior styling
d Innenraumgestaltung *f*
f aménagement *m* intérieur
n interieurinrichting
p stylistyka wnętrza

2938 interlock ball
d Riegelkugel *f*
f bille *f* de verrou
n vergrendelkogel
p kulka zatrzasku

2939 interlocking
d Sperre *f*
f blocage *m*
n vergrendeling
p blokada

2940 interlocking device
d Vorriegelungssystem *n*
f système *m* de verrouillage
n vergrendelingssysteem
p system blokady

2941 interlock plug
d Riegelstopfen *m*
f bouchon *m* de verrou
n vergrendelplug
p korek blokujący

2942 interlock spring
  d Riegelfeder *f*
  f ressort *m* de verrou
  n vergrendelveer
  p sprężyna zatrzasku

2943 intermediate ball bearing
  d Zwischenlager *n*
  f roulement *m* à billes intermédiaires
  n tussenkogellager
  p łożysko kulkowe pośrednie

2944 intermediate band brake
  d mittlere Bandbremse *f*
  f frein *m* central à bande
  n tussenliggende bandrem
  p środkowy hamulec taśmowy

2945 intermediate bearing carrier
  d Zwischenlagerfassung *f*
  f support *m* de roulement intermédiaire
  n tussenlagerdrager
  p obsada łożyska pośredniego

2946 intermediate bearing flexible carrier
  d elastischer Ring *m* der Wellenlageraufhängung
  f support *m* élastique de roulement intermédiaire
  n elastische ring voor aslagerbok
  p elastyczna obsada łożyska

2947 intermediate cross member
  d Zwischentraverse *f*
  f traverse *f* intermédiaire
  n tussenliggende dwarsbalk
  p belka poprzeczna pośrednia

2948 intermediate disc
  d Zwischenscheibe *f*
  f disque *m* intermédiaire
  n tussenschijf
  p tarcza pośrednia

2949 intermediate layer
  d Zwischenlage *f*; Zwischenschicht *f*
  f couche *f* intermédiaire
  n tussenlaag
  p warstwa pośrednia; międzywarstwa

2950 intermediate lever
  d Zwischenhebel *m*
  f levier *m* intermédiaire
  n tussenhefboom
  p dźwignia pośrednia

2951 intermediate shaft
  d Hilfswelle *f*

  f arbre *m* intermédiaire
  n tussenas
  p wał pośredni

2952 intermediate speed clutch
  d Kupplung *f* des zweiten Ganges
  f embrayage *m* de deuxième vitesse
  n koppeling van de tweede versnelling
  p sprzęgło biegu pośredniego

2953 intermediate tap
  d Mittelschneider *m*
  f taraud *m* intermédiaire
  n tussensnijder
  p drugi gwintownik wstępny

2954 intermediate washer
  d Zwischenring *m*
  f anneau *m* intermédiaire
  n tussenring
  p pierścień pośredni

2955 intermittent wiper
  d intermittierender Scheibenwischer *m*
  f essuie-glace *f* à impulsions
  n ruitenwisser met intervalschakeling
  p wycieraczka z włączeniem przerywanym

2956 internal combustion engine; fuel engine
  d Verbrennungsmotor *m*
  f moteur *m* à combustion
  n verbrandingsmotor
  p silnik spalinowy

2957 internal contact friction gearing
  d Innenreibradgetriebe *n*
  f transmission *f* à friction avec contact intérieur
  n frictieoverbrenging met inwendige aandrijving
  p przekładnia z wewnętrznym sprzężeniem ciernym

2958 internal geared oil pump
  d innenverzahnte Ölpumpe *f*
  f pompe *f* à huile à denture intérieure
  n oliepomp met inwendige vertanding
  p pompa oleju o wewnętrznym zazębieniu

2959 internal gearing
  d Innenverzahnung *f*
  f denture *f* intérieure
  n binnenvertanding
  p uzębienie wewnętrzne

  * **internal gear wheel → 232**

2960 internal grinder
  d Innenschleifmaschine *f*

f rectificatrice *f* d'intérieurs
n inwendig-slijpmachine
p szlifierka do otworów

2961 **internal grinding**
d Innenschleifen *n*
f rectification *f* intérieure
n inwendig slijpen
p szlifowanie otworów

2962 **internal marking**
d innere Markierung *f*
f identifications *fpl* intérieures
n markering aan de binnenzijde
p oznaczenie wewnętrzne

2963 **internal thread**
d Innengewinde *n*
f filet *m* intérieur
n binnenschroefdraad
p gwint wewnętrzny

2964 **inverse current**
d Sperrstrom *m*
f courant *m* inversé
n sperstroom; omkeerstroom
p prąd wsteczny

2965 **involute spline joint**
d Zahnwellenverbindung *f* mit
   Evolventenflanken
f cannelures *fpl* à flancs en développante
n evolvente spieverbinding
p połączenie wielowypustowe ewolwentowe

2966 **involute tooth system**
d Evolventenverzahnung *f*
f engrenage *m* à développante de cercle
n evolvente vertanding
p zazębienie ewolwentowe

* **iodine bulb** → 2576

2967 **iron**
d Eisen *n*
f fer *m*
n ijzer
p żelazo

* **iron crowbar** → 2459

2968 **irregular ignition**
d unregelmässige Zündung *f*
f allumage *m* irrégulier
n onregelmatige ontsteking
p nieregularny zapłon

2969 **isolating switch**
d Trennschalter *m*
f coupe-circuit *m*
n scheidingsschakelaar; sectieschakelaar
p odłącznik

2970 **isomer**
d Isomer *n*
f isomère *m*
n isomeer
p izomer

2971 **isothermal compression**
d isothermische Verdichtung *f*
f compression *f* isothermique
n isothermische compressie
p sprężanie izotermiczne

J

2972 jack
   d Heber *m*
   f cric *m* élévateur
   n krik
   p lewarek

2973 jacket
   d Mantel *m*
   f enveloppe *f*
   n mantel; omhulling; ommanteling
   p osłona; płaszcz

2974 jacking point
   d Heberabstützung *f*
   f point *m* de cric
   n kriksteunpunt
   p miejsce przyłożenia dźwignika

2975 jack lead
   d Wagenheberleitung *f*
   f tuyauterie *f* de cric
   n krikvoorloop
   p przewód podnośnika

2976 jack plate
   d Wagenheberstützplatte *f*
   f plaque *f* d'appui de cric
   n krikplaat
   p płyta oporowa podnośnika

2977 jack pump
   d Wagenheberpumpe *f*
   f pompe *f* de cric
   n hydraulische krik
   p podnośnik hydrauliczny

2978 jack screw
   d Hebebock *m*
   f vérin *m*
   n krikspindel
   p dźwignik śrubowy zespołowy

2979 jammed clutch
   d festsitzende Kupplung *f*
   f embrayage *m* grippé
   n niet-vrijkomende koppeling
   p sprzęgło zakleszczające się

2980 jamming roller
   d Klemmrolle *f*
   f rouleau *m* de coincement
   n klemrol
   p wałek klinujący

* jam nut → 3124

2981 jar
   d Bohrstange *f*
   f coulisse *f* de perforateur
   n boorstaaf
   p drąg wiertniczy

* jaw clutch → 1648

2982 jeep
   d Jagdwagen *m*
   f voiture *f* de chasse
   n terreinauto
   p łazik

2983 jet blocs
   d Zerstäuberblock *m*
   f corps *m* de pulvérisateur
   n verstuiverblok
   p korpus rozpylacza

2984 jet chamber
   d Düsenkammer *f*
   f chambre *f* du gicleur
   n sproeierkamer
   p komora dyszowa

2985 jig borer
   d Bohrmaschine *f* mit Schablone
   f foreuse *f* à gabarit
   n boormachine met mal
   p wiertarka współrzędnościowa

2986 jim crow
   d drehbarer Werkzeughalter *m*
   f porte-outil *m* orientable
   n draaibare gereedschapshouder
   p uchwyt narzędziowy ruchomy

2987 job chart
   d Arbeitskarte *f*
   f fiche *f* de travail
   n werkkaart
   p karta pracy

2988 joiner's hammer
   d Schreinerhammer *m*
   f marteau *m* de menuisier
   n timmermanshamer
   p młotek stolarski

2989 joining iron
   d Fugeneisen *n*

   f fer *m* à joints
   n verbindingsijzer
   p żelazo spoinowe

2990 joint greese
   d Gelenkfett *n*
   f graisse *f* pour articulation
   n gewrichtsvet
   p tłuszcz do przegubu

2991 joint sleeve
   d Gelenkstulpe *f*
   f gaîne *f* d'étanchéité
   n verbindingsmof
   p pierścień uszczelniający przegubu

2992 joint transmission
   d Kardanantrieb *m*
   f transmission *f* à cardan
   n cardanaandrijving
   p napęd Cardana

   * **journal** → 3740

   * **journal bearing** → 4416

   * **jumper cable** → 422

K

2993 key
   d Keil *m*
   f clavette *f*
   n spie
   p klin; wpust

2994 key and spline joints
   d Keilpassfeder *f* und Keilwellenverbindungen *fpl*
   f clavetages *mpl* et cannelures *fpl*
   n spieverbindingen en vertandingen
   p klinowy wpust pasowany i połączenie wałem wielowypustowym

2995 key blanks
   d Schlüsselrohlinge *mpl*
   f embauches *fpl* de clés
   n onbewerkte sleutels
   p klucze nieobrobione

2996 key cabinet
   d Schlüsselschrank *m*
   f armoire *f* de clé
   n sleutelkast
   p szafka na klucze

2997 key chuck
   d Zahnkranzbohrfutter *n*
   f mandrin *m* à clé
   n boorhouder met sleutel
   p uchwyt wiertarski zaciskany kluczem

2998 key cutting machine
   d Schlüsselmaschine *f*
   f machine *f* à fraiser des clés
   n sleutelfreesmachine
   p frezarka do kluczy

2999 key gauge
   (tool for a quick demonstration of lever key bit forms)
   d Schlüssellehre *f*
   f jauge *f* pour clés
   n sleutelmal
   p wzorzec kluczy

3000 key joint
   d Passfederverbindung *f*
   f assemblage *m* à clavette; clavetage *m*

   n spieverbinding
   p połączenie klinowe

3001 key tower
   d Schlüsselturm *m*
   f tour *m* de clés
   n sleuteltoren
   p wieża kluczy

3002 key vice
   (rotatable and vertical adjustment enables treatment of key without shifting)
   d Schlüsselschraubstock *m*
   f étau *m* de clés
   n sleutelklem
   p imadło do kluczy

3003 keyway tool
   d Nutstahl *m*
   f outil *m* de rainurage
   n spiebaanstaal
   p nóż nacinak do rowków

3004 key working center
   d Schlüsselarbeitszentrum *n*
   f centre *m* de travail
   n sleutelwerkplek
   p ośrodek roboczy kluczy

3005 kick down contact
   d Kickdownkontakt *m*
   f contact *m* de kick-down
   n kickdown-contact
   p styk pedału

3006 kick down valve
   d Kickdownventil *n*
   f valve *f* à tiroir de kick-down
   n kickdown-klep
   p zawór wymuszonego przełączania

3007 kinematic viscosity
   d kinematische Viskosität *f*
   f viscosité *f* cinématique
   n kinematische viscositeit
   p lepkość kinematyczna

3008 kinetic energy
   d kinetische Energie *f*
   f énergie *f* cinétique
   n kinetische energie; bewegingsenergie
   p energia kinetyczna

* **king pin bush → 4925**

3009 king pin side inclination angle
   d Spreizwinkel *m*

f angle *m* d'inclination latérale des pivots de
fusée
n fuseehellingshoek
p kąt pochylenia sworznia zwrotnicy

3010 knee
 d Kniestück *n*; Knie *n*
 f coudé *m*
 n kniestuk; leidingsbocht
 p kolanko

3011 knife tool
 d Seitenstahl *m*
 f outil *m* couteau
 n zijbeitel
 p nóż boczny odsadzony

3012 knob
 d Knopf *m*
 f bouton *m*
 n knop
 p guzik

 * **knocking** → **1531**

3013 knuckle arm
 d Achsenschenkel *m*
 f fusée *f* d'essieu
 n fusee
 p zwrotnica

3014 knuckle joint
 d Kreuzgelenk *n*
 f joint *m* à croisillon
 n kruiskoppeling
 p przegub krzyżakowy

3015 knurled nut
 d gerändte Mutter *f*
 f écrou *m* moleté
 n kartelmoer; geribde moer
 p nakrętka radełkowana

L

3016 label; sticker
  d Aufklebezettel *m*; Etikett *n*
  f étiquette *f*
  n sticker
  p etykieta; nalepka

3017 labyrinth seal
  d Labyrinthdichtung *f*
  f joint-labyrinthe *m*
  n labyrintafdichting
  p uszczelnienie labiryntowe

3018 lack of spark
  d Funkmangel *m*
  f manque *m* d'étincelle
  n lek van ontstekingsvork
  p brak iskry

3019 laminar flow
  d laminare Strömung *f*
  f écoulement *m* laminaire
  n laminaire stroming
  p przepływ laminarny; przepływ warstwowy

3020 laminated glass; multilayer glass
  d Verbundglas *n*; Schichtglas *n*;
     Mehrschichtenglas *n*
  f verre *m* feuilleté; verre *m* collé
  n gelaagd glas
  p szkło wielowarstwowe

3021 lamp
  d Leuchte *f*
  f lampe *f*
  n lamp; licht
  p lampa

3022 lamp base
  d Lampensockel *m*
  f socle *m* de lampe
  n lampvoet
  p podstawa lampy

3023 lamp bulb screen
  d Lampenschirm *m*
  f blindé *m* d'ampoule de lampe
  n lampscherm
  p osłona żarówki

3024 landing gear
  d Abstellvorrichtung *f*; Stützwinde *f*

  f appui *m* de semi-remorque
  n steunpoot van oplegger
  p podpora naczepy

3025 lap belt; strap belt
  d Leibgurt *m*; Beckengurt *m*
  f ceinture *f* abdominale
  n heupgordel
  p pas biodrowy

3026 lap joint
  d Überlappungsverbindung *f*
  f assemblage *m* à recouvrement
  n overlappingsverbinding
  p połączenie zakładkowe; złącze na zakładkę

3027 lap welded joint
  d Überläppstoss *m*
  f soudure *f* à clin
  n overlaplas
  p połączenie zakładkowe

3028 large scale series production
  d Grosserienbau *m*
  f fabrication *f* en grande série
  n serieproductie
  p produkcja seryjna

3029 latch
  d Drehmaschine *f*; Drehbank *f*
  f tour *m*
  n draaibank
  p tokarka

3030 latch plate
  d Schnalle *f*
  f boucle *f* de fermeture
  n gesp van veiligheidsgordel
  p klamra pasów bezpieczeństwa

  * **latch spring → 3988**

3031 lateral block profile
  d Seitenblockprofil *n*
  f profil *m* à blocs latéraux
  n dwarsprofiel
  p kształtownik boczny

  * **lateral electrode → 4474**

3032 lateral panel
  d Seitenwand *f*
  f cloison *m* latérale
  n zijschot
  p przegroda boczna

3033 lateral stability
  d Seitenstabilität *f*; Querstabilität *f*

f stabilité *f* latérale
n dwarsstabiliteit
p stabilność poprzeczna

3034 lateral vent
d Seitenlüftung *f*
f reniflard *m* latéral
n ventilatiegat aan de zijkant; ontluchting aan de zijkant
p wywietrznik boczny

3035 lateral whip
d Querschlag *m*
f faux-rond *m*
n dwarsslag
p bicie poprzeczne

3036 latex
d Gummimilch *f*
f latex *m*
n natuurrubber
p latex naturalny; mleczko kauczukowe

3037 lathe center
d Drehmaschinenspitze *f*
f pointe *f* de tour
n draaibankcenter
p kieł tokarski

3038 lathe file
d Drehbankfeile *f*
f lime *f* de tour
n draaibankvijl
p pilnik do pracy na tokarkach

3039 lathe tool
d Drehbankgerät *n*
f outil *m* de tour
n draaibankgereedschap
p nóż tokarski

* **laying-on trowel → 2166**

* **lead → 1220**

3040 lead
d Blei *n*
f plomb *m*
n lood
p ołów

3041 lead accumulator
d Bleiakkumulator *m*
f accumulateur *m* au plomb
n loodaccu
p akumulator ołowiowy

* **leaded petrol → 2019**

* **lead free gasoline → 5267**

3042 leading shoe; primary shoe
d auflaufende Bremsbacke *f*; Primärbacke *f*
f mâchoire *f* comprimée; segment *m* comprimé
n oplopende remschoen; primaire remschoen
p szczęka współbieżna

3043 lead of thread
d Gewindesteigung *f*
f pas *m* du filet; course *f* de filetage
n spoed van schroefdraad
p skok gwintu

3044 leaf spring
d Blattfeder *f*
f ressort *m* à lames
n bladveer
p resor piórowy

3045 leaf spring bolt
d Federbolzen *m*
f axe *m* d'articulation de ressort
n veerbout
p sworzeń resoru

* **leaf spring bracket → 2374**

3046 leaf spring clip
d Federbügel *m*
f étrier *m* de ressort
n veerbeugel; veerstrop
p jarzmo resoru

3047 leaf spring seat; spring perch
d Federsitz *m*; Federsattel *m*
f patin *m* de ressort
n veerpad; veerzitting
p siodło resoru

3048 leaf spring support
d Blattfederstütze *f*
f support *m* de ressort à lames
n bladveersteun
p wspornik resoru piórowego

3049 leaf spring wedge
d Federkeil *m*
f cale *f* inférieure de ressort
n veerspie
p klin wsporny resoru

3050 leak
d Undichtheit *f*
f manque *m* d'étanchéité
n lekkage
p nieszczelność

3051 leakage
- *d* Undichtheitsverlust *m*
- *f* fuite *f*
- *n* lekkage
- *p* wyciek

3052 leakage detector
- *d* Stromentweichungssucher *m*
- *f* cherche-pertes *m* de courant
- *n* lekzoeker
- *p* detektor upływu prądu

3053 leakage fuel
- *d* Leckkraftstoff *m*
- *f* carburant *m* de fuite
- *n* lekbrandstof
- *p* nadmiar paliwa

3054 leak detector
- *d* Leckdetektor *m*
- *f* détecteur *m* de fuites
- *n* lekdetector
- *p* wykrywacz nieszczelności

3055 leak off adapter
- *d* Unterdruckstutzen *m*
- *f* raccord *m* des fuites de combustible
- *n* onderdrukpasstuk
- *p* króciec przewodu nadmiarowego

3056 leakproofing material
- *d* Dichtungsmittel *n*
- *f* enduit *m* hermétique
- *n* afdichtmiddel; afdichtvloeistof
- *p* środek uszczelniający; dodatek uszczelniający

3057 leak tester for petrol engines
  (tool to pinpoint any loss of compression)
- *d* Dichtigkeitsprüfgerät *n* für Benzinmotoren
- *f* contrôleur *m* d'étanchéité de moteur à essence
- *n* lektester voor benzinemotoren
- *p* manometr do sprawdzania stopnia sprężania w cylindrach silników benzynowych

3058 lean mixture control unit
- *d* Verminderungsdose *f*
- *f* capsule *f* d'appauvrissement
- *n* regeleenheid van mager mengsel
- *p* blok sterujący mieszanki ubogiej

3059 leather
- *d* Leder *n*
- *f* cuir *m*
- *n* leder
- *p* skóra

3060 leather washer
- *d* Lederunterlegscheibe *f*
- *f* rondelle *f* en cuir
- *n* lederen onderlegring
- *p* podkładka skórzana

3061 left door
- *d* linke Tür *f*
- *f* porte *f* gauche
- *n* linker portier
- *p* drzwi lewe

3062 left hand drive
- *d* links angeordnete Lenkung *f*
- *f* direction *f* à gauche
- *n* linkse besturing
- *p* układ kierowniczy lewostronny

3063 left handed screw
- *d* linksgängige Schraube *f*
- *f* vis *f* à filet gauche
- *n* linksdraaiende moer
- *p* śruba lewoskrętna

3064 legroom
- *d* Beinraum *m*; Beinfreiheit *f*
- *f* place *f* pour les jambes
- *n* beenruimte
- *p* przestrzeń dla nóg

3065 lens
- *d* Streuscheibe *f*
- *f* glace *f* diffusante
- *n* koplichtglas
- *p* szyba reflektora

3066 let *v* the clutch slip
- *d* die Kupplung schleifen lassen
- *f* permettre aux accouplements de se faire
- *n* de koppeling laten slippen
- *p* wyłączać sprzężenie

3067 level
- *d* Libelle *f*
- *f* niveau *m*
- *n* niveau
- *p* poziom

3068 levelling pole
- *d* Nivellierlatte *f*; Nivellierstab *m*
- *f* mire *f* de nivellement
- *n* nivelleerlat; meetstok
- *p* łata niwelacyjna

3069 levelling valve
- *d* Nivellierventil *n*
- *f* soupape *f* de correction d'assiette
- *n* hoogteregelklep; niveauregelaar
- *p* zawór poziomujący; zawór stabilizacyjny

3070 levelling valve of front axle
d Nivellierventil *n* der Vorderachse
f soupape *f* de nivellement d'essieu avant
n hoogteregelklep van vooras; niveauregelaar van vooras
p zawór poziomujący osi przedniej

3071 level plug joint
d Dichtung *f* für Niveaustopfen
f joint *m* de vis bouchon de niveau
n niveaustopafdichtingsring
p uszczelka wkrętu wlewu

3072 lever
d Hebel *m*
f levier *m*
n hefboom
p dźwignia

3073 lever action
d Hebelwirkung *f*
f fonctionnement *m* de levier
n hefboomwerking
p mechanizm działania dźwigni

3074 lever actuated friction clutch
d handschaltbare Reibungskupplung *f*
f embrayage *m* à friction à levier
n frictiekoppeling met hefboombediening
p sprzęgło cierne przełączane ręcznie

3075 lever bracket
d Hebelstütze *f*
f support *m* de levier
n hefboomsteun
p wspornik dźwigni

3076 lever cover
d Deckel *m* für Hebel
f chapeau *m* de levier
n dekplaat van hefboom
p pokrywa dźwigni

3077 lever train
d Baugruppe *f* des Gestänge
f ensemble *m* de leviers
n stangenstelsel
p zespół dźwigni

3078 L head engine
d seitengesteuerter Motor *m*
f moteur *m* à soupapes latérales
n zijklepmotor
p silnik dolnozaworowy

3079 liability insurance
d Kraftfahrhaftpflichtversicherung *f*
f assurance *f* responsabilité
n aansprakelijkheidsverzekering
p ubezpieczenie od odpowiedzialności cywilnej

* **licence plate → 3482**

3080 licence production
d Lizenzherstellung *f*
f fabrication *f* sous licence
n productie onder licentie
p produkcja licencyjna; produkcja na licencji

3081 lift control handle
d Steuerzugknopf *m* der Niveauregulierung
f poignée *f* de réglage
n bedieningsknop van niveauregeling
p gałka sterownicza

3082 lifting bracket
d Hebeöse *f*; Aufhängeöse *f*
f œillet *m* de levage
n hijsoog
p ucho zawieszenia

3083 lifting screw
d Druckschraube *f*; Abziehschraube *f*
f vis *f* d'arrêt; vis *f* de arrachement
n drukbout
p śruba dociskowa

3084 lifting screws
d Hebeschrauben *fpl*
f vis *fpl* de soulèvement
n hefschroeven
p śruby podnośne; śruby podniesieniowe

3085 lifting slings; lifting tackle
d Hebeschlinge *f*
f boucle *f* de levage
n hijsblok
p wciągnik

* **lifting tackle → 3085**

3086 lifting truck
d Hubtransportkarre *f*
f chariot *m* élévateur à main
n vorkheftruck
p wózek podnośnikowy

3087 light alloy
d Leichtmetallegierung *f*
f alliage *m* léger
n lichtmetaallegering
p stop lekki

* **lighting → 2821**

3088 lighting switch
d Lichtschalter *m*
f interrupteur *m* d'éclairage
n lichtschakelaar
p wyłącznik świateł

3089 limitation cam
d Begrenzungsnocken *m*
f came *f* de limitation
n begrenzingsnok
p krzywka ograniczająca

3090 limited slip differential
d Sperrdifferential *n*
f différentiel *m* autobloquant
n differentieel met gedeeltelijke blokkering;
 sperdifferentieel met gedeeltelijke blokkering
p samoblokujący mechanizm różnicowy

3091 limited visibility
d beschränkte Sichtweite *f*; Sichtbehinderung *f*
f visibilité *f* restreinte
n slecht zicht
p ograniczona widoczność

3092 limit gauge
d Grenzlehre *f*
f calibre *m* de limites
n grenskaliber
p sprawdzian graniczny

3093 limit switch
d Grenzschalter *m*
f interrupteur *m* de fin de course
n eindschakelaar
p łącznik ograniczający; ogranicznik

3094 lining
d Auskleidung *f*
f revêtement *m*
n bekleding
p wykładzina; okładzina

3095 link
d Glied *n*
f élément *m*
n schakel; element
p ogniwo; człon

3096 linkage ball joint
d Kugelzapfengelenk *n*
f rotule *f* de timonerie
n stuurkogel
p przegub kulowy

3097 link motion slider
d Kulissengleitstein *m*

f poussoir *m* coulissant
n geleidbaan
p ślizgacz kulisy

3098 lipped bush
d Lippenring *m*
f bague *f* à levres
n mof met lip
p pierścień uszczelniający o kształcie litery C

3099 liquefied petroleum gas
d Autogas *n*
f gaz *m* de pétrole liquide
n autogas
p gaz płynny

3100 liquid gas
d Flüssiggas *n*
f gaz *m* liquide
n vloeibaar gas
p gaz ciekły

3101 load capacity
d Nutzlast *f*; Ladefähigkeit *f*
f charge *f* utile
n draagvermogen; laadvermogen
p ładowność

3102 loading gauge
d Ladelehre *f*
f indicateur *m* de charge
n ladingaanwijzer
p skrajnik ładunkowy

3103 loading platform
d Laderampe *f*; Ladebuhne *f*
f rampe *f* de chargement
n laadperron
p pomost załadowczy

3104 loading state
d Beladungszustand *m*
f condition *f* de chargement
n belastingstoestand
p stan obciążenia

3105 load rating
d Tragfähigkeit *f* von Reifen
f charge *f* limite du pneu de voiture
n draagvermogen van autoband
p nośność opony

3106 load tester
(tool to test the tension between terminals)
d Batteriespannungsprüfer *m*
f contrôleur *m* de tension de batterie
n accuspanningsmeter
p próbnik naładowania akumulatora

3107 local transport
  d örtlicher Transport *m*
  f transport *m* local
  n plaatselijk vervoer
  p przewóz lokalny; transport lokalny

3108 lock
  d Schloss *n*
  f serrure *f*
  n slot
  p zamek

3109 lock ball sleeve
  d Verriegelungshülse *f*
  f douille *f* de verrouillage
  n blokkeringsbus
  p tulejka zatrzasku

3110 lock button
  d Sperrknopf *m*
  f bouton *m* de verrouillage
  n vergrendelknop
  p przycisk ustalający

3111 lock cam
  d Verschlussnocken *m*
  f came *f* de serrure
  n vergrendelnok
  p krzywka zamka

3112 lock cylinder
  d Schlosszylinder *m*
  f cylindre *m* de serrure
  n slotcilinder
  p cylinder zamka

3113 lock grip welding clamp
  d Schweissergripzange *f*
  f pince *f* pour soudure
  n vastklembare lastang
  p szczypce spawalnicze

3114 locking arm
  d Verriegelungsnocken *m*
  f came *f* de verrouillage
  n vergrendelnok
  p krzywka blokująca

3115 locking ball
  d Verschlusskugel *f*
  f bille *f* de verrouillage
  n vergrendelkogel
  p kula zatrzasku

3116 locking compound
  d Sicherungsmittel *n*
  f produit *m* de scellement
  n borgmiddel
  p środek zabezpieczający

3117 locking lever
  d Verriegelungshebel *m*
  f levier *m* de verrouillage
  n vergrendelhefboom
  p dźwignia blokująca

3118 locking pawl
  d Sperrklinke *f*
  f rochet *m*
  n blokkeerpal
  p zapadka

3119 locking pawl lever
  d Sperrhebel *m*
  f levier *m* de rochet
  n afsluithefboom
  p dźwignia zapadki

3120 locking screw
  d Fangschraube *f*; Klemmschraube *f*
  f vis *f* d'arrêt
  n klemschroef
  p śruba zabezpieczająca

3121 locking seat; anchorage
  d Verankerung *f*
  f ancrage *m*
  n verankering
  p gniazdo osadcze

3122 locking spring
  d Verriegelungsfeder *f*
  f ressort *m* de verrouillage
  n vergrendelingsveer
  p sprężyna zabezpieczająca

3123 locking wheel nut
  d Felgenschloss *n*
  f antivol *m* de roue
  n wielslot
  p zamek zabezpieczający koło przed kradzieżą

3124 lock nut; jam nut
  d Gegenmutter *f*
  f contre-écrou *m*
  n zelfborgende moer
  p przeciwnakrętka

3125 lock pin
  d Sperraste *f*
  f boulon *m* de blocage
  n blokkeerpen; blokkeerstift
  p kołek ustalający; czop ryglujący

3126 lock plate
  d Klemmplatte *f*
  f plaque *f* de serrage

n borgplaat; slotplaat
p płytka ustalająca

3127 lock ring
d Verschlussring *m*
f jonc *m* de verrouillage
n borgring
p pierścień sprężynujący zabezpieczający

3128 locksmith
d Schlosser *m*
f serrurier *m*
n slotenmaker
p ślusarz

3129 locksmith's hammer
d Schlosserhammer *m*
f marteau *m* de serrurier
n slotenmakershamer
p młotek ślusarski

3130 lock wire
d Sicherungsdraht *m*; Sicherheitsdraht *m*
f fil *m* d'arrêt
n borgdraad
p drut bezpiecznikowy

3131 long dome dresser
d Profilgegenhalter *m*, längs gewölbt
f batte *f* bombée en long
n plethamer in de lengte gebold
p młotek blacharski zgięty

3132 longitudinal engine
d Längsmotor *m*
f moteur *m* longitudinal
n motor in langsrichting
p silnik usytuowany wzdłużnie

3133 longitudinal floor beam
d Fussbodenlängsträger *m*
f longeron *m* de plancher
n overlangse vloerbalk
p wzdłużna belka podłogi

3134 longitudinal joint
d Längenverbindung *f*
f assemblage *m* en long
n lengteverbinding
p połączenie wzdłużne; połączenie na długość

3135 longitudinal oscillation
d Nickbewegung *f*
f tangage *m*
n dompbeweging
p wahanie wzdłużne

3136 longitudinal stability
d Längsstabiliät *f*

f tenue *f* de cap
n langsstabiliteit
p stateczność podłużna

3137 long reach drifts
d langer Splintreiber *m*
f chasse-goupilles *mpl* longs
n lange doorslagen
p długie wybijaki

3138 long reach socket
d langer Steckschlüssel *m*
f douille *f* longue
n lange dop
p długi klucz nasadowy

3139 long snipe nose pliers
d Langbeckzange *f* mit spitz zulaufenden flachen Backen
f pince *f* à becs longs affilés
n tang met lange platte spits uitlopende bekken
p szczypce płaskie wydłużone

3140 long stem nozzle
d lange Lochdüse *f*
f injecteur *m* long
n verstuiver met lange neus
p rozpylacz otworkowy wydłużony

3141 long stroke engine
d Langhubmotor *m*
f moteur *m* à longue course
n langeslagmotor
p silnik długoskokowy

3142 long wheelbase truck
d Langradstandwagen *m*
f camion *m* rallongé
n vrachtwagen met lange wielbasis
p samochód z dużym rozstawem kół

3143 loop scavenge
d Umkehrspülung *f*
f balayage *m* à courants ascendants
n omkeerspoeling
p przepłukiwanie pętlicowe

3144 loose connection
d Wackelkontakt *m*
f contact *m* intermittent
n los contact
p styk chwiejny

3145 loose pulley
d Leerlaufrolle *f*
f poulie *f* folie
n losse katrol
p koło pasowe jałowe; koło pasowe luźne

3146 **loss of compression**
   d Verdichtungsverlust *m*
   f perte *f* de compression
   n compressieverlies
   p strata przy sprężaniu

3147 **loss of gloss**
   d Glanzverlust *f*
   f perte *f* de brillant
   n glansverlies
   p utrata połysku

3148 **loss of pressure**
   d Druckverlust *m*
   f perte *f* de pression
   n drukverlies
   p strata ciśnienia

3149 **loud speaker**
   d Lautsprecher *m*
   f haut-parleur *m*
   n luidspreker
   p głośnik

3150 **louver**
   d Luftschlitz *m*
   f persienne *f*
   n luchtspleet
   p szczelina wentylacyjna

* **low beam** → 237

3151 **low bed semitrailer**
   d Tiefladeauflieger *m*
   f semi-remorque *f* surbaissée
   n dieplader
   p naczepa niskopodłogowa

3152 **lower hinge**
   d unteres Scharnier *n*
   f charnière *f* inférieure
   n onderste scharnier
   p zawiasa dolna

3153 **lower supporting flange seal**
   d Zylinderfussdichtung *f*
   f joint *m* de base de cylindre
   n cilindervoetafdichting
   p dolny uszczelniacz tulei cylindra

3154 **lower suspension arm**
   d Unterarm *m*
   f bras *m* inférieur
   n onderste wieldraagarm
   p wahacz najniższy

3155 **low gear; bottom gear**
   d erster Gang *m*

   f première vitesse *f*
   n eerste versnelling
   p pierwszy bieg

3156 **low grade**
   d minderwertig
   f de moindre valeur
   n minderwaardig
   p niskowartościowy; małowartościowy

3157 **low maintenance**
   d wartungsarm
   f nécessitant peu d'entretien
   n onderhoudsarm
   p nie wymagający większej konserwacji

3158 **low pressure chamber**
   d Niederdruckkammer *f*
   f chambre *f* à basse pression
   n lagedrukkamer
   p przestrzeń niskiego ciśnienia

3159 **low pressure pomp**
   d Niederdruckpumpe *f*
   f pompe *f* à basse pression
   n lagedrukpomp
   p pompa niskiego ciśnienia

3160 **low pressure tyre**
   d Niederdruckreifen *m*
   f pneumatique *m* à basse pression
   n lagedrukband
   p opona niskociśnieniowa

3161 **low range of gears**
   d Geländegang *m*
   f vitesse *f* tous terrains
   n zeer lage versnelling geschikt voor rijden in ruw terrein
   p bieg terenowy skrzynki przekładniowej

3162 **low revolutions**
   d niedrige Touren *fpl*
   f basses fréquences *fpl*
   n lage toeren
   p niskie obroty

3163 **low speed downshift valve**
   d Niedergeschwindigkeitsrückschaltventil *n*
   f soupape *f* de rétrogradation à basse vitesse
   n terugschakelklep voor lage snelheden
   p zawór zwrotnego przełączenia małej prędkości

3164 **low tension cable**
   d Niederspannungskabel *m*
   f câble *m* à basse tension

n laagspanningskabel
p kabel niskiego napięcia

3165 low tension ignition
d Niederspannungszündung *f*
f allumage *m* à basse tension
n laagspanningsontsteking
p niskie napięcie zapłonu

* **low voltage winding** → 3897

3166 lubricating ability
d Schmierfähigkeit *f*
f pouvoir *m* lubrifiant
n smerend vermogen
p smarność

3167 lubricating felt; greasing pad
d Schmierfilz *m*; Schmierkissen *n*
f feutre *m* lubrifiant
n smeervilt
p poduszka smarująca

3168 lubricating film
d Schmierfilm *m*; Ölfilm *m*
f film *m* lubrifiant
n smeerfilm
p warstewka smaru

3169 lubricating oil
d Schmieröl *n*
f huile *f* de graissage
n smeerolie
p olej smarowy

3170 lubricating point
d Schmierstelle *f*
f point *m* à graisser
n smeerpunt
p punkt smarowania

3171 lubrication
d Schmierung *f*
f graissage *m*
n smering
p smarowanie

3172 lubrication chart
d Schmierplan *m*
f schéma *m* de graissage
n smeringstabel
p karta smarowania

3173 lug and post cleaner
d Pol- und Klemmenreiniger *m*
f nettoyeur *m* de cosses et bornes
n borstel voor accupool en -klem
p szczotka do czyszczenia styków i biegunów

3174 lug disconnector
d Polklemmenabzieher *m*
f arrache-casse *m* extracteur
n accupooltrekker
p ściągacz do zacisków akumulatorów

3175 luggage grid
d Gepäckbrücke *f*
f porte-bagages *m*
n dakdrager; imperiaal; bagagerek op auto
p bagażnik dachu

3176 luggage net
d Gepäcknetz *n*
f filet *m* à bagages
n bagagenet
p siatka bagażnika

3177 luggage rail
d Gepäckgalerie *f*
f galerie *f* à bagages
n dakimperiaal
p krata bagażowa na dachu

3178 luggage trailer
d Gepäckanhänger *m*
f remorque *f* pour bagages
n bagage-aanhanger
p przyczepa bagażowa

3179 lug socket
(socket for lower ball joints)
d Kappe *f* mit Stiften
f douille *f* à ergots
n dop met stiften
p nasadka z kołkami

3180 lug socket
(socket used for rear shock absorber nuts)
d verzahnte Kappe *f*
f douille *f* à ergots
n vertande dop
p klucz do nakrętek zawieszeń amortyzatorów

3181 luminous intensity
d Lichtstärke *f*
f intensité *f* lumineuse
n lichtsterkte
p natężenie światła

3182 luminous paint
d Leuchtfarbe *f*; Tagesleuchtfarbe *f*
f peinture *f* lumineuse; peinture *f* phosphorescente
n lichtgevende lak
p farba świecąca

* **lump hammer** → 1094

M

* **machine elements** → 3184

3183 machine linked line
d Transferstrasse *f*
f ligne *f* des machines-outils
n geautomatiseerde productielijn; machinestraat
p linia produkcyjna

3184 machine parts; machine elements
d Maschinenelemente *npl*
f éléments *mpl* des machines
n machineonderdelen
p elementy maszyn; części maszyn

3185 machine tool
d Werkzeugmaschine *f*
f machines-outil *f*
n gereedschapswerktuigen
p obrabiarka do wytwarzania narzędzi

3186 machining
d Bearbeitung *f*
f usinage *m*
n machinale bewerking
p obróbka

3187 machining center
d Bearbeitungszentrum *n*
f centre *m* d'usinage
n bewerkingsstation
p centrum obróbkowe

3188 Mach number
d Machzahl *f*
f nombre *m* de Mach
n getal van Mach
p liczba Macha

3189 magnet
d Magnet *m*
f aimant *m*
n magneet
p magnes

3190 magnetic chuck
d Magnetspannfutter *n*
f mandrin *m* magnétique
n elektromagnetische boorhouder
p uchwyt magnetyczny

3191 magnetic closure
d Magnetverschluss *m*

f fermeture *f* magnétique
n magnetische sluiting
p zamknięcie magnetyczne

3192 magnetic clutch
d magnetische Kupplung *f*
f embrayage *m* magnétique
n magneetkoppeling
p sprzęgło elektromagnetyczne

3193 magnetic core
d Magnetkern *m*
f noyau *m* magnétique
n magneetkern
p rdzeń magnetyczny

3194 magnetic drain plug
d Stopfen *m* mit Magneteinsatz
f bouchon *m* à aimant
n aftapplug met magneet; aftapstop met magneet
p wkręt z wkładem magnetycznym

3195 magnetic field
d Magnetfeld *n*
f champ *m* magnétique
n magnetisch veld; magneetveld
p pole magnetyczne

3196 magnetic material
d Magnetwerkstoff *m*
f matière *f* magnétique
n magnetische stof
p materiał magnetyczny

3197 magnetic particle coupling
d Magnetpulverkupplung *f*
f embrayage *m* magnétique à poudre métallique
n magnetische vloeistofkoppeling
p sprzęgło magnetyczne hydrauliczne

3198 magnetic pole
d Magnetpol *m*
f pôle *m* magnétique
n magnetisch veld
p biegun magnetyczny

3199 magnetic retriever
d Magnetsucher *m*
f doigt *m* magnétique
n flexibele magneet
p doszukiwacz magnetyczny

3200 magnetic sensor
d magnetischer Geber *m*
f capteur *m* magnétique
n magnetische meetsonde
p czujnik magnetyczny

3201 magnetic separator
d Magnetscheider *m*
f séparateur *m* magnétique
n magnetische scheider
p oddzielacz magnetycny; separator
   magnetyczny

3202 magnetic switch
d Magnetschalter *m*
f interrupteur *m* magnétique
n magneetschakelaar
p wyłącznik magnetyczny

3203 magnetism
d Magnetismus *n*
f magnétisme *m*
n magnetisme
p magnetyzm

3204 magneto
d Magnetzünder *m*
f magnéto *m*
n ontstekingsmagneet
p iskrownik

3205 magneto distributor
d Magnetverteiler *m*
f distributeur *m* de magnéto
n magneetverdeler
p rozrząd magnetyczny

3206 magnet plate
d Elektromagnetplatte *f*
f plaque *f* d'électroaimant
n elektromagneetplaat
p płytka elektromagnetyczna

3207 magnifying glass
d Lupe *f*
f loupe *f*
n vergrootglas
p lupa; szkło powiększające

3208 main bearing
d Hauptlager *n*
f roulement *m* principal
n hoofdlager
p łożysko główne

3209 main bearing diameter
d Durchmesser *m* der Hauptzapfen
f alésage *m* des tourillons
n hoofdlagerdiameter
p średnica czopu głównego

3210 main brake circuit
d Hauptbremskreis *m*

f circuit *m* secondaire de freinage
n hoofdremcircuit
p obwód hamulca głównego

3211 main connector
d Hauptstecker *m*
f contact *m* à fiche principal
n hoofdstekker
p wtyczka główna

3212 main cut-off valve
d Hauptabsperrventil *n*
f vanne *f* d'arret; clapet *m* de fermeture à main
n hoofdafsluiter
p główny zawór odcinający; główny zawór
   zamykający

3213 main filter
d Hauptfilter *m*
f filtre *m* principal
n fijnfilter; hoofdfilter
p filtr główny

3214 main flow
d Hauptströmung *f*
f courant *m* principal
n hoofdstroming
p ruch główny cieczy

3215 main frame
d Hauptrahmen *m*
f châssis *m*
n hoofdchassis
p rama główna

3216 main group
d Hauptgruppe *f*
f groupe *m* principal
n hoofdgroep
p grupa główna

3217 main jet
d Hauptdüse *f*
f gicleur *m* principal
n hoofdsproeier
p dysza główna

3218 main journal
d Hauptlagerzapfen *m*
f tourillon *m*
n hoofdlagertap
p czop główny

3219 main key
d Hauptschlüssel *m*
f clé *f* principale
n hoofdsleutel
p klucz główny

3220 main leaf
d Hauptfederblatt *n*
f lame *f* de ressort
n hoofdveerblad
p pióro główne resoru

3221 main oil passage
d Ölhauptkanal *m*
f canal *m* principal d'huile
n hoofdoliekanaal
p główny kanał oleju

3222 main shaft
d Hauptwelle *f*
f arbre *m* principal
n hoofdas
p wał główny

3223 main shaft bearing
d Hauptwellenlager *n*
f palier *m* d'arbre principal
n hoofdaslager
p łożysko wału głównego

3224 main shaft seal ring
d Dichtring *m* der Hauptwelle
f bague *f* d'étanchéité d'arbre principal
n hoofdasafdichting; hoofdaskeerring
p uszczelniacz wału głównego

3225 main silencer
d Hauptauspufftopf *m*
f pot *m* principal
n hoofddemper
p główny tłumik dźwięków

3226 maintenance
d Instandhaltung *f*
f entretien *m*
n onderhoud
p konserwacja; utrzymanie

3227 mainteance tool set
d Werkzeugwandschrank *m* für allgemeine Instandsetzung
f composition *f* "mécanique entretien"
n onderhoudsgereedschapsset
p zestaw narzędzi do konserwacji

3228 main valve
d Hauptventil *n*
f soupape *f* principale
n hoofdklep
p zawór główny

3229 major repair
d Generalreparatur *f*

f révision *f* générale
n algehele revisie
p naprawa główna

3230 major road
d Vorrangstrasse *f*
f route *f* à priorité
n voorrangsweg
p droga z pierwszeństwem przejazdu

3231 makeshift
d provisorisch
f provisoire
n tijdelijk
p prowizoryczny

3232 malfunction
d Anomalie *f* der Wirkungsweise
f anomalie *m* de fonctionnement
n storing
p działanie nieprawidłowe

3233 malleable cast iron
d Temperguss *m*
f fonte *f* malléable
n tempergietijzer
p żeliwo ciągliwe

3234 mallet
d Holzhammer *m*
f maillet *m*
n houten hamer
p młotek drewniany

3235 manganese
d Mangan *n*
f manganèse *m*
n mangaan
p mangan

3236 manhole
d Mannloch *n*
f trou *m* d'homme
n mangat
p właz

3237 manifold
d Auspuffkrümmer *m*
f collecteur *m*
n uitlaatspruitstuk
p przewód rurowy rozgałęziony

3238 manifold pressure
d Leitungsdruck *m*
f pression *f* dans la tuyauterie
n luchtdruk in inlaatspruitstuk
p ciśnienie przewodu

* **manipulator** → 3578

3239 manual
d handbedient
f manuel
n handmatig
p ręczny

3240 manual injection advance device
d handverstellbarer Spritzversteller *m*
f variateur *m* manuel d'avance
n inspuitverstelling met de hand
p przestawiacz wyprzedzenia wtryskiwania

3241 manual lubrication
d Handschmierung *f*
f lubrication *f* manuelle
n smering met de hand
p smarowanie ręczne

3242 manual pump
d Handpumpe *f*
f pompe *f* à main
n handpomp
p pompa ręczna

3243 manual valve
d Einstellventil *n*
f clapet *m* de surpression
n regelbare klep
p zawór nastawny

3244 map pocket
d Ablagefach *m*
f récepteur *m* de cartes
n kaartenvak
p kieszeń na mapy

3245 map reading lamp
d Fahrgastraumleuchte *f*
f lampe *f* pour la lecture de la carte
n kaartleeslamp
p lampka do czytania mapy

3246 marker lamp
d Begrenzungslichter *m*
f feu *m* d'encombrement
n markeerlicht
p światło pozycyjne; światło gabarytowe

3247 marking
d Markierung *f*
f identifications *fpl*
n markering; merkteken
p oznaczenie

3248 marking sealant
d Markierungskitt *m*

f produit *m* adhésif de marqage
n markeerkit
p kit do znakowania

3249 masking paper cover
d Abdeckpapier *n*
f papier *m* adhésif de protection
n afplakpapier
p papier maskujący; papier osłaniający

3250 massive bar tyre
d Grobstollenreifen *m*
f pneu *m* à crampons massifs
n massieve band
p opona masywna

3251 master batch
d Grundmischung *f*
f mélange *m* mère
n hoofdpartij
p przedmieszka

3252 master cylinder piston
d Hauptbremszylinderkolben *m*
f piston *m* de maître-cylindre
n zuiger van de hoofdremcilinder
p tłok pompy hamulca

3253 master cylinder push rod
d Kolbenstange *f* der Ausrückpumpe
f poussoir *m* de piston de pompe
n zuigerstang van de ontkoppelingspomp
p tłoczysko pompy wyłączającej; drąg tłokowy
pompy wyłączającej

* **master gauge** → 4061

3254 master switch
d Meisterschalter *m*
f contrôleur *m* indirect
n hoofdschakelaar
p wyłącznik główny

3255 material
d Material *n*; Werkstoff *m*
f matériel *m*
n materiaal
p materiał

3256 material damage
d Sachschaden *m*
f dommage *m* matériel
n materiaalschade
p zniszczenie materiału

3257 material defect
d Materialfehler *m*

f défaut *m* de matière
n materiaaldefect
p wada materiału

3258 material fatigue
d Materialermüdung *f*
f fatigue *f* du métal
n materiaalmoeheid
p zmęczenie materiałowe

3259 mat lacquer
d Mattlack *m*
f vernis *m* terne
n matte lak
p lakier matowy

3260 mat sheen
d Mattglanz *m*
f éclat *m* terne
n doffe glans; matte glans
p połysk matowy

3261 matting
d Mattieren *n*
f dépolissage *m*
n matteren
p matowanie

3262 mattock
d Rodehacke *f*; Karst *m*
f pioche *f*
n piekhouweel
p motyka; kilof

3263 maul; beetle
d Schlegel *m*
f maillet *m*
n moker; beukhamer
p pobijak

3264 maximal
d maximal
f maximal
n maximaal
p maksymalny

3265 maximal load of the roof
d Maximaldachbelastung *f*
f poids *m* maxi sur barres de toit
n maximale dakbelasting
p maksymalne obciążenie dachu

3266 maximal position
d maximale Stellung *f*
f hauteur *f* maximale
n hoogste stand
p wysokość maksymalna; pozycja
     maksymalna

3267 maximal revolutions
d Maximaldrehzahl *f*
f fréquence *f* maximale
n maximale toeren
p obroty maksymalne

3268 maximum ball pressure
d Höchstkugeldruck *m*
f poids *m* maxi sur flèche
n maximale kogeldruk
p maksymalny nacisk kulki

3269 maximum power
d Höchstleistung *f*
f puissance *f* maximum
n maximumvermogen
p moc maksymalna

3270 maximum retardation
d Maximalabbremsung *f*
f décélération *f* de maxi
n maximaal haalbare remvertraging
p hamowność maksymalna

3271 maximum speed; speed limit
d Höchstgeschwindigkeit *f*
f vitesse *f* maximum
n maximumsnelheid
p ograniczenie szybkości jazdy

3272 maximum speed limitation range
d Endregelung *f*
f régime *m* de limitation de vitesse maximum
n traject van de maximumtoerenafstelling
p zakres ograniczania prędkości biegu

3273 maximum speed spring; stiff spring
d Endregelfeder *f*
f ressort *m* de vitesse maximum
n regelveer voor maximumsnelheid
p sprężyna dopuszczalnej prędkości biegu

3274 maximum tolerance
d Grenzspiel *n*
f tolérance *f* maximale
n maximumtolerantie
p tolerancja maksymalna

3275 maximum torque
d Höchstdrehmoment *n*
f couple *m* maximal
n maximaal draaimoment
p największy moment obrotowy

3276 maximum torque speed
d Höchstmomentdrehzahl *f*
f vitesse *f* de rotation au couple maximal

n motortoerental bij maximumkoppel
p prędkość obrotowa największego momentu

3277 McPherson separator
d Druckwerkzeug *n* für McPherson
f écarteur *m* McPherson
n drukgereedschap McPherson
p aparat McPherson

3278 McPherson strut suspension
d McPherson-Aufhängung *f*
f système *m* McPherson
n McPherson-wielophanging
p układ zawieszenia McPhersona

3279 mean effective pressure
d mittlerer Arbeitsdruck *m*
f pression *f* moyenne effective
n gemiddelde effectieve druk
p średnie ciśnienie efektywne; średnie ciśnienie użyteczne

3280 mean pressure
d Mitteldruck *m*
f pression *f* moyenne
n gemiddelde druk
p ciśnienie średnie

3281 mean velocity
d durchschnittliche Strömungsgeschwindigkeit *f*
f vitesse *f* moyenne d'écoulement
n gemiddelde stroomsnelheid
p średnia prędkość pizepływu

3282 measure
d messen
f mesurer
n meten
p mierzyć

3283 measuring glass
d Messglas *n*
f verre *m* gradué
n meetglas
p szkiełko pomiarowe

3284 measuring instrument
d Messgerät *n*
f instrument *m* de mesure
n meetinstrument
p przyrząd pomiarowy

3285 measuring tape
d Bandmass *n*; Massband *n*
f ruban *m* de mesure
n maatband; bandmaat; meetlint
p taśma miernicza; przymiar taśmowy

3286 measuring trammel
d Messzirkel *m*
f compas *m* de mesure
n meetpasser
p cyrkiel pomiarowy

3287 mechanic
d Handwerker *m*
f ouvrier *m*
n monteur
p monter; mechanik

3288 mechanical
d mechanisch
f mécanique
n mechanisch
p mechaniczny

3289 mechanical advantage
d Hebelübersetzung *f*
f rapport *m* de levier
n hefboomoverbrenging
p przekładnia dźwigniowa

3290 mechanical clutch operation
d mechanischer Kupplungsausrücker *m*
f commande *f* mécanique de débrayage
n mechanische bediening van koppeling
p mechaniczny wyprzęgnik sprzęgła

3291 mechanical efficiency
d mechanischer Wirkungsgrad *m*
f rendement *m* mécanique
n mechanisch rendement
p sprawność mechaniczna

3292 mechanical friction
d mechanische Reibung *f*
f frottement *m* mécanique
n mechanische wrijving
p tarcie mechaniczne

3293 mechanical governor; centrifugal governor
d Fliehkraftregler *m*
f régulateur *m* mécanique
n centrifugaalregelaar
p regulator mechaniczny; regulator odśrodkowy

3294 mechanical joint
d Schraubverband *n*
f boulonnage *m*
n boutverbinding; schroefverband
p połączenie mechaniczne

3295 mechanical metering
d mechanische Dosierung *f*
f dosage *m* mécanique

n mechanische dosering
p dozowanie mechaniczne

3296 mechanical wear
d mechanischer Verschleiss *m*
f usure *f* mécanique
n mechanische slijtage
p zużycie mechaniczne

3297 mechanic's general purpose attaché case
d Kundendienstkoffer *m*
f valise *f* universelle
n monteurskoffer
p skrzynia mechanika do obsługi klientów

3298 mechanic's wall cabinet
d Werkzeugwandschrank *m* für Automechaniker
f armoire *f* "mécanique auto"
n monteurgereedschapswand
p szafa narzędziowa dla mechanika

3299 mechanization
d Mechanisierung *f*
f mécanisation *f*
n mechanisering
p mechanizacja

3300 medium piston speed
d mittlere Kolbengeschwindigkeit *f*
f vitesse *f* moyenne de piston
n gemiddelde zuigersnelheid
p średnia prędkośc tłoka

3301 megger; insulation tester
d Isolationsmessapparat *m*; Isolationsprüfer *m*
f appareil *m* à essayer l'isolement; testeur *m* d'isolation
n isolatietestapparaat
p próbnik izolacji; miernik oporności izolacyjnej

3302 melt *v*
d schmelzen
f fuser
n smelten
p wytapiać

3303 melting pot
d Schmelztiegel *m*
f creuset *m*
n smeltkroes
p tygiel do topienia

* membrane → 1540

3304 metachromotype
d Abziehbild *n*

f décalcomanie *f*
n metachroomletter
p kalkomania

3305 metal
d Metall *n*
f métal *m*
n metaal
p metal

3306 metal bed catalytic converter
d Metallkatalysator *m*
f catalyseur *m* à métal
n katalysator met metalen drager; metaaldraagkatalysator
p katalizator metaliczny

3307 metal conduit
d Metallschlauch *m*
f tuyau *m* métallique
n metalen slang
p przewód metalowy; wąż metalowy

3308 metal cutting saw blade
d Metallsägeblatt *n*
f lame *f* de scie à métaux
n metaalzaagblad
p brzeszczot piły do metalu

3309 metal fatigue
d Metallermüdung *f*
f fatigue *f* du métal
n metaalmoeheid
p zmęczenie metalu

3310 metal filler
d Metallspachtel *m*
f enduit *m* pour métal
n metaalplamuur
p szpachla metalowa

3311 metal joint
d Metallgelenk *n*
f articulation *f* métallique
n metalen verbinding
p przegub metalowy

3312 metal lacquer
d Metallack *m*
f vernis *m* à métaux
n metaallak
p lakier do metalu

3313 metallic coating
d Metalliklackschicht *f*
f peinture *f* métallisée
n metallic laklaag
p powłoka metaliczna

3314 **metallic packing**
  d Metalldichtung *f*
  f garniture *f* métallique
  n metaalpakking
  p uszczelka metalowa

3315 **metallic sheen**
  d Metallglanz *m*
  f éclat *m* métallique
  n metaalglans
  p połysk metaliczny

3316 **metal slitting saw**
  d Metallkreissäge *f*
  f scie *f* circulaire pour couper métaux
  n metaalcirkelzaag
  p piła tarczowa do cięcia metalu

3317 **metal spraying**
  d Metallspritzen *n*
  f application *f* de métal par projection
  n metalliseren door spuiten
  p opryskiwanie lakierem metalicznym

3318 **metering device**
  d Dosiergerät *n*
  f appareil *m* de dosage
  n doseerinrichting
  p dawkownik

3319 **metering helix**
  d Regulierkante *f*; Steuerkante *f*
  f hélice *f* de réglage
  n helix van brandstofinspuitpomp
  p krawędź sterująca

3320 **metering needle; tapered needle**
  d Dosiernadel *f*
  f aiguille *f* de dosage
  n sproeiernaald
  p igła dozująca; igła dawkująca

3321 **methanol; methyl alcohol**
  d Methanol *n*; Methylalkohol *m*
  f méthanol *m*; alcool *m* méthylique
  n methanol
  p metanol; alkohol metylowy

  * **methyl alcohol** → 3321

3322 **metric system**
  d metrisches System *n*
  f système *m* métrique
  n metriek stelsel
  p system metryczny

3323 **metric thread**
  d Millimetergewinde *n*; metrisches Gewinde *n*

  f filet *m* métrique
  n metrische schroefdraad
  p gwint metryczny

3324 **micrometer**
  (instrument to measure cutting depth in keys)
  d Schnelltaster *m*
  f micromètre *m*
  n micrometer
  p szybki taster

3325 **microprocessor**
  d Prozessrechner *m*
  f calculateur *m*
  n microprocessor
  p mikroprocesor

3326 **microprocessor digital stroboscope**
  d Stroboskoplampe *f* für Benzinmotor
  f lampe *f* stroboscopique digitale à
     microprocesseur
  n digitale afstellamp met microprocessor
  p mikroprocesowy stroboskop cyfrowy

3327 **middle position**
  d Zentralstellung *f*
  f hauteur *f* intermédiaire
  n middelste stand
  p pozycja środkowa

3328 **mild steel; structural steel**
  d Flusseisen *n*
  f acier *m* doux; fer *m* fondu
  n zachtstaal
  p stal miękka

3329 **mileage between services**
  d Fahrstrecke *f* zwischen Durchsichten
  f parcours *m* entre deux entretiens
  n kilometers tussen onderhoudsbeurten
  p przebieg międzyobsługowy

  * **mileage indicator** → 3495

3330 **milled bayonet screwdriver blades**
  d gefräste Bajonettklingen *fpl*
  f lames *fpl* baïonnettes fraisées
  n bewerkte bajonetklingen
  p frezowane brzeszczoty Bayonetta

3331 **milled edge**
  d gezackter Rand *m*
  f morsure *f*
  n kartelrand
  p krawędź radełkowana

3332 **milled file**
  d Metallraspel *f*

f lime f fraisée
n metaalrasp
p pilnik frezowany

3333 milling
d Fräsen n
f fraisage m
n frezen
p frezowanie

3334 milling cutter
d Fräser m
f fraise f
n frees
p frez

3335 milling machine
d Fräsmaschine f
f machine f à fraiser
n freesbank; freesmachine
p frezarka

* milling table → 1421

3336 mineral oil
d Mineralöl n
f huile f minérale
n aardolie
p olej mineralny

3337 miniblock spring
d Miniblockfeder f
f ressort m minibloc
n miniblokveer
p sprężyna minibloku

3338 minibus
d Kleinbus m
f microbus m
n busje
p mikrobus

3339 minimal
d minimal
f minimal
n minimaal
p minimalny

3340 minimal position
d minimale Stellung f
f hauteur f minimale
n laagste stand
p wysokość najniższa; pozycja najniższa

3341 minimum continuous speed
d niedrigste Dauerbetriebsdrehzahl f
f régime m minimal en service continu

n laagst mogelijke toerental
p najmniejsza prędkość obrotowa

3342 minimum radius
d Mindesthalbmesser m
f rayon m minimum
n minimum draaicirkel
p najmniejszy promień; promień minimalny

3343 mini pipe bender
d Minirohrbiegezange f
f minicintreuse f
n minibuigtang
p giętarka do rurek

3344 mini pipe cutter
(instrument to cut copper and light alloy pipes)
d kleiner Rohrschneider m
f mini-coupe-tubes f
n minipijpsnijder
p obcinak do rurek

3345 minor road
d Seitenweg m
f voie f latérale
n secundaire weg
p droga boczna

3346 mirror
d Spiegel m
f miroir m
n spiegel
p lusterko

3347 misalignment
d Fluchtungsfehler m
f désalignement m
n foute uitlijning
p niewspółosiowość; nieprostoliniowość

3348 mishap
d Panne f
f ennui m; panne f
n pech; storing
p awaria

3349 mitre block
d Gehrungsblock m
f boîte f à onglets
n verstekblok
p opór uciosowy

3350 mitre joint
d Gehrverbindung f
f assemblage m à onglets
n verstekverbinding
p połączenie kątowe na ucios

3351 mitre square
d Gehrdreieck *n*
f équerre *f* à onglet
n verstekhaak
p kątownik stały

3352 mixing can
d Mischkanne *f*
f mélangeur *m*
n mengkan
p bańka do mieszania

3353 mixing chamber
d Mischkammer *f*
f chambre *f* de carburation
n mengkamer
p komora mieszalna

3354 mixing switch
d Mischschalter *m*
f commutateur *m* mélangeur
n mengschakelaar
p przełącznik układu mieszania

3355 mixture
d Mischung *f*; Gemisch *n*
f mélange *m*
n mengsel
p mieszanka

3356 mixture correction
d Gemischkorrektion *f*
f correction *f* du mélange
n mengselcorrectie
p korekcja mieszanki

3357 mixture formation
d Gemischbildung *f*
f formation *f* du mélange
n mengselvorming
p wytwarzanie mieszanki

3358 mixture regulator
d Gemischregler *m*
f correcteur *m* du mélange
n mengselregelaar
p regulator mieszanki

3359 mobile crane
d Mobilkran *m*; Autokran *m*
f grue *f* automotrice
n verrijdbare kraan; kraanwagen
p dźwig samojezdny

3360 mobile frame
d Gleitrahmen *m*
f cadre *m* mobile
n uitneembaar frame
p rama ślizgowa

3361 mobile shop; shoptruck
d Verkaufswagen *m*
f camion-magasin *m*
n winkelwagen; winkelauto
p samochód handlowy

3362 mode button
d Funktionknopf *m*
f bouton *m* de fonctionnement
n functieknop
p guzik funkcyjny

3363 model
d Modell *n*
f modèle *m*
n model
p model

3364 modulated output fan; proportional flow fan
d Ventilator *m* mit veränderlichem Luftdurchsatz
f ventilateur *m* à débit variable
n ventilateur met regelbare luchtopbrengst
p wentylator nastawny

3365 modulator
d Modulator *m*
f modulateur *m*
n modulator
p modulator

3366 modulator valve
d Modulatorventil *n*
f soupape *f* de modulateur
n modulatieklep
p zawór modulacyjny

3367 module
d Modul *m*
f module *m*
n module; moduul
p moduł

3368 modulus of elasticity
d Elastizitätsmodul *m*
f module *m* d'élasticité
n elasticiteitsmodulus
p moduł sprężystości

3369 moisture
d Feuchtigkeit *f*
f humidité *f*
n vocht
p wilgoć

3370 **momentary power**
  d Kurzleistung *f*
  f puissance *f* instantanée
  n voor korte duur afgegeven vermogen
  p moc chwilowa

3371 **moment of friction**
  d Reibungsmoment *n*
  f moment *m* de frottement
  n wrijvingsmoment
  p moment tarcia

3372 **moment of inertia**
  d Schwungmoment *n*
  f moment *m* d'inertie
  n traagheidsmoment
  p moment bezwładności

3373 **monocoque chassis**
  d monocoque Fahrwerk *n*
  f monocoque *f*
  n monocoque-chassis
  p konstrukcja skorupowa podwozia

3374 **monopoint fuel injection**
  d Zentraleinspritzung *f*; Benzineinspritzung *f* mit einer einzigen Düse
  f injection *f* d'essence monopoint
  n monopoint benzine-inspuiting
  p wtryskiwanie centralne

3375 **monotone exterior colour**
  d einfache Lackfarbe *f*
  f teinte *f* de laque unique
  n enkelvoudige lakkleur
  p pojedynczy kolor lakieru

3376 **Morse taper**
  d Morsekegel *m*
  f cône *m* morse
  n Morse-kegel
  p stożek Morse'a

3377 **motive force**
  d Triebkraft *f*
  f force *f* mouvante
  n drijfkracht
  p siła napędowa

3378 **motor harness**
  d Kabelbündel *n* für Motor
  f câblerie *m* de moteur
  n kabelbundel voor de motor
  p zespół przewodów silnika

3379 **motor highway; motor road; motor way**
  d Autobahn *f*
  f autoroute *f*
  n autosnelweg
  p autostrada

* **motor road** → 3379

3380 **motor vehicle**
  d Kraftfahrzeug *n*
  f véhicule *m* moteur
  n motorvoertuig
  p pojazd mechaniczny

3381 **motor vehicle insurance**
  d Kraftfahrzeugversicherung *f*
  f assurance *f* automobile
  n autoverzekering
  p ubezpieczenie samochodu

* **motor way** → 3379

3382 **motor way guide planks**
  d Leitplanken *fpl*
  f bandes *fpl* de guidage pour routes
  n vangrails
  p poręcze drogowe

3383 **moulding**
  d Zierleiste *f*
  f moulure *f*
  n sierlijst
  p listwa ozdobna

3384 **mount** *v*
  d befestigen
  f fixer
  n bevestigen
  p mocować

3385 **mounting bracket**
  d Tragstütze *f*; Befestigungsschelle *f*
  f support *m*; bride *f* de fixation
  n bevestigingsbeugel
  p wspornik montażowy

3386 **mounting cover plate**
  d Befestigungsplatte *f*
  f support *m* de montage
  n bevestigingsplaat
  p płyta mocująca

3387 **mounting eye**
  d Befestigungsöse *f*
  f patte *f* de fixation
  n bevestigingsoog
  p ucho mocujące

3388 **mounting flange**
  d Befestigungsflansch *m*

f collerette *f* de fixation
n bevestigingsflens
p kołnierz mocujący

3389 movable part
d beweglicher Teil *m*
f partie *f* mobile
n beweegbaar deel
p część ruchoma

3390 mud and snow tyre
d Winterreifen *m*
f pneu *m* hiver
n geprofileerde band
p opona zimowa

3391 mudguard; rear wing
d Hinterkottflügel *m*
f aile *f* arrière; garde-boue *m* arrière
n achterspatscherm
p błotnik tylny

* **mudguard cover** → 3392

3392 mudguard flap; mudguard cover
d Schmutzfänger *m*
f bavette *f*
n spatlap
p fartuch błotnika

3393 multibush type flexible coupling
d elastische Buchsenringkupplung *f*
f joint *m* à bagues élastiques
n elastische koppeling meerbustype
p wielotulejowy przegub elastyczny

3394 multicircuit switch
d Serienschalter *m*
f contacteur *m* en série
n serieschakelaar
p przełącznik dwugrupowy

3395 multicore cable
d vieladriges Kabel *n*
f câble *m* multifiliaire
n meeraderige kabel
p kabel wielożyłowy

3396 multicylinder engine
d Mehrzylindermotor *m*
f moteur *m* multicylindre
n meercilindermotor
p silnik wielocylindrowy

3397 multicylinder injection pump
d Reiheneinspritzpumpe *f*
f pompe *f* d'injection en ligne

n meercilinderinspuitpomp; lijninspuitpomp
p wielosekcyjna pompa wtryskowa

* **multidisc clutch** → 3403

3398 multifuel engine
d Vielstoffmotor *m*
f moteur *m* à combustibles multiples
n meerbrandstofmotor
p silnik wielopaliwowy

3399 multigated solid friction
d Grenzreibung *f*
f frottement *m* mixte
n marginale wrijving
p tarcie graniczne

3400 multigrade oil
d Mehrbereichöl *n*
f huile *f* "multigrade"
n multigrade-olie
p olej uniwersalny

3401 multihole nozzle
d Mehrlochdüse *f*
f injecteur *m* à trous
n meergatverstuiver
p rozpylacz wielootworkowy

* **multilayer glass** → 3020

* **multimeter** → 1054

3402 multipitch roof; foldable roof
d Faltdach *n*
f toit *m* repliable
n opvouwbaar dak
p dach składany

3403 multiple clutch; multidisc clutch
d Lamellenkupplung *f*
f embrayage *m* multidisque
n lamellenkoppeling; meervoudige koppeling
p sprzęgło wielotarczowe

3404 multiple disc brake
d Mehrscheibenbremse *f*
f frein *m* multidisque
n meerdere-schijfrem
p hamulec wielotarczowy

3405 multiple range carburetter
d Registervergaser *m*
f carburateur *m* à registre
n registercarburateur
p gaźnik wieloprzelotowy; gaźnik o stopniowym działaniu

3406 multiple start thread
d mehrgängiges Gewinde *n*
f filet *m* multiple
n meervoudige schroefdraad
p gwint wielokrotny; gwint wielozwojny

3407 multiple strand chain
d Mehrfachkette *f*
f chaîne *f* multiple
n meervoudige ketting
p łańcuch drabinkowy wielokrotny

3408 multiplying gear; overdrive
d Multiplikator *m*
f surmultiplicateur *m*
n multiplicator
p przekładnia przyśpieszjąca

3409 multipoint injection
d Anzaugkanaleinspritzung *f*;
  Mehrfacheinspritzung *f*
f injection *f* multiple
n multipoint benzine-inspuiting
p wtrysk wielopunktowy

3410 multipurpose additive
d Detergentdispersant *m*
f additif *m* multifonctionnel
n universeel additief
p dodatek wielofunkcyjny

3411 multipurpose vehicle
d Mehrzweckwagen *m*
f camion *m* universel; véhicule *m* universel
n auto voor meerdere toepassingen
p samochód uniwersalny

3412 multistage
d mehrstufig
f gradué
n meertraps
p wielostopniowy

3413 multivalve engine
d Mehrventilenmotor *m*
f moteur *m* à plusieurs soupapes
n meerklepsmotor
p silnik wielozaworowy

3414 municipal transport
d städtischer Verkehr *m*
f transport *m* urbain
n stedelijk transport
p transport miejski

3415 municipal vehicle
d Kommunaldienstfahrzeug *n*

f voiture *f* de voirie
n voertuig voor openbare diensten
p pojazd komunalny

3416 mushroom tappet
d Pilzstössel *m*
f poussoir *m* à sabot
n paddestoelvormige klepstoter
p popychacz talerzykowy

N

3417 nail bag
d Geldtasche *f* für Nägel
f sacoche *f* à clous portemarteau
n spijkerzak
p worek na gwoździe

3418 nail nippers
d Nagelzange *f*
f tenaille *f*
n spijkerbuigtang
p obcęgi do gwoździ

3419 name plate
d Fabrikschild *n*
f plaque *f* de fabricant
n firmaplaat; merkaanduiding
p tabliczka firmowa

3420 narrow
d schmal
f étroit
n smal
p wąski

3421 narrow domed single spoon
d kurzes, schmales Löffeleisen *n*
f palette *f* single bombée étroite
n smalle lepel, gebold
p wąska łyżka zgięta

3422 narrowing of the road
d Fahrbahnverengung *f*
f chaussée *f* rétrécie
n wegversmalling
p zwężenie jezdni

3423 narrow spoon
d Löffeleisen *n* schmal
f cuillère *f* étroite
n smalle uitdeuklepel
p łyżka wąska

3424 nationality label; nationality plate
d Nationalitätszeichen *n*
f plaque *f* de nationalité
n nationaliteitsteken; nationaliteitsplaat
p znak przynależności państwowej

* **nationality plate → 3424**

3425 natural frequency
d Eigenschwundigkeitszahl *f*

f nombre *m* d'oscillations propres
n eigenfrequentie
p częstotliwość własna; częstotliwość drgań własnych

3426 neck journal
d Halszapfen *m*
f tourillon *m* intermédiaire
n halstap
p czop poprzeczny środkowy

3427 needle
d Nadel *f*
f aiguille *f*
n naald
p igła

3428 needle cage
d Nadelkäfig *m*
f cage *f* à aiguilles
n naaldkooi van kogelkringloopbesturing
p koszyczek igieł

3429 needle carburetter
d Nadeldüsenvergaser *m*
f carburateur *m* à aiguille
n naaldcarburateur
p gaźnik iglicowy

3430 needle file
d Nadelfeile *f*
f lime *f* aiguille
n naaldvijl
p pilnik igiełkowy

3431 needle roller
d Nadelrolle *f*
f rouleau *m* aiguille
n naaldrol
p wałeczek igiełkowy

3432 needle roller bearing
d Nadellager *n*
f roulement *m* à aiguilles
n naaldlager
p łożysko igiełkowe

3433 needle sleeve
d Nadelhülse *f*
f douille *f* d'aiguilles
n naaldhuls
p łożysko igiełkowe bez pierścienia wewnętrznego

3434 needle thrust bearing
d Drucknadellager *n*
f butée *f* à aiguilles
n axiaal naaldlager
p skośne łożysko igiełkowe

3435 needle valve
d Nadelventil *n*
f soupape *f* à aiguille; valve *f* à aiguille
n naaldklep
p zawór iglicowy

3436 needle valve seat
d Nadelventilsitz *m*
f siège *m* du pointeau
n vlotternaaldzitting
p gniazdo rozpylacza

3437 negative plate
d Minusplatte *f*
f plaque *f* négative
n accuminuspool
p płyta ujemna akumulatora

3438 negative pole
d negativer Pol *m*
f pôle *m* négatif
n minklem; minpool
p biegun ujemny

3439 nest of tubes
d Rohrbundel *n*
f faisceau *m* tubulaire
n buizenbundel; pijpenbundel
p wiązka rur

3440 net of distribution
d Verteilernetz *n*
f réseau *m* de distribution
n distributienet
p sieć rozdzielcza

3441 neutral axis
d neutrale Achse *f*
f ligne *f* neutre
n neutrale as
p oś obojętna

3442 neutral grounding
d Nullpunkterdung *f*
f mise *f* au neutre
n aarding van de nulleiding
p uziemienie punktu zerowego

3443 neutral position
d Nullstellung *f*
f position *f* de repos
n nulstand
p położenie zerowe; położenie obojętne

3444 neutral wire
d Nulleiter *m*
f conducteur *m* neutre; neutre *m*

n nulleider
p przewód zerowy

3445 new car registration
d Neuzulassung *f*
f immatriculation *f* de voitures neuves
n registratie van nieuwe auto's
p nowa rejestracja samochodu

* **nibbing cutter** → **3446**

3446 nibbing machine; nibbing cutter
d Knabbelschere *f*
f grignoteuse *f*
n knabbelschaar
p nożyce wibracyjne

3447 nickel
d Nickel *n*
f nickel *m*
n nikkel
p nikiel

3448 nickel iron storage battery
d Nickeleisenakkumulator *m*
f accumulateur *m* au ferro-nickel
n nikkel-ijzeraccu
p akumulator żelazowo-niklowy

3449 nickel plated
d vernickelt
f nickelé
n vernikkeld
p niklowany

3450 nickel steel
d Nickelstahl *m*
f acier *m* au nickel
n nikkelstaal
p stal niklowa

3451 nippers
d Beisszange *f*
f pince *f*
n knijptang
p szczypce do cięcia drutu

3452 nipple
d Nippel *m*
f raccord *m* fileté
n nippel
p złączka skrętna

3453 nitrocellulose lacquer
d Nitrozelluloselack *m*
f vernis *m* nitrocellulosique
n nitrocelluloselak
p lakier nitrocelulozowy; nitrolakier

3454 nitrogen
  d Stickstoff *m*
  f azote *m*
  n stikstof
  p azot

3455 noise detector
  d Geräuschdetektor *m*
  f détecteur *m* de bruit
  n lawaaidetector
  p wykrywacz hałasu

3456 noiseless
  d geräuschlos
  f sans bruit
  n geruisloos
  p bezszumowy; bezszmerowy

3457 nominal capacity
  d Nennkapazität *f*
  f capacité *f* nominale
  n nominale capaciteit
  p pojemność znamionowa

3458 nominal value
  d Nennwert *m*
  f valeur *f* nominale
  n nominale waarde
  p wartość znamionowa; wartość
       nominalna

3459 nominal voltage
  d Nennspannung *f*
  f tension *f* nominale
  n nominale spanning
  p napięcie znamionowe

3460 nonconductor
  d Nichtleiter *m*
  f non-conducteur *m*
  n isolator
  p nieprzewodnik

3461 nonlocking retractor safety belt
  d nichtsperrender Sicherheitsgurt *m*
  f ceinture *f* de sécurité statique
  n niet-automatische veiligheidsgordel
  p nieautomatyczny pas bezpieczeństwa

3462 nonpolluting engine
  d umweltfreundlicher Motor *m*
  f moteur *m* moins polluant
  n motor met gunstige uitlaatgasemissie
  p motor nie zanieczyszczający

3463 nonreturn valve
  d Rückschlagventil *n*

  f soupape *f* d'arrêt; soupape *f* de retenue
  n eenrichtingsklep
  p zawór jednokierunkowy

3464 nonselective catalytic reduction
  d nichtselektive Abgasreduktion *f*
  f purification *f* des gaz d'échappement
  n ongeregelde uitlaatgasreiniging
  p redukcja katalityczna nieselektywna

3465 nonskid pattern
  d Gleitschutzmuster *n*
  f sculpture *f* antidérapante
  n antislipprofiel
  p profil przeciwślizgowy

3466 no parking
  d Parkverbot *n*
  f stationnement *m* interdit
  n verboden te parkeren
  p zakaz postoju; zakaz parkowania

3467 noppy tread
  d Noppenprofil *n*
  f sculpture *f* à tétons
  n noppy-loopvlak
  p bieżnik z ostrogą przeciwpoślizgową

3468 normal pitch of rack
  d Normalteilung *f* der Zahnstange
  f pas *m* normal de crémaillère
  n normaalsteek van een tandstang
  p normalna podziałka zębatki

3469 normal position
  d normale Stellung *f*
  f position *f* normale de route
  n normale stand
  p wysokość normalna; pozycja normalna

3470 normal section of rack
  d Normalschnitt *m* durch Zahnstange
  f section *f* normale de crémaillère
  n normaaldoorsnede van een tandstang
  p przekrój normalny zębatki

3471 no U turns
  d Wendeverbot *n*
  f interdiction *f* de faire demi-tour
  n keerverbod
  p zakaz zawracania

3472 no waiting
  d Halteverbot *n*
  f interdiction *f* de s'arrêter
  n stopverbod
  p zakaz postoju

3473 nozzle bore
d Düsenbohrung *f*
f orifice *m* calibré de gicleur
n diameter van inspuitopening
p otwór rozpylacza

3474 nozzle for pins
d Mundstück *n* für Nagel
f buse *f* pour clous
n mondstuk voor spijkers
p otwór do kołków

3475 nozzle fouling
d Düsenverstopfung *f*
f encrassement *m* du gicleur
n verstopte spuitneus
p zatkanie dyszy

3476 nozzle holder
d Düsenhalter *m*
f porte-injecteur *m*
n verstuiverhouder
p korpus wtryskiwacza

3477 nozzle needle
d Düsennadel *f*
f aiguille *f* d'injecteur
n verstuivernaald
p iglica wtryskiwacza

3478 nozzle plate
d Düsenplatte *f*
f plaque *f* d'injecteur
n verstuiverplaat
p płyta rozpylacza

3479 nozzle valve spring
d Druckfeder *f*
f ressort *m* de tarage
n drukveer van verstuiver
p sprężyna rozpylacza

3480 numbering
d Numerierung *f*
f numérotage *m*
n nummering
p numeracja

3481 number of teeth
d Zähnezahl *f*
f nombre *m* des dents
n tandaantal
p liczba zębów

3482 number plate; licence plate
d Kennzeichenschild *n*; Nummerschild *n*
f plaque *f* d'immatriculation; plaque *f* de
licence
n kentekenplaat
p tablica rejestracyjna

3483 number plate light
d Nummerschildleuchte *f*
f feu *m* de plaque de licence
n kentekenplaatverlichting
p oświetlenie tablicy rejestracyjnej

3484 nut
d Mutter *f*
f écrou *m*
n moer
p nakrętka

3485 nut clip
d Clipsmutter *f*
f écrou *m* clip
n klem van moer
p nakrętka klipsowa

3486 nut for pinion shaft
d Mutter *f* für Antriebsritzel
f écrou *m* de pignon d'attaque
n tandwielasmoer
p śruba zębnika

3487 nut runner
d Mutteranzieher *m*
f clé *f* de serrage d'écrous
n moeraanzetter
p wkrętak mechaniczny do nakrętek

3488 nut spinner
d Gelenkgriff *m*
f poignée *f* articulée
n handgreep met kniegewricht
p rękojeść przegubowa

3489 nut splitter
d Mutternsprenger *m*
f casse-écrou *m*
n moersplijter
p rozbijacz nakrętek

3490 nut with grooves
d Nutmutter *f*
f écrou *m* à encoches
n kruisgleufmoer
p nakrętka okrągła rowkowa

3491 nylon cover
d Kunststoffhülle *f*
f hausse *f* en plastique
n kunststof hoes
p pokrowiec nylonowy

O

3492 octagonal chisel
d achtkantiger Meissel *m*
f burin *m* octagonal
n achthoekige koudbeitel
p przecinak ośmiokątny

3493 octagonal cross cut chisel
d achtkantiger Kreuzmeissel *m*
f bédane *f* octagonale
n achthoekige ritsbeitel
p wycinak ośmiokątny

3494 octane number
d Oktanzahl *f*
f indice *f* d'octane
n octaangetal
p liczba oktanowa

3495 odometer; mileage indicator
d Kilometerzähler *m*
f odomètre *m*
n kilometerteller
p licznik kilometrów

3496 odor component
d Geruchsstoff *m*; Riechstoff *m*
f odorifère *m*
n bestanddeel van geur
p substancja zapachowa

3497 odor threshold
d Geruchsschwelle *f*
f seuil *m* d'odeur
n geurgrenswaarde
p próg zapachu

3498 odourless
d geruchlos
f inodore
n reukloos
p bezwonny

3499 offroad vehicle
d Riesenfahrzeug *n*
f camion *m* gros porteur
n terreinvoertuig
p samochód pozadrogowy

3500 offset adjusting wrench
d doppeltversetzter Aufsteckschlüssel *m*

f clé *f* de réglage contrecoudée
n dubbel gebogen ringsleutel
p klucz nasadowy odsadzony

3501 offset radius
d Lenkrollradius *m*
f rayon *m* de flottement de roue directrice
n schuurstraal
p promień zataczania

3502 offset screwdriver
d rechtwinkliger Schraubendreher *m*
f tournevis *m* coudé lame
n haakse schroevendraaier
p wkrętak kątowy

3503 offset socket wrench
d Pfeifenkopfschlüssel *m*
f clé *f* à pipe
n pijpsleutel
p fajkowy klucz głowicowy

3504 ohm meter
d Ohmmeter *n*
f ohmmètre *m*
n ohmmeter
p omomierz

3505 oil
d Öl *n*
f huile *f*
n olie
p olej

3506 oil absorbent
d Ölentfernungsmittel *n*
f détachant *m* d'huile
n olieabsorbeermiddel
p środek do usuwania oleju

3507 oil baffle; oil deflector
d Ölleitblech *n*
f déflecteur *m* d'huile
n oliekeerschot in carterpan
p odrzutnik oleju

3508 oil bath
d Ölbad *n*
f bain *m* d'huile
n oliebad
p kąpiel olejowa

3509 oil can
d Ölkanne *f*
f burette *f* à huile
n oliekan; oliespuit
p olejarka

3510 oil change
d Ölwechsel m
f changement m d'huile
n olieverversing
p wymiana oleju

3511 oil circulation
d Ölumlauf m
f circulation f d'huile
n oliecirculatie
p obieg oleju

3512 oil cleaner
d Ölreiniger m
f épurateur m d'huile
n oliereinigingsmiddel
p urządzenie do oczyszczania oleju; aparat do regeneracji oleju

3513 oil consumption
d Ölverbrauch m
f consommation f d'huile
n olieverbruik
p zużycie oleju

3514 oil content
d Ölgehalt m
f teneur f d'huile
n oliegehalte
p zawartość oleju

3515 oil cooler
d Ölkuhler m
f réfrigérant m d'huile
n oliekoeler
p chłodnica oleju

* oil deflector → 3507

3516 oil dilution
d Ölverdünnung f
f dilution f de l'huile
n dilutie van olie
p rozrzedzenie oleju

3517 oil dipper
d Ölschöpfer m
f cuillère f à huile
n olielikker
p czerpak oleju

3518 oil dipstick; oil level dipstick
d Ölkontrollstab m; Ölmessstab m
f jauge f d'huile
n oliepeilstok; oliemeetstaaf
p prętowy wskaźnik poziomu oleju; olejowskaz prętowy

3519 oil drain hole
d Ölauslassöffnung f
f orifice m de vidange d'huile
n olieaftapgat
p otwór spustowy oleju

3520 oil drip
d Tropföler m
f graisseur m à goutte
n oliedruppelaar
p smarownica kroplowa; olejarka kroplowa

3521 oil duct; oil passage
d Ölkanal m
f canal m d'huile
n oliekanaal
p kanał oleju

3522 oiler
d Druckschmiertopf m; Öler m
f graisseur m à compte-gouttes
n doorsmeerder; oliespuit
p smarownica olejowa

3523 oil filler cap
d Ölverschlussschraube f
f bouchon m de remplissage d'huile
n olievuldop
p pokrywa wlewu oleju

3524 oil filler hole
d Ölfüllöffnung f
f orifice m de remplissage d'huile
n olievulopening
p gniazdo wlewu oleju

3525 oil filler tube
d Ölfüllstutzen m
f tubulure f de remplissage d'huile
n olievulopening
p króciec wlewu oleju

3526 oil film wedge
d Keilspalt m; Schmierkeil m
f coin m d'huile
n oliewig
p klin smarowy; klin olejowy

3527 oil filter
d Ölfilter m
f filtre m d'huile
n oliefilter
p filtr oleju

3528 oil filter spanner
d Schlüssel m für Ölfilter
f clé f pour filtre à huile

n sleutel voor oliefilter
p klucz do filtru oleju

3529 oil flexible hose
d Ölschlauch *m*
f tuyau *m* armé d'huile
n flexibele olieslang
p giętki przewód oleju

* **oil flinger → 3562**

3530 oil funnel
d Öleinfülltrichter *m*
f entonnoir *m* à huile
n olietrechter
p lejek do napełniania oleju

3531 oil groove
d Ölnut *f*
f rainure *f* de graissage
n oliegroef
p rowek oleju

3532 oil hole
d Ölloch *n*; Schmierloch *n*
f trou *m* d'huile
n oliegaatje
p otwór smarowy

3533 oil leak
d Lecköl *n*
f fuite *f* d'huile
n lekolie
p przeciek oleju

3534 oil leakage detector
d Ölundichtheitsanzeige *f*
f indicateur *m* de fuite d'huile
n olielekkageverklikker
p detektor przecieku oleju

3535 oil level
d Ölstand *m*
f niveau *m* d'huile
n oliepeil
p poziom oleju

* **oil level dipstick → 3518**

3536 oil nozzle
d Öldüse *f*
f gicleur *m* d'huile; jet *m* d'huile
n oliesproeier
p dysza olejowa

3537 oil pan
d Ölwanne *f*
f carter *m* inférieur

n oliecarter
p miska olejowa

3538 oil pan drain plug
d Ablassschraube *f* der Ölwanne
f bouchon *m* de vidange de carter inférieur
n aftapplug van oliecarter
p korek spustowy miski olejowej

* **oil passage → 3521**

3539 oil pipe
d Ölleitung *f*
f tube *m* de graissage
n olieleiding
p przewód olejowy

3540 oil plug wrench
d Verschlusskappe *f* zum Erneuern des Schmieröls
f douille *f* de vidange
n dop voor olieverversen
p klucz do korka olejowego

3541 oil pocket
d Ölkammer *f*
f poche *f* d'huile
n olieholte
p zbiornik oleju

3542 oil pressure
d Öldruck *m*
f pression *f* d'huile
n oliedruk
p ciśnienie oleju

3543 oil pressure gauge
d Öldruckmesser *m*
f manomètre *m* d'huile
n oliedrukmeter
p manometr olejowy

* **oil pressure indicator → 3549**

3544 oil pressure relief valve
d Ölüberdruckventil *n*
f soupape *f* de suppression d'huile
n olieoverloopklep
p zawór przelewowy

3545 oil pressure relief valve ball
d Kugel *f* des Überströmventils
f bille *f* de clapet de décharge
n klepkogel van oliedrukveiligheid
p miseczka zaworu przelewowego

3546 oil pressure relief valve spring
d Feder *f* des Überströmventils

f ressort *m* de clapet de décharge
n overloopklepveer
p sprężyna zaworu przelewowego

3547 oil pressure sensor
d Fühler *m* des Öldruckmessers
f monocontact *m* de pression d'huile
n sensor van oliedrukmeter
p czujnik ciśnienia oleju

3548 oil pressure switch
d Öldruckschalter *m*
f contact *m* de pression d'huile
n oliedrukschakelaar
p przełącznik ciśnienia oleju

3549 oil pressure warning light; oil pressure indicator
d Öldruckkontrolleuchte *f*
f témoin *m* de pression d'huile
n oliedrukcontrolelicht; verklikkerlicht voor oliedruk
p światło kontrolne ciśnienia oleju

3550 oil pump
d Ölpumpe *f*
f pompe *f* à huile
n oliepomp
p pompa oleju

3551 oil pump delivery
d Ölpumpenförderung *f*
f débit *m* de pompe à huile
n opbrengst van de oliepomp
p wydajność pompy oleju

3552 oil pump driving gear shaft
d Antriebswelle *f* der Ölpumpe
f arbre *m* de commande de pompe à huile
n oliepompaandrijfas met tandwiel
p wał napędowy pompy olejowej

3553 oil pump gears
d Ölpumpenzahnräder *npl*
f pignons *mpl* de pompe à huile
n tandwielen van oliepomp
p koła zębate pompy oleju

3554 oil pump housing
d Ölpumpengehäuse *n*
f corps *m* de pompe à huile
n oliepomplichaam; oliepomphuis
p kadłub pompy oleju

3555 oil pump outlet pipe
d Öldruckpumpenleitung *f*
f tubulure *f* de refoulement de pompe à huile

n oliepompdrukleiding
p przewód tłoczny pompy oleju

* **oil reservoir → 3567**

3556 oil resistance
d Ölbeständigkeit *f*
f résistance *f* à huile
n oliebestendigheid
p olejoodporność

3557 oil ring expander
d Spreizfeder *f*; Expanderfeder *f*
f segment *m* élastique
n expanderveer
p sprężyna rozpierająca; sprężyna rozprężająca

3558 oil ring sensor
d Öldrucksensor *m*
f capteur *m* de pression d'huile
n oliedruksensor
p sensor ciśnienia oleju; czujnik ciśnienia oleju

3559 oil seal
d Ölabdichtung *f*
f disque *m* de retenue d'huile
n olieafdichting
p uszczelnienie olejowe

3560 oil seal extractor
(tool for easy removal of concealed seals)
d Ölabstreifringabzieher *m*
f extracteur *m* des bagues d'étanchéité
n keerringtrekker
p ściągacz do usuwania uszczelnień olejowych

3561 oil separator of crankcase ventilating system
d Ölseparator *m* der Entlüftung des Kurbelgehäuses
f déshuileur *m* de ventilation de carter
n carterontluchting
p odolejacz układu odpowietrzania skrzyni korbowej

3562 oil slinger; oil flinger
d Ölschleuderring *m*
f anneau *m* de retour d'huile; anneau *m* de rejet d'huile
n oliekeerring; oliespatring
p odrzutnik oleju

3563 oil sludge
d Ölschlamm *m*
f boue *f* d'huile
n oliedrab
p osad oleju

3564 oil starvation
 d Ölmangel *m*
 f manque *m* d'huile
 n gebrek aan olie
 p niedostateczny dopływ oleju

3565 oil stone
 d Ölstein *m*
 f pierre *f* à huile
 n olieslijpsteen
 p osełka

3566 oil sump gasket
 d Ölwannendichtung *f*
 f joint *m* de carter d'huile
 n oliecarterafdichting
 p uszczelnienie miski olejowej

3567 oil tank; oil reservoir
 d Ölbehälter *m*
 f réservoir *m* d'huile
 n olietank
 p zbiornik oleju

3568 oil thermostat
 d Öltemperaturregler *m*
 f thermostat *m* pour l'huile
 n olietemperatuurregelaar
 p regulator temperatury oleju

3569 oil trap
 d Entöler *m*
 f déshuileur *m*
 n olieafscheider
 p oddzielacz oleju

3570 oil trough
 d Ölmulde *f*
 f cuvette *f* à huile
 n oliepan
 p rynienka olejowa

3571 one tube hydraulic damper
 d hydraulischer Einrohrstossdämpfer *m*
 f amortisseur *m* monotube hydraulique
 n hydraulische schokdemper met enkele
    buis
 p jednorurowy amortyzator hydrauliczny

3572 one way clutch
 d einseitige Kupplung *f*
 f embrayage *m* d'un côté
 n vrijwielkoppeling
 p sprzęgło jednokierunkowe

3573 one way traffic
 d Einrichtungsverkehr *m*

 f trafic *m* à sens unique
 n eenrichtingsverkeer
 p ruch jednokierunkowy

3574 open front press
 d Einständerpresse *f*
 f presse *f* à bâti en col-de-cygne
 n pers met open voorkant
 p prasa jednostojakowa

3575 open guideways
 d offene Führungen *fpl*
 f glissières *fpl* ouvertes
 n open geleidingen
 p prowadnice otwarte

3576 operating cam
 d Steuernocken *m*
 f came *f* de commande
 n aandrijfnok
 p krzywka sterująca

3577 operating costs
 d Betriebskosten *mpl*
 f frais *mpl* d'exploitation
 n bedrijfskosten
 p koszty produkcji

3578 operating device; manipulator
 d Bedienungselement *n*
 f organe *m* de commande
 n bedieningsorgaan
 p element obsługi

3579 operating force
 d Betätigungskraft *f*
 f force *f* de commande
 n bedieningskracht
 p siła napędowa

3580 operating fork ball end
 d Kugelbolzen *m* der Ausrückgabel
 f rotule *f* de fourchette
 n kogelbout van ontkoppelingvork
 p sworzeń kulowy widełek

3581 operating lever
 d Steuerhebel *m*
 f levier *m* de commande
 n bedieningshendel
 p dźwignia sterownicza

3582 operating lever articulation
 d Ausrückhebelgelenk *n*
 f articulation *f* de fourchette de débrayage
 n verbinding met de ontkoppelingsvork
 p przegub widełek wyłączających

3583 operating principle
  d Betriebsweise f
  f fonctionnement m
  n werkingswijze
  p sposób działania

* operating rod → 4087

3584 operation life
  d Benutzungsdauer f
  f durée f de service
  n levensduur
  p okres użytkowania

3585 operation mode switch
  d Kennfeldschalter m
  f commutateur m de ronde de fonctionnement
  n schakelaar van besturingssysteem
  p przełącznik zakresów pracy

3586 operation range; radius of action
  d Aktionsradius m
  f autonomie f
  n actieradius
  p zasięg

3587 opposed milling
  d Gegenlauffräsen n
  f fraiser à contre-sens
  n tegenloopfrezen
  p frezowanie przeciwbieżne

3588 opposite profile
  d Gegenstück n
  f profil m contraire
  n tegenprofiel
  p zarys przeciwny

3589 optical unit
  d Leuchteinheit f
  f bloc m optique
  n verlichtingseenheid
  p blok optyczny

3590 option
  d Option f
  f option f
  n optie
  p opcja

3591 order
  d Auftrag m
  f commande f
  n opdracht
  p zlecenie; zamówienie

3592 orifice
  d Blende f

  f porte f de travail
  n opening
  p otwór

3593 original equipment
  d Erstausrüstung f
  f première monte f
  n standaarduitrusting
  p wyposażenie standartowe; wyposażenie oryginalne

3594 original part
  d Originalteil n
  f pièce f originale
  n origineel onderdeel
  p część oryginalna

3595 oscillating axle
  d Pendelachse f
  f essieu m oscillant
  n schommelas
  p oś łamana (niezależnego zawieszenia tylnych kół)

3596 oscillating lever
  d Schwinghebel m
  f levier m oscillant
  n schommelarm
  p dźwignia swobodna

3597 oscillating reflector
  d Pendelrückstrahler m
  f catadioptre m oscillant
  n oscillerende reflector
  p reflektor wahadłowy

* oscillation → 5360

3598 oscillation moment
  d Schwingmoment n
  f moment m de renversement
  n kantelmoment
  p moment drgań

3599 outdoor temperature gauge
  d Aussenthermometer n
  f thermomètre m extérieure
  n buitentemperatuurmeter
  p termometr zewnętrzny

* outer compass → 841

3600 outer conical disc; sliding conical disc
  d aussere Kegelscheibe f; verschiebbare Kegelscheibe f
  f flasque m extérieur; flasque m mobile
  n buitenste conische schijf
  p tarcza stożkowa; tarcza przesuwna

3601 outer race
- *d* Kugelkappe *f*
- *f* calotte *f*
- *n* buitenkogelschaal
- *p* czasza

3602 outer rear view mirror
- *d* Aussenrückblickspiegel *m*
- *f* rétroviseur *m* extérieur
- *n* buitenspiegel
- *p* lusterko wsteczne

3603 outer scavenger rotor
- *d* Aussenrotor *m*
- *f* pignon *m* extérieur; pignon *m* rotor
- *n* buitenrotor
- *p* wirnik zewnętrzny

3604 outer valve spring
- *d* aussere Ventilfeder *f*
- *f* ressort *m* extérieur de soupape
- *n* buitenste klepveer
- *p* zewnętrzna sprężyna zaworu

3605 outflow plate
- *d* Abflussplatte *f*
- *f* plaque *f* d'écoulement
- *n* afvoerplaat
- *p* płyta spustowa

3606 outlet closing
- *d* Auslassende *f*
- *f* fermeture *f* d'échappement
- *n* einde uitlaten
- *p* koniec wylotu

3607 outlet hose
- *d* Abflussleitung *f*
- *f* conduit *m* de décharge
- *n* afvoerslang
- *p* przewód odpływowy

* **outline → 1270**

3608 output diode
- *d* Ausgangdiode *f*
- *f* diode *f* de sortie
- *n* vermogensdiode; hoofddiode
- *p* elektroda wyjściowa

3609 output flange
- *d* Abtriebsflansch *m*
- *f* flasque *m* de sortie
- *n* uitvoerflens
- *p* kołnierz napędzany

3610 output gear
- *d* Abtriebszahnradpaar *n*
- *f* couple *m* d'engrenage secondaire
- *n* secundaire tandwielen
- *p* przekładnia wyjściowa; przekładnia zdawcza

3611 output link
- *d* Abtriebsglied *n*
- *f* élément *m* mené
- *n* aangedreven element
- *p* człon bierny; człon napędzany

3612 output pulley coil spring
- *d* Schraubenfeder *f* der Abtriebsscheibe
- *f* ressort *m* hélicoïdal de poulie réceptrice
- *n* schroefveer van aandrijfrol
- *p* sprężyna walcowa rozsuwnego koła pasowego

3613 output shaft
- *d* Abtriebswelle *f*
- *f* arbre *m* récepteur
- *n* uitgangsas
- *p* wał zdawczy

3614 output shaft flange
- *d* Abtriebswellenflansch *m*
- *f* flasque *m* d'arbre récepteur
- *n* flens van de uitgangsas
- *p* kołnierz wału zdawczego

3615 outside diameter of turning circle
- *d* Wendekreisdurchmesser *m*
- *f* diamètre *m* de braquage extérieur
- *n* buitendiameter van de draaicirkel
- *p* zewnętrzna średnica zawracania

3616 oval cylinder
- *d* ovaler Zylinder *m*
- *f* cylindre *m* ovalisé
- *n* onronde cilinder
- *p* cylinder owalny

3617 oval file
- *d* Vogelzunge *f*; Vogelzungenfeile *f*
- *f* lime *f* feuille de sauge
- *n* ovale vijl
- *p* pilnik soczewkowy

* **oval head screw → 3979**

3618 overall
- *d* Arbeitsanzug *m*
- *f* combinaison *f* de travail
- *n* overall
- *p* kombinezon roboczy

3619 overall length
- *d* Gesamtlänge *f*
- *f* longueur *f* totale

n totale lengte
p długość całkowita

3620 overcooled
 d übergekühlt
 f surrefroidi
 n overgekoeld
 p przechłodzony

* **overdrive → 3408**

3621 overdrive control knob
 d Schnellgangengetriebeknopfschalter m
 f commutateur m à poussoir de
    surmultiplication
 n overdrive-bedieningsknop
 p wyłącznik wciskowy nadbiegu

3622 overdrive kick down switch
 d Kickdownschnellgangschalter m
 f commutateur m kick-down de
    surmultiplicateur
 n kickdown-schakelaar van overdrive
 p wyłącznik krańcowy nadbiegu

3623 overdrive warning light
 d Schnellgangkontrollanzeiger m
 f lampe f témoin de surmultiplication
 n controlelicht voor ingeschakelde
    overdrive
 p wskaźnik włączenia nadbiegu

3624 overflow pipe
 d Überlaufrohr n
 f tube m de trop-plein
 n overlooppijp
 p rura przelewowa

3625 overflow plug
 d Überlaufstopfen m
 f bouchon m de trop-plein
 n niveauplug
 p zatyczka przelewowa

3626 overflow valve
 d Überlaufventil n
 f clapet m de décharge
 n overloopklep
 p zawór przelewowy

3627 overhang length
 d Überhanglänge f
 f longueur f de porte-à-faux
 n overhanglengte
 p długość zwisu

* **overhaul → 2904**

3628 overhead camshaft
 d obenliegende Nockenwelle f
 f arbre m à cames par le haut
 n bovenliggende nokkenas
 p wałek rozrządczy górny

3629 overhead camshaft engine
 d Motor m mit obenliegender Nockenwelle
 f moteur m à arbre à cames et tête
 n motor met een bovenliggende nokkenas
 p silnik o wale rozrządu w głowicy

3630 overhead valve
 d hängendes Ventil n
 f soupape f en tête
 n kopklep
 p zawór górny

3631 overhead valve engine
 d obengesteuerter Motor m
 f moteur m à soupapes en tête
 n kopklepmotor
 p silnik górnozaworowy

3632 overheat v
 d überhitzen
 f surchauffer
 n oververhitten
 p przegrzewać

3633 overlapping
 d Überschneidung f
 f recoupement m
 n overlap
 p łączenie na zakładkę

3634 overload
 d Überlastung f
 f surcharge f
 n overbelasting
 p przeciążenie

* **overload leaf → 2650**

3635 overload spring; supplementary spring
 d Zusatzfeder f; Hilfsfeder f
 f ressort m compensateur; ressort m auxiliaire
 n hulpveer
 p resor dodatkowy

3636 overpressure; excess pressure
 d Überdruck m
 f surpression f
 n overdruk
 p nadciśnienie

3637 overrun brake
 d Auflaufbremse f

f frein *m* d'échouement
n mechanische oplooprem
p mechaniczny hamulec najazdowy

3638 oversize
d Übergrosse *f*
f surdimension *f*
n overmaat
p nadwymiar; zwiększony wymiar

3639 oversize piston
d Übermasskolben *m*
f piston *m* surdimensionné
n overmaatzuiger
p tłok o zwiększonej średnicy

3640 oversquare engine
d Motor *m* mit überquadratischen Dimensionen
f moteur *m* supercarré
n overvierkante motor; korteslagmotor
p silnik nadkwadratowy

3641 oversteering
d Übersteuerung *f*
f surdirection *f*
n overstuur
p nadsterowność

3642 overstress
d Überbeanspruchung *f*
f surcharge *f*
n overbelasten
p przeciążenie

3643 overtake *v*
d überholen
f dépasser
n inhalen
p wyprzedzać

3644 oxidation
d Oxidation *f*
f oxydation *f*
n oxidatie
p utlenianie

3645 oxygen
d Sauerstoff *m*
f oxygène *m*
n zuurstof
p tlen

P

3646 packing; gasket
d Abdichtung *f*; Dichtung *f*
f garniture *f*; joint *m*
n pakking
p uszczelka

3647 packing antirattle
d Geräuschdämpfscheibe *f*
f cale *f* d'insonorisation
n antidreunstrip
p podkładka tłumika szumu

3648 packing cloth
d Packleinwand *f*
f toile *f* d'emballage
n paklinnen
p płótno pakowe

* packing ring → 4323

3649 padded
d gepolstert
f rembourré
n gepolsterd
p wyściełany

3650 padlock
d Hangschloss *n*
f cadenas *m*
n hangslot
p kłódka

3651 pad saw
d Stichsäge *f*
f scie *f* à guichet
n sleutelgatzaag
p piła otwornica

3652 paint cleaner
d Lackpflegemittel *n*
f nettoyeur *m* de peinture
n lakreiniger
p oczyszczacz lakieru

3653 paint drying installation
d Lacktrocknungslage *f*
f installation *f* de séchage de peinture
n verfdrooginstallatie
p urządzenie do suszenia lakieru

3654 painter
d Anstreicher *m*

f peintre *m*
n spuiter
p malarz; lakiernik

3655 painting booth; paint spray booth
d Anstrichspritzkabine *f*
f cabine *f* de peinture
n spuitcabine
p komora natryskowa

3656 paint mixer
d Farbmixer *m*
f mélangeur *m* de peinture
n verfmenger
p mikser do farby

3657 paint scraper
d Farbkratzer *m*
f grattoir *m* de peintre
n verfkrabber
p skrobak do farby

3658 paint shop
d Lackiererei *f*
f atelier *m* de peinture
n spuiterij
p lakiernia

3659 paint spray
d Lacksprühdose *f*
f laque *f* du pistoleur aérosol
n lak uit spuitbus
p lakier w aerozolu rozpryskowym

* paint spray booth → 3655

3660 paint spray respirator
d Farbspritzermaske *f*
f masque *m* pour air
n verfspuitmasker
p respirator

3661 paint storage
d Lager *n* von Lackiererei
f magasin *m* de pistolage
n magazijn van spuiterij
p magazyn lakierni

3662 paint thickness gauge
d Lackdickemesser *m*
f gauge *m* d'épaisseur de laque
n lakdiktemeter
p przyrząd do pomiaru grubości lakieru

3663 paint tin
d Farbbüchse *f*
f bidon *m* de couleur

n verfbus
p puszka farby

3664 panel frame
d Pressrahmen *m*
f cadre *m* à panneau
n bakchassis
p rama płytowa tłoczona

3665 panel structure body
d Zellenaufbau *m*
f carrosserie *f* en caisson; carrosserie *f* à
  panneaux
n kooiloze carrosserie; carrosserie zonder
  kooiconstructie
p nadwozie członowe

3666 Panhard rod
d Panhardstab *m*
f barre *f* de Panhard
n Panhard-stang
p drążek Panharda

3667 paper element
d Papiereinsatz *m*
f élément *m* en papier
n papieren element
p wkład papierowy

3668 paper roll
d Papierwalze *f*
f rouleau *m* de papier
n papierrol
p rolka papieru

3669 parabolic reflector
d parabolischer Scheinwerferspiegel *m*
f réflecteur *m* parabolique
n parabolische reflector
p paraboliczny odbłyśnik reflektora

3670 paraffin
d Kerosin *n*
f kérosène *m*
n kerosine
p nafta

3671 parallel connection
d Parallelschaltung *f*
f connexion *f* en parallèle
n parallelschakeling
p połączenie równoległe

3672 parallel key
d Passfeder *f*
f clavette *f* parallèle
n prismaspie; passpie
p wpust pryzmatyczny

3673 parker screw; self-tapping screw
d Blechschraube *f*
f vis *f* autofileteuse
n parker; plaatschroef
p wkręt samogwintujący

3674 parking brake
d Feststellbremse *f*
f frein *m* de parcage
n parkeerrem
p hamulec postojowy

3675 parking brake cable pliers
d Zange *f* für Parkbremsenzug
f pince *f* pour le câble de frein à main
n tang voor handremkabel
p szczypce do linki hamulca postojowego

3676 parking brake lever
d Handbremshebel *m*
f levier *m* de frein à main
n handremhendel
p dźwignia hamulca postojowego

3677 parking brake lever pull rod
d Zugstange *f* der Feststellbremse
f tringle *f* de frein de stationnement
n trekstang van de handrem
p cięgno dźwigni hamulca ręcznego

3678 parking brake valve
d Feststellbremsventil *n*
f robinet *m* de frein de stationnement
n parkeerremklep
p zawór nastawczy hamulca postojowego

3679 parking lamp
d Parklicht *n*
f feu *m* de stationnement
n parkeerlicht
p światła postojowe

3680 parking meter
d Parkuhr *f*
f parcomètre *m*
n parkeermeter
p parkometr

3681 parking ratchet
d Parksperre *f*
f ratchet *m* de verrouillage
n parkeerblokkering
p ryglownik postojowy

3682 partial load
d Teillast *f*
f charge *f* partielle

n deellast
p obciążenie częściowe

3683 parting tool; grooving tool
d Grabstichel m; Plattenstahl m
f ciseau m de tourneur
n afsteekbeitel
p nóż przecinak

3684 partition panel
d Trennwand f
f cloison m
n tussenwand; tussenschot
p ściana działowa; przegroda

3685 partition panel frame
d Trennwandrahmen m
f cadre m de cloison de séparation
n omranding van scheidingsplaat
p rama przegrody

3686 partition window glass
d Trennscheibe f
f vitre f de séparation
n scheidingsruit
p szyba rozdzielcza

3687 parts list
d Stückliste f
f liste f des pièces
n onderdelenlijst
p wykaz części

3688 passenger
d Fahrgast m
f passager m
n passagier
p pasażer

3689 passenger compartment
d Abteil n für Fahrgäste
f compartiment m réservé aux voyageurs
n passagierscompartiment
p przedział pasażerski

3690 passenger insurance
d Insassenversicherung f
f assurance f occupants
n inzittendenverzekering
p ubezpieczenie pasażerów

3691 passenger semitrailer
d Autobussattelauflieger m
f semi-remorque f à passagers
n opleggercombinatie voor
  personenvervoer
p naczepa autobusowa

3692 passenger trailer
d Autobusanhängerwagen m
f remorque f à passagers
n aanhanger voor personenvervoer
p przyczepa autobusowa

3693 passing prohibited
d Überholverbot n
f interdiction f de dépasser
n inhaalverbod
p zakaz wyprzedzania

3694 passive safety
d passive Unfallsicherheit f
f sécurité f passive
n passieve veiligheid
p bezpieczeństwo bierne

3695 paste
d Paste f
f pâte f
n pasta
p pasta

3696 patch
d Metallblech n
f lame f
n metalen plaatje
p płytka metalowa

3697 patent drawing
d Patentzeichnung f
f dessin m de brevet
n octrooitekening
p rysunek patentowy

3698 patented system
d patentiertes System n
f système m breveté
n gepatenteerd systeem
p układ patentowany

3699 patent key
d Sicherheitsschlüssel m
f clé f de sûreté
n veiligheidssleutel
p klucz patentowy

3700 pattern plate
d Modellplatte f
f plaque-modèle f
n modelplaat
p płyta modelowa

3701 paver's pinching bar
d Pflasterbrechstange f
f pince f de paveur

n stratenmakersklemstang
p łom

3702 pawl
d Schaltklinke *f*
f cliquet *m*
n pal
p zapadka

3703 payload
d Nutzlast *f*; Zuladung *f*
f charge *f* payante
n nuttige last
p ładunek użyteczny

3704 pearlescent paint
d Perlenlack *m*
f laque *f* nacrée
n parelmoerlak
p lakier perłowy

3705 pecking hammer
d Pickhammer *m*
f marteau *m* à pince
n aftastershamer
p młotek dekarski

3706 pedal
d Pedal *n*
f pédale *f*
n pedaal
p pedał

3707 pedal bracket; pedal support
d Fusshebelbock *m*
f support *m* de pédale
n pedaalsteun
p wspornik pedału

3708 pedal lever bracket
d Pedalhebellagerbock *m*
f support *m* de levier de pédale
n assupport voor hefboom van het gaspedaal
p wspornik dźwigni pedału

3709 pedal lever seal
d Kupplungspedalwandabdichtung *f*
f joint *m* de passage de pédale d'embrayage
n afdichting van pedaalhefboom
p przelotka pedału sprzęgła

* **pedal return spring** → 21

3710 pedal spindle
d Pedalbolzen *m*
f axe *m* de pédale
n pedaalas
p wrzeciono pedału

* **pedal support** → 3707

3711 peg arm
d Lenkfingerhebel *m*
f levier *m* de goujon
n stuurvingerhefboom
p dźwignia palca ramienia wewnętrznego mechanizmu kierowniczego

3712 penetrating oil
d Eindringöl *n*
f huile *f* pénétrante
n kruipolie
p olej grafitowany

3713 perforated disc wheel
d Lochscheibenrad *n*
f roue *f* à voile ajouré
n schijfwiel met openingen
p tarcza koła z otworami

3714 performance test
d Funktionstest *m*
f contrôle *m* de fonctionnement
n functietest
p próba eksploatacyjna

3715 periodical service
d regelmassige Wartungsarbeiten *fpl*
f entretien *m* périodique
n periodiek onderhoud
p obsługa okresowa

3716 permanent joint
d unlösbare Verbindung *f*
f assemblage *m* non-démontable
n permanente verbinding
p połączenie nierozłączne

3717 permanently elastic sealer
d konstant elastischer Kitt *m*
f mastic *m* restant élastique
n elastisch blijvende kit
p kit stale elastyczny

3718 permanent magnet motor
d Dauermagnetmotor *m*
f moteur *m* avec aimant permanent
n permanente magneetmotor
p silnik o magnesie trwałym

3719 permissible clearance; admissible clearance
d zulässiges Spiel *n*
f jeu *m* admissible
n toelaatbare speling
p luz dopuszczalny

3720 permissible load
d zulässige Last *f*

f charge *f* admissible
n toelaatbare belasting
p obciążenie dopuszczalne

3721 personal code
d persönliche Kode *f*
f code *m* confidentiel
n persoonlijke code
p kod osobisty

* **petrol** → **2449**

* **petrol engine** → **2450**

* **petrol filter** → **2451**

* **petrol gauge** → **2397**

3722 petrol heating
d Benzinheizung *f*
f chauffage *m* à essence
n benzineverwarming
p ogrzewanie benzynowe

3723 petrol hose
d Benzinschlauch *m*
f tuyau *m* à essence
n benzineslang
p wąż benzyny

3724 petrol injection
d Benzineinspritzung *f*
f injection *f* d'essence
n benzine-inspuiting
p zasilanie wtryskowe benzyną

3725 petrol injection engine
d Benzineinspritzmotor *m*
f moteur *m* à injection d'essence
n benzine-injectiemotor
p benzynowy silnik wtryskowy

3726 petrol pipe
d Benzinleitung *f*
f conduite *f* d'essence
n benzineleiding
p przewód benzyny

3727 phase
d Phase *f*
f phase *f*
n fase
p faza

3728 phase difference
d Unterschied *m* der Phase

f différence *f* de phase
n faseverschil
p przesunięcie fazowe

3729 Phillips screw; cross recessed head screw
d Kreuzschlitzschraube *f*
f vis *f* de crosse
n kruiskopschroef
p wkręt z łbem z gniazdkiem krzyżowym

3730 photoelectric cell
d Photozelle *f*
f cellule *f* photoélectrique
n fotocel; foto-elektrische cel
p komórka fotoelektryczna

3731 photoelectric effect
d Photoeffekt *m*
f effet *m* photoélectrique
n foto-elektrisch effect
p zjawisko fotoelektryczne

3732 photoelectric sensor
d Photozellenfühler *m*
f capteur *m* photoélectrique
n foto-elektrische sensor
p nadajnik fotoelektryczny

3733 pigment
d Pigment *n*
f pigment *m*
n pigment
p pigment

3734 pillar
d Pfosten *m*
f pied *m*
n dakstijl
p słupek dachu

3735 pillar bolt
d Kragenschraube *f*; Bundschraube *f*
f boulon *m* à épaulement; vis *f* à collet
n borstbout
p śruba kołnierzowa

3736 pillar shield
d Säulenschild *n*
f revêtement *m* de pied
n scherm van kolom
p nakładka słupka

3737 pilot bearing
d Führungslager *n*
f palier *m* guide
n centreerlager
p łożysko środkujące

3738 pilot line
  d Steuerleitung *f*
  f ligne *f* hydraulique commandée
  n hydraulische aanstuurleiding
  p przewód sterowniczy

3739 pilot pin
  d Zentrierstift *m*
  f goujon *m* de centrage
  n centreerstift
  p kołek środkujący

  *   **pilot sleeve** → **958**

3740 pin; journal
  d Zapfen *m*; Bolzen *m*
  f pivot *m*; goujon *m*; tourillon *m*
  n pen; stift
  p czop; kołek

3741 pin insulation
  d Zündbolzenisolierung *f*
  f isolement *m* de tige
  n isolatie
  p izolacja rdzenia

3742 pinion setting
  d Ritzeleinstellung *f*
  f réglage *m* de la distance des pignons
  n afstelling van rondsel
  p nastawianie zębnika

3743 pinion shaft
  d Ritzwelle *f*
  f arbre *m* pignon
  n rondselas
  p wał z kolem zębatym

3744 pin punch
  d Durchschlag *m*
  f chasse-goupilles *m*
  n doorslag
  p wybijak

3745 pintle
  d Spritzzapfen *m*
  f téton *m* d'aiguille d'injection
  n spuittap
  p czopik iglicy

3746 pintle hook
  d Anhängeröse *f*
  f anneau *m* de remorque
  n scharnierende haak
  p zaczep przyczepy

3747 pintle nozzle
  d Zapfendüse *f*
  f injecteur *m* à téton
  n tapverstuiver
  p rozpylacz czopikowy

3748 pintle strangler
  d Drosselzapfen *m*
  f embase *f* de téton
  n smoortap
  p dławik czopika

3749 pipe
  d Rohr *n*
  f tuyau *m*
  n pijp; buis
  p rura

3750 pipe bender
  d Rohrbiegezange *f*
  f pince *f* à cintrer
  n buigtang
  p giętarka rurowa

3751 pipe bundle
  d Leitungsbündel *n*
  f faisceau *m*; tube *m*
  n leidingbundel
  p wiązka przewodów

3752 pipe clamp
  d Rohrschelle *f*
  f collier *m* de tuyau
  n pijpbeugel
  p zacisk rury; opaska rury

3753 pipe coupling
  d Muffe *f*
  f raccord *m* de tubes
  n mof; sok
  p połączenie rurowe; złączka rury

3754 pipe dresser
  d Instandsetzungswerkzeug *n* für Rohrkante
  f reformeur *m* de tube
  n pijprandherstelgereedschap
  p przyrząd do wyrównywania końcówek rur

3755 pipe expander
  d Auftrumpfwerkzeug *n*
  f expandeur *m* de tube
  n optrompgereedschap
  p naprężacz do rur

3756 pipe fitting press
  d Doppelbördelgerät *n*
  f presse *f* à collets
  n persgereedschap voor kragen
  p przyrząd do formowania złączek rurowych

3757 pipe joint separator
(tool for easy insertion between pipe walls)
d Zerlegungslöffel *m*
f cuillère *f* décolle tube
n demontagelepel
p przyrząd do rozdzielania rur

3758 pipeline
d Rohrleitung *f*
f conduite *f*
n pijpleiding
p przewód rurowy

3759 pipe nipple
d Rohrnippel *n*
f raccord *m* de tuyau
n pijpnippel
p złączka rurowa

3760 pipe support
d Halteblechn für Leitung
f support *m* de tube
n pijpsteun
p podtrzymka przewodu rurowego

3761 pipe U bolt
d Bügelschraube *f*
f manille *f* de tuyau
n U-bout
p strzemię rury

3762 pipe wrench
d Rohrzange *f*
f pince *f* multiprise
n pijptang
p szczypce do rur; obcęgi do rur

3763 piston
d Kolben *m*
f piston *m*
n zuiger
p tłok

3764 piston bottom
d Kolbenboden *m*
f dessus *m* de piston; face *f* de piston
n zuigerbodem
p denko tłoka

3765 piston crown; piston head
d Kolbenkopf *m*
f tête *f* de piston
n zuigerkop
p główka tłoka

3766 piston diameter
d Kolbendurchmesser *m*

f alésage *m* de piston
n zuigerdiameter
p średnica tłoka

3767 piston freezing
d Kolbenfresser *m*
f grippage *m* du piston
n vretende zuiger
p chłodzenie tłoku

3768 piston gear
d Kolbenzahnkranz *m*
f roue *f* dentée de piston
n rotorringwiel
p koło zębate tłoka

* piston head → 3765

3769 piston internal space
d Kolbenzelle *f*
f espace *m* résiduel
n inwendige ruimte in zuiger
p przestrzeń wewnętrzna tłoka

3770 piston joint
d Dichtung *f* für Kolben
f joint *m* de piston
n zuigerafdichting
p uszczelnienie tłoka

* piston knock → 3784

3771 piston pin; gudgeon pin
d Kolbenbolzen *m*
f axe *m* de piston
n zuigerpen
p sworzeń tłokowy

3772 piston pin boss
d Kolbenbolzennabe *f*
f bossage *m* de piston
n zuigerpenboring
p piasta sworznia tłokowego

3773 piston play
d Kolbenspiel *n*
f jeu *m* de piston
n zuigerspeling
p luz tłoka

3774 piston ring
d Kolbenring *m*
f segment *m* de piston
n zuigerring
p pierścień tłokowy

3775 piston ring clamp
d Kolbenringspannband *n*

f   collier *m* à segments
n   zuigerveerklem
p   zacisk pierścienia tłokowego

3776 piston ring compressor
d   Kolbenringspannband *n*
f   collier *m* à segments
n   zuigerveerspanband; zuigerveerklemband
p   ściskacz pierścieni tłokowych

3777 piston ring flutter
d   Kolbenringflattern *n*
f   mouvement *m* du segment de piston
n   onstabiele trilling van zuigerveer
p   wibracja pierścienia tłokowego

3778 piston ring gap
d   Stossspiel *n*
f   coupure *f* de segment
n   slotspeling van zuigerveer
p   luz pierścienia tłokowego

3779 piston ring groove; ring groove
d   Kolbenringnut *f*
f   gorge *f* de segment de piston
n   zuigerveergroef
p   rowek pierścienia tłokowego

3780 piston ring pliers
d   Kolbenringaufleger *m*
f   pince *f* à segments
n   zuigerveertang
p   szczypce do pierścieni tłokowych

3781 piston ring tightener
d   Kolbenmontagegerät *n*; Kolbenringspanner *m*
f   collier *m* à segments
n   zuigerveerspanband; zuigerveerklemband
p   zaciskacz pierścienia tłokowego

3782 piston rod
d   Kolbenstange *f*
f   tige *f* du piston
n   zuigerstang
p   tłoczysko

3783 piston skirt
d   Kolbenmantel *m*
f   corps *m* de piston
n   zuigermantel
p   płaszcz tłoka

3784 piston slap; piston knock
d   Kolbenschlagen *n*; Kolbenklopfen *n*
f   claquement *m* de piston
n   klapperen van zuiger
p   stukanie tłoka

3785 piston spring
d   Kolbenfeder *f*
f   ressort *m* de piston
n   zuigerveer
p   sprężyna tłoka

3786 piston stroke
d   Kolbenhub *m*
f   course *f* de piston
n   zuigerslag
p   skok tłoka

3787 piston temperature
d   Kolbentemperatur *f*
f   température *f* de piston
n   zuigertemperatuur
p   temperatura tłoka

3788 pitch diameter
d   Teilkreisdurchmesser *m*
f   diamètre *m* du cercle primitif
n   steekcirkeldiameter
p   średnica toczna

3789 pitching
d   Bremsnicken *n*
f   tangage *m*
n   dompbeweging
p   wahania wzdłużne

3790 pivoting point
d   Drehpunkt *m*
f   point *m* de rotation
n   scharnierpunt
p   punkt obrotu

*   **pivot pin → 2679**

*   **plain bearing → 4539**

3791 plain washer
d   Unterlegscheibe *f*
f   rondelle *f*
n   ronde vulring
p   podkładka okrągła

3792 plan
d   Projekt *n*
f   projet *m*
n   ontwerp; project
p   projekt

3793 plancher
d   Wagenboden *m*
f   plancher *m*
n   wagenbodem
p   podłoga samochodu

3794 **planetary gear**
d Planetengetriebe n
f train m planétaire
n planetair tandwieloverbrenging
p przekładnia planetarna; przekładnia obiegowa

* **planetary interaxial differential** → 2015

3795 **planet long pinion**
d Planetenrad n lang
f satellite m long
n lange satelliet
p długi satelit

3796 **plane truss**
d ebenes Fachwerk n
f ferme f plane
n vlak vakwerk
p kratownica płaska

3797 **planet short pinion**
d Planetenrad n kurz
f satellite m court
n korte satelliet
p krótki satelit

3798 **planet wheel**
d Planetenrad n; Umlaufrad n
f pignon m satellite
n planeetwiel
p satelit

3799 **planet wheel carrier**
d Umlaufradträger m
f porte-satellites m
n planeetwieldrager
p koszyk satelitów

3800 **planet wheel sliding cup**
d Gleitschale f des Planetenrades
f coupelle f de satellite
n glijring van de satellieten
p płytka ślizgowa satelita

3801 **planishing hammer**
d Treibhammer m
f marteau m à aplaner
n vlakhamer
p młotek klepak

3802 **plan outline**
d Draufsicht f
f vue f de dessus
n bovenaanzicht
p obrys poziomy

3803 **plastic bush**
d Kunststoffbüchse f

f douille f en matière plastique
n kunststof huls
p tuleja plastykowa

3804 **plastic cleaner**
d Kunststoffreinigungsmittel n
f rénovateur m plastique
n kunststofreiniger
p środek czyszczący do plastyku

3805 **plastic handle**
d Plastikgriff m
f manche f plastique
n kunststof handgreep
p uchwyt plastykowy

3806 **plastic parts**
d Kunststoffteile mpl
f éléments mpl en matière plastique
n kunststof delen
p części plastykowe

3807 **plate active material**
d wirksame Masse f der Platte
f matière f active de plaque
n actieve massa in accu
p czynna masa płyty

3808 **plate connecting pin**
d Plattenverbinder m
f barrette f de liaison des plaques
n poolbrug
p śruba złączna plyt

3809 **platinum**
d Platinum n
f platine f
n platina
p platyna

3810 **plexiglass**
d Plexiglas n
f verre m acrylique
n plexiglas
p pleksiglas

3811 **plug**
d Stecker m
f fiche f
n stekker
p wtyczka

3812 **plug channel**
d Stopfenkanal m
f conduit m de bouchon
n groef van stop
p kanał korka

3813 plug fitting
d Stopfen *m*
f bouchon *m* d'obturation
n plug; stop
p zatyczka; korek

3814 plug gauge
d Lehrdorn *m*
f jauge *f* tampon
n penkaliber; gatkaliber
p sprawdzian tłoczkowy; sprawdzian
   trzpieniowy

3815 plug screw
d Verschlussstopfen *m*
f vis *f* d'obturation
n afsluitdop met schroefdraad
p wkręt zaślepiający

3816 plug socket
d Steckdose *f*
f prise *f* de courant
n stekkerdoos
p gniazdo wtykowe

3817 plunger
d Plunger *m*
f plongeur *m*
n plunjer
p tłoczek

3818 plunger return spring
d Kolbenfeder *f*
f ressort *m* de rappel de plongeur
n plunjerveer
p sprężyna tłoczka

3819 plunger type fuel feed pump
d Kolbenförderpumpe *f*
f pompe *f* d'alimentation à piston
n plunjeropvoerpomp
p tłoczkowa pompa zasilająca

3820 ply separation
d Lagentrennung *f*
f décollement *m* des nappes
n koordlaagseparatie
p rozwarstwienie

3821 pneumatic brake system
d Druckluftbremsanlage *f*
f dispositif *m* de freinage à air comprimé
n luchtdrukreminrichting
p pneumatyczny układ hamulcowy

3822 pneumatic cutter for bonded windows
d Pressluftmesser *m* für verleimte
   Fensterscheiben

f couteau *m* pneumatique pour les glaces collées
n persluchtmes voor gelijmde ruiten
p przecinak pneumatyczny do usuwania szyb
   klejonych

3823 pneumatic rammer
d Druckluftstampfer *m*
f fouloir *m* pneumatique
n persluchtstamper
p ubijak pneumatyczny

3824 pneumatic regulator
d Unterdruckversteller *m*
f régulateur *m* pneumatique
n onderdrukregelaar van stroomverdeler
p regulator podciśnieniowy; regulator
   pneumatyczny

3825 pneumatic spring
d Luftfeder *f*
f ressort *m* pneumatique
n luchtveer
p resor pneumatyczny

**3826 pneumatic suspension; air spring
    suspension; air suspension**
d Luftfederaufhängung *f*
f suspension *f* pneumatique
n pneumatische ophanging
p zawieszenie pneumatyczne

3827 pneumatic valve; air valve
d pneumatisches Ventil *n*
f soupape *f* pneumatique
n pneumatische klep
p zawór pneumatyczny

3828 polarizable
d polarisierbar
f polarisable
n polariseerbaar
p polaryzowany

3829 polarization
d Polarisierung *f*
f polarisation *f*
n polarisatie
p polaryzacja

3830 pole shoe
d Polschuh *m*; Feldpol *m*
f corne *f* polaire; masse *f* polaire
n poolschoen
p nabiegunnik

3831 polishing hammer
d Polierhammer *m*

f   marteau *m* postillon
n   polijsthamer
p   gładzik

3832 polishing red
d   Läppaste *f*
f   pâte *f* à roder
n   slijppasta
p   pasta do docierania

3833 polluant emission
d   Schadstoffausstoss *m*
f   émission *f* de polluants
n   schadelijke uitlaatgasemissie
p   szkodliwa emisja spalin

3834 polyester lacquer
d   Polyesterlack *m*
f   vernis *m* de polyester
n   polyester lak
p   lakier poliestrowy

3835 polyester thinner
d   Polyesterverdünnung *f*
f   diluant *m* polyester
n   polyester verdunning
p   rozcieńczalnik do wyrobów poliestrowych

3836 poor mixture
d   armes Gemisch *n*; mageres Gemisch *n*
f   mélange *m* pauvre
n   mager mengsel; arm mengsel
p   mieszanka uboga

3837 portable lamp; hand lamp
d   Handlampe *f*; Arbeitslampe *f*
f   lampe *f* baladeuse
n   handlamp; looplamp
p   lampa przenośna; lampa ręczna

3838 portable press with hand pump
d   drehbare Presse *f* mit Handpumpe
f   presse *f* portable à pompe manuelle
n   draagbare pers met handpomp
p   prasa przenośna z ręczną pompą

3839 portable tripod stand
d   tragbares Arbeitsgerät *n*
f   établi *m* pliant portatif
n   draagbare driepoot
p   przenośne urządzenie robocze

3840 position
d   Stellung *f*
f   position *f*
n   positie
p   pozycja

3841 positive plate unit
d   Plusplatteverbinder *m*
f   barrette *f* de plaques positives
n   plusplaatverbindingsklem
p   mostek biegunowy zespołu płyt dodatnich

3842 potentiometer
d   Potentiometer *m*
f   potentiomètre *m*
n   potentiometer
p   potencjometr

3843 power balance
d   Leistungsbilanz *f*
f   bilan *m* de puissance
n   vermogenbalans
p   bilans mocy

3844 power cable
d   Stromzuführungskabel *n*
f   câble *m* d'alimentation
n   stroomtoevoerkabel
p   kabel zasilający

3845 power consumption
d   Leistungsaufnahme *f*
f   puissance *f* absorbée
n   opgenomen vermogen
p   pobór mocy

3846 power cylinder bracket
d   Servomotorauflager *n*
f   support *m* de vérin
n   steun voor servomotor
p   wspornik siłownika

3847 power factor
d   Leistungsfaktor *m*
f   coefficient *m* de puissance
n   vermogenfactor
p   współczynnik mocy

3848 power grooved pliers
d   Zangen *fpl* mit Führungsnuten
f   pinces *fpl* à cran demi-lune
n   tang met groef
p   obcęgi z rowkiem prowadzącym

3849 power lever
d   Servomotorhebel *m*
f   levier *m* de servo
n   servomotorarm
p   dźwignia siłownika

3850 power screw; translation screw
d   Bewegungsschraube *f*; Spindel *f*
f   vis *f* de mouvement

n bewegingsschroef
p śruba napędowa

3851 power screw thread
d Bewegungsgewinde n
f filet m du mouvement
n bewegingsschroefdraad
p gwint napędowy

* **power steering** → **4411**

3852 power stroke; expansion stroke
d Arbeitshub m
f temps m moteur
n arbeidsslag
p suw pracy; suw rozprężania

3853 power transmission; transmission
d Kraftübertragung f
f transmission f de puissance
n krachtoverbrenging
p przenoszenie napędu

3854 precautionary measures
d Sicherheitsmassnahmen fpl
f précautions fpl
n voorzorgsmaatregelen
p środki zapobiegawcze

3855 precision tweezer
d Präzisionspinzetten fpl
f brucelles fpl de professionnels
n precisiepincet
p pinceta precyzyjna

3856 precombustion chamber
d Vorkammer f
f préchambre f
n voorkamer
p komora wstępna

3857 precompressed mixture
d vorverdichtetes Gemisch n
f mélange m précomprimé
n voorgecomprimeerd mengsel
p mieszanka wstępnie sprężona

3858 preliminary filter
d Grobfilter m
f préfiltre m
n voorfilter
p filtr wstępnego oczyszczania

3859 preselective gearbox electronic control
d elektronische Vorwählergetriebesteuerung f
f commande f électronique de boîte présélective
n elektronische voorkeuzeschakelaar van

versnellingsbak
p elektroniczne sterowanie skrzynki
   preselektywnej

3860 preselector relay
d Vorwahlrelais n
f relais m de présélection
n relais van keuzeschakelaar
p przekaźnik selektora

3861 preserving
d Unterbringen n
f entretien m préventif
n preventief onderhoud
p konserwacja zapobiegawcza

3862 preset torque handle
d automatisch ausschaltender Schlüssel m
f poignée f de serrage préréglée
n handgreep met vast moment
p klucz dynamometryczny

3863 press v down
d anziehen; anpressen
f serrer
n aandrukken
p dociskać

3864 pressure
d Druck m
f pression f
n druk
p ciśnienie

3865 pressure angle
d Eingriffswinkel m
f angle m d'engagement; angle m de pression
n aangrijpingshoek
p kąt przyporu

3866 pressure boost valve
d Drucksteigerungsventil n
f soupape f de surcharge
n drukopvoerklep
p zawór podwyższający ciśnienie

3867 pressure chamber
d Druckkammer f
f chambre f de pressage
n drukkamer
p komora ciśnienia

3868 pressure control valve; delivery valve
d Druckventil n
f soupape f de pression
n drukklep
p zawór tłoczny; zawór ciśnieniowy

3869 **pressure diagram**
  d Flächenpressungsverlauf *m*
  f diagramme *m* des pressions
  n drukdiagram
  p wykres ciśnień

3870 **pressure die casting**
  d Druckgiessen *n*
  f coulée *f* sous pression
  n spuitgieten; persgieten
  p odlewanie ciśnieniowe

3871 **pressure dowel**
  d Druckstift *m*; Druckzapfen *m*
  f téton *m* de pression; goujon *m* de pression
  n drukstift
  p kołek oporowy

3872 **pressure drop warning light**
  d Differenzdruckwarnlampe *f*
  f lampe *f* témoin de chute de pression
  n drukverminderingcontrolelicht
  p lampka sygnalizacyjna zaniku ciśnienia

3873 **pressure feed**
  d Druckzuführung *f*
  f alimentation *f* par pression
  n toevoer onder druk
  p zasilanie pod ciśnieniem

3874 **pressure force**
  d Anpresskraft *f*
  f force *f* de pression
  n drukkracht
  p siła nacisku; siła docisku; napór

3875 **pressure head**
  d Druckhohe *f*
  f hauteur *f* représentative
  n drukhoogte
  p wysokość ciśnienia

3876 **pressure limitation**
  d Druckbegrenzung *f*
  f limitation *f* de pression
  n drukbegrenzing
  p ograniczenie ciśnienia

3877 **pressure plate; thrust plate**
  d Druckscheibe *f*
  f plateau *m* de pression
  n drukplaat
  p płytka oporowa

3878 **pressure reducing valve**
  d Druckminderventil *n*
  f soupape *f* de réduction

  n reduceerafsluiter; reduceerklep
  p zawór redukcyjny

3879 **pressure regulating valve**
  d Druckregelventil *n*
  f soupape *f* de reglage de pression
  n drukregelklep
  p zawór regulacyjny ciśnienia

3880 **pressure regulator**
  d Druckregler *m*
  f régulateur *m* de pression
  n drukregelaar
  p regulator ciśnienia

3881 **pressure ring**
  d Verdichtungsring *m*
  f segment *m* de compression
  n compressieveer van zuiger
  p sprężyna tłokowa pierścienia

3882 **pressure rise**
  d Druckzunahme *f*
  f accroissement *m* de la pression
  n drukstijging
  p wzrost ciśnienia

3883 **pressure sensor**
  d Druckfühler *m*
  f manocontact *m*
  n druksensor
  p czujnik ciśnienia

3884 **pressure spring**
  d Druckfeder *f*
  f ressort *m* de pression
  n drukveer
  p sprężyna dociskowa

3885 **pressure switch**
  d Druckschalter *m*
  f contacteur *m* de pression
  n drukschakelaar
  p wyłącznik przyciskowy

3886 **pressure tapping connector**
  d Verbindungsstück *n* für Öldruckkontrolle
  f raccord *m* pour prise de pression d'huile
  n zelftappend koppelstuk
  p element łączący kontroli ciśnienia oleju

3887 **pressure test**
  d Druckprüfung *f*
  f essai *m* sous pression
  n druktest
  p próba na ciśnienie

3888 **pressure transmitter**
  d Druckgeber *m*

f transmetteur *m* de pression
n drukseingever
p czujnik ciśnienia

3889 pressure valve
   d Druckventil *n*
   f clapet *m* de pression
   n persklep
   p zawór ciśnieniowy

3890 pressure washer
   d Anpressscheibe *f*
   f rondelle *f* d'appui
   n drukring; steunring
   p podkładka dociskowa

3891 pretreatment
   d Vorbearbeitung *f*
   f dégrossissage *m*; ébauchage *m*
   n voorbehandeling
   p obróbka wstępna

3892 primary circuit
   d Hauptstromkreis *m*
   f circuit *m* primaire
   n primaire stroomkring
   p obwód pierwotny

3893 primary head
   d Setzkopf *m*
   f tête *f* première
   n zetkop
   p łeb nitu

3894 primary piston
   d Primärkolben *m*
   f piston *m* primaire
   n primaire pompzuiger
   p tłok pierwotny

3895 primary shaft
   d Primärwelle *f*
   f arbre *m* primaire
   n primaire as
   p wał pierwotny

   * **primary shoe** → 3042

3896 primary voltage
   d Primärspannung *f*
   f tension *f* primaire
   n primaire spanning
   p napięcie pierwotne

3897 primary winding; low voltage winding
   d Primärwicklung *f*
   f enroulement *m* primaire

n primaire wikkeling
p uzwojenie pierwotne

   * **priming colour** → 3899

3898 priming lever
   d Handantriebshebel *m*
   f culbuteur *m* de commande manuelle
   n handhefboom
   p dźwignia ręczna

3899 priming paint; priming colour
   d Grundierfarbe *m*
   f peinture *f* de fond
   n grondverf; vullende grondlak
   p farba podkładowa; farba gruntowa

3900 printed circuit
   d gedruckte Schaltung *f*
   f circuit *m* imprimé
   n gedrukte schakeling
   p obwód drukowany

3901 prismatic cutter
   d Prismenfräser *m*
   f fraise *f* prismatique
   n prismatische beitel
   p frez pryzmatyczny

3902 processing
   d Bearbeitung *f*
   f usinage *m*
   n bewerking
   p obróbka

3903 profile
   d Profil *n*
   f profil *m*; dessin *m*
   n profiel
   p profil; zarys

3904 profile cylinder
   d Profilzylinder *m*
   f cylindre *m* profilé
   n profielcilinder
   p cylinder profilowany

   * **profile grinding** → 2307

3905 profile width
   d Profilbreite *f*
   f largeur *f* du profil
   n profielbreedte
   p szerokość profilu

3906 programming of code
   d Kodeprogrammierung *f*

f introduction *f* du code
n programmeren van een code
p programowanie kodu

3907 propane
d Propan *n*
f propane *m*
n propaan
p propan

3908 prop brace
d Stützwinkel *m*
f tirant *m* de béquille
n steunbalk
p kątownik podporowy

3909 propeller shaft retarder
d Gelenkwellendauerbremse *f*
f ralentisseur *m* d'arbre de transmission
n transmissiereelrem
p zwalniacz wału pędnego

3910 propeller shaft tunnel
d Gelenkwellentunnel *m*
f tunnel *m* d'arbre de transmission
n cardantunnel
p tunel wału pędnego

* **proportional flow fan** → **3364**

3911 proportioning valve
d Kompensator *m*; Ausgleicher *m*
f compensateur *m*
n compensatieklep
p kompensator

3912 protecting layer
d Schutzschicht *f*
f couche *f* de protection
n beschermende laag
p warstwa ochronna

3913 protecting plate
d Schutzunterlage *f*
f plaquette *f* de protection
n beschermingsplaat
p płytka ochronna

3914 protection paint
d Schutzlack *m*
f vernis *m* de protection
n beschermende lak
p lakier ochronny

3915 protective cap
d Schutzring *m*
f anneau *m* de protection

n beschermring
p pierścień ochronny

3916 protective clothing
d Schutzkleidung *f*
f vêtement *m* protecteur
n beschermkleding
p ubranie ochronne

3917 protective coating
d Schutzüberzug *m*
f revêtement *m* de protection
n beschermlaag
p powłoka ochronna

3918 protective rubber
d Gummiabschirmung *f*
f protecteur *m* caoutchouc
n beschermingsrubber
p guma zabezpieczająca

3919 prototype
d Prototyp *m*
f prototype *m*
n prototype
p prototyp

3920 protractor attachment
d Winkelschlüssel *m*
f clé *f* de serrage angulaire
n hoekverdraaiingssleutel
p klucz kątowy

3921 public road
d öffentliche Strasse *f*
f voie *f* publique
n openbare weg
p droga publiczna

3922 puff control
d Rauchbegrenzer *m*
f limiteur *m* de fumée
n rookbegrenzer
p ogranicznik spalin

3923 pulley
d Riemenscheibe *f*
f poulie *f*
n poelie; riemschijf
p koło pasowe

3924 pulley nut
d Mutter *f* für Riemenscheibe
f écrou *m* de poulie
n poeliemoer
p nakrętka koła pasowego

3925 pulley remover
d Antriebsscheibeabzieher *m*

f extracteur *m* de poulies
n krukaspoelietrekker
p ściągacz koła pasowego wału korbowego

3926 pulley stay
    (tool to hold alternator or water pump belt
    pulleys while removing retaining screw)
d Schlüssel *m* zum Festhalten der
    Riemenscheiben
f clé *f* pour immobiliser les poulies
n poeliesleutel
p klucz do unieruchamiania koła pasowego

3927 pull jack
d Zugrammklotz *m*
f vérin *m* tireur
n trekram
p cylinder do wyciągania

3928 pull lever
d Mitnehmerhebel *m*
f levier *m* d'engrènement
n meeneemhefboom
p dźwignia zabieraka

3929 pulse motor; stepping motor
d Schrittmotor *m*
f moteur *m* pas à pas
n stappenmotor
p silnik skokowy

3930 pump
d Pumpe *f*
f pompe *f*
n pomp
p pompa

3931 pump capacity
d Förderstrom *m*
f débit *m* de la pompe
n pompvermogen
p wydajność pompy

3932 pump casing
d Pumpengehäuse *n*
f carter *m* de pompe
n pomphuis
p obudowa pompy

3933 pump connector
d Pumpenanschluss *m*
f raccord *m* souple de pompe
n pompaansluitstuk
p łącznik pompy

3934 pump cover
d Pumpendeckel *m*

f couvercle *m* de pompe
n pompdeksel
p pokrywa pompy

3935 pump cylinder
d Pumpenzylinder *m*
f cylindre *m* de pompe
n pompcilinder
p cylinderek pompy

3936 pump drive coupling
d verstellbare Kupplung *f*
f accouplement *m* de pompe pour calage
n verstelbare koppeling van pompaandrijving
p sprzęgło nastawne

3937 pumping chamber
d Saugraum *m*
f chambre *f* de refoulement
n zuigruimte
p komora pompowania

3938 pump jet
d Pumpendüse *f*
f gicleur *m* de pompe
n acceleratiesproeier; pompsproeier
p pompowtryskiwacz

3939 pump working space
d Pumpenarbeitsraum *m*
f espace *m* de travail de pompe
n werkruimte voor de pomp
p przestrzeń robocza pompy

3940 punch
d Dorn *m*
f mandrin *m*
n centerpons; drevel
p przebijak

3941 punching disc
d Schlagräder *npl*
f disque *m* poinçon
n ponsschijf
p tarcze udarowe

3942 puncture indicator
d Reifenwächter *m*
f avertisseur *m* de crevaison
n verklikker van lekke band
p lampka kontroli ciśnienia w ogumieniu

3943 push button
d Druckknopf *m*
f bouton *m* poussoir
n drukknop
p przycisk guzikowy

3944 push jack
d Hydraulikzylinder *m*
f vérin *m* pousseur
n persram
p cylinder do pchania

3945 push rod
d Stösselstange *f*
f tige *f* de poussoir
n stoterstang
p drążek popychacza

3946 push rod boot
d Kolbengestängeschutzbalg *m*
f soufflet *m* de tige de poussoir
n beschermbalg voor de zuigerstang
p osłona ochronna tłoczyska

3947 push rod cover
d Stösselstangenverkleidung *f*
f couvercle *m* de tige de poussoir
n deksel van lichterstang
p osłona dźwigni popychacza

3948 push rod fork
d Zugstangengabel *f*
f fourchette *f* de tige
n kleplichterstangvork
p widełki cięgna

3949 putty
d Kitt *m*
f mastic *m*
n kit; plamuur
p kit; szpachlówka

3950 putty knife
d Spachtel *f*
f couteau *m* à mastiquer
n plamuurmes
p szpachlówka

Q

3951 quality control
   d Qualitätskontrolle *f*
   f contrôle *m* de qualité
   n kwaliteitscontrole
   p kontrola jakości

3952 quarter elliptic spring
   d viertelelliptische Feder *f*
   f demi-cantilever *m*
   n kwart-elliptische bladveer
   p resor ćwierćepileptyczny

   * **quartz iodine bulb** → **2576**

R

3953 rack pinion
 d Zahnstangenritzel *n*
 f pignon *m* de direction
 n rondsel
 p zębnik wału kierownicy

3954 radial packing strip
 d radiale Dichtleiste *f*
 f baguette *f* d'étanchement radiale
 n radiale afdichting
 p listwa uszczelnienia promieniowego

3955 radial piston motor
 d Radialkolbenmotor *m*
 f moteur *m* hydraulique à piston radial
 n radiale hydraulische zuigermotor
 p silnik hydrauliczny wielotłokowy
   promieniowy

3956 radial piston pump
 d Radialkolbenpumpe *f*
 f pompe *f* à piston radial
 n radiale zuigerpomp
 p pompa wielotłokowa promieniowa

3957 radiator baffle plate
 d Kühlerspritzblech *n*
 f protège-radiateur *m*
 n radiateurkeerschot
 p blacha przegrodowa chłodnicy

3958 radiator bonnet
 d Kühlernabe *f*
 f couvre-radiateur *m*
 n radiateurhoes
 p przesłona chłodnicy

3959 radiator cap; radiator locking cap
 d Kühlerverschraubung *f*
 f bouchon *m* de radiateur
 n radiateurdop
 p korek chłodnicy

3960 radiator cleaner
 d Kühlerreinigungsmittel *n*
 f produit *m* de nettoyage pour radiateur
 n radiateurspoelmiddel
 p środek czyszczący chłodnicę

3961 radiator coil
 d Kühlschlange *f*

 f serpentin *m* de radiateur
 n koelslang
 p wężownica chłodząca

3962 radiator core
 d Kühlnetz *n*
 f corps *m* de radiateur
 n radiateurframe
 p rdzeń chłodnicy

3963 radiator filler cap
 d Einfüllverschluss *m*
 f bouchon *m* de remplissage de radiateur
 n radiateurdop
 p pokrywa wlewu chłodnicy

3964 radiator frame
 d Kühlergehäuse *n*
 f cadre *m* de radiateur
 n radiateurhuis
 p obudowa chłodnicy

3965 radiator grille
 d Kühlerschutzgitter *n*
 f grille *f* de radiateur
 n radiateurgrille
 p osłona chłodnicy

3966 radiator hose clip
 d Schlauchverbinder *m*
 f collier *m* de serrage
 n radiateurslangklem
 p opaska zaciskowa węża chłodnicy

* radiator locking cap → 3959

3967 radiator mounting
 d Kühlerträger *m*
 f support *m* de radiateur
 n bevestigingsbeugel voor de radiateur
 p wspornik chłodnicy

3968 radiator outlet connection
 d Kühlerauslaufstützen *m*
 f tubulure *f* de sortie d'eau
 n afvoerpijp van radiateur
 p rura odprowadzająca chłodnicy

3969 radiator roller blind; radiator shutter
 d Kühlerjalousie *f*; Kühlerabdeckung *f*
 f rideau *m* de radiateur
 n radiateurrolhoes; radiateurjaloezie
 p roletowa przesłona chłodnicy

3970 radiator screen
 d Kühlerattrappe *f*
 f calandre *f* de radiateur

n radiateurgrille
p osłona chłodnicy

* **radiator shutter** → 3969

3971 radiator strut
d Kühlerstrebe *f*
f tirant *m* de radiateur
n radiateursteun
p wspornik chłodnicy

3972 radiator tank
d Wasserkasten *m* des Kühlers
f réservoir *m* de radiateur
n radiateurtank
p zbiornik chłodnicy

3973 radiator tube
d Kühlrohr *n*
f tube *m* de refroidissement
n koelbuis; koelpijp
p rura chłodząca

3974 radio console
d Autoradiokonsole *f*
f console *f* de radio
n autoradioconsole
p konsola radiowa

3975 radio installation
d Radioeinbausatz *m*
f emplacement *m* autoradio
n inbouwradio
p radio wbudowane

3976 radio interference
d Radiostörung *f*
f parasitage *m*
n radiostoring
p zakłócenie radiowe

3977 radius dolly
d Handfaust *f* für Radien
f tas *m* à rayon
n straalzettas
p kowadełko promieniowe

* **radius of action** → 3586

3978 railway crossing
d Bahnübergang *m*
f passage *m* à niveau
n onbewaakte spoorwegovergang
p przejazd kolejowy niestrzeżony

**3979 raised countersunk head screw; oval head
screw**
d Linsenschraube *f*; Linsenkopfschraube *f*

f vis *f* à tête fendue fraisée bombée
n schroef met verzonken lenskop
p wkręt o soczewkowym łbie stożkowym

3980 rake angle
d Spanwinkel *m*
f angle *m* d'attaque
n spanhoek
p kąt natarcia

3981 random test
d Stichprobe *f*
f contrôle *m* au hasard
n steekproef
p próba wyrywkowa

3982 range selector; drive selector
d Wählschalter *m*
f sélecteur *m* de vitesses
n keuzeschakelaar
p wybierak zakresów pracy

3983 rasp; gratel
d Raspel *f*
f râpe *f*
n raspvijl
p tarnik

3984 ratchet flare nut wrench
(tool used for self gripping and release in
opposite direction)
d offener Ringschlüssel *m*
f clé *f* à tuyauter à cliquet
n open ratelsleutel
p klucz z mechanizmem zapadkowym

3985 ratchet gearing
d Klinkenschaltwerk *n*
f mécanisme *m* par rachet et cliquet
n palwerk
p mechanizm zapadkowy

3986 ratchet handle
d Klinkengriff *m*
f poignée *f* à cliquet
n handvat van klinksperwerk
p uchwyt zapadkowy

3987 ratchet rod
d Klinkenstange *f*
f tige *f* à cliquet
n klinksperwerkstang
p drążek zapadkowy

3988 ratchet spring; latch spring
d Klinkenfeder *f*
f ressort *m* de cliquet

n klinksperwerkveer
p sprężyna zapadki

3989 ratchet wheel
d Sperrad *n*
f roue *f* à rachet
n palwiel
p koło zapadkowe

3990 rated current
d Nennstrom *m*
f courant *m* nominal
n nominale stroom
p prąd znamionowy

3991 rated power
d Nennleistung *f*
f puissance *f* nominale
n nominaal vermogen
p moc znamionowa

3992 rated power engine speed
d Nennleistungdrehzahl *f*
f vitesse *f* de rotation à puissance nominale
n toerental bij het nominaal vermogen
p prędkość obrotowa mocy znamionowej

3993 rate of combustion
d Brenngeschwindigkeit *f*
f rapidité *f* de la combustion
n verbrandingssnelheid
p szybkość spalania

3994 rate of flow
d Strömungsgeschwindigkeit *f*
f vitesse *f* de courant
n doorstroomsnelheid
p natężenie przepływu

3995 ratio of compression
d Verdichtungsverhältnis *n*
f degré *m* de compression
n compressieverhouding
p stopień sprężania

3996 ratio of mixture
d Mischungsverhältnis *n*
f dosage *m* de mélange
n mengverhouding
p skład mieszaniny; skład mieszanki

3997 ratio of transmission
d Übersetzungsverhältnis *n*
f rapport *m* d'engrenage
n overbrengingsverhouding
p stosunek przełożenia

3998 raw material
d Rohstoff *m*

f matière *f* première
n onbewerkt materiaal
p surowiec

3999 reaction time
d Schreckzeit *f*
f durée *f* de réaction
n reactietijd
p czas trwania reakcji

4000 reamed bolt
d Passschraube *f*
f boulon *m* de mécanique
n pasbout
p śruba pasowana

* **reamer** → 760

4001 reamer
(tool to clean contact surfaces and terminals)
d Polklemmenfräser *m*
f fraise *f*
n frees
p rozwiertak

4002 rear arm positioning tool
(tool which allows correct positioning of the arm)
d Werkzeug *n* zum Einbauen des hinteren Tragarms
f outil *m* de positionnement du bras arrière
n gereedschap voor het plaatsen van de achterdraagarm
p przyrząd do pozycjonowania i ustalania tylnych ramion nośnych

4003 rear axle
d Hinterachse *f*
f pont *m* arrière
n achteras
p oś tylna

4004 rear axle beam
d Hinterachskörper *m*
f corps *m* de l'essieu arrière
n achterste aslichaam
p korpus osi tylnej

4005 rear axle casing
d Hinterachsbrücke *f*
f pont *m* arrière
n achterbrug
p most tylny

4006 rear axle cover
d Hinterachsbrückendeckel *m*
f couvercle *m* de pont arrière

n achterbrugdeksel
p pokrywa tylnego mostu

4007 rear axle housing
d Hinterachsgehäuse *n*
f carter *m* de pont arrière
n achterashuis
p obudowa tylnego mostu; pochwa tylnego mostu

4008 rear axle stabilizer
d Hinterachsschubstange *f*
f bielle *f* stabilisatrice d'essieu arrière
n achterasstabilisator
p drążek reakcyjny tylnego zawieszenia

4009 rear axle subframe
d Lagerung *f* für Hinterachse
f faux châssis *m* de train arrière
n subframe voor de achteras
p obudowa tylnej osi

4010 rear axle tube
d Hinterachsrohr *n*
f tube *m* de pont arrière
n achterasbuis
p osłona półosi tylnego mostu

4011 rear bearing extractor
d Lagerabzieher *m* für Hinternaben
f extracteur *m* de roulement de fusée arrière
n lagertrekker voor achternaven
p ściągacz łożyska tylnego

4012 rear brake actuator
d Hinterbremsenhydrozylinder *m*
f récepteur *m* de frein arrière
n werkcilinder van achterrem
p siłownik tylnego hamulca

4013 rear brake lever
d hinterer Bremshebel *m*
f levier *m* de frein arrière
n achterremhefboom
p tylna dźwignia hamulcowa

4014 rear bumper
d hinterer Stossfänger *m*
f pare-chocs *m* arrière
n achterbumper
p zderzak tylny

4015 rear clutch actuator
d Hinterkupplungshydrozylinder *m*
f récepteur *m* d'embrayage arrière
n werkcilinder van achterste koppeling
p siłownik tylnego sprzęgła

4016 rear cover
d Hinterdeckel *m*
f couvercle *m* arrière
n achterdeksel
p pokrywa tylna

4017 rear cross member
d hintere Querstrebe *f*
f traverse *f* arrière
n achterste dwarsbalk
p poprzeczka tylna

4018 rear door
d Hintertür *f*
f porte *f* arrière
n achterportier
p drzwi tylne

*   rear drive → 4037

4019 rear engine support
d hinteres Motorlager *n*
f support *m* arrière de moteur
n achterste motorsteun
p tylne łoże silnika

4020 rear flaps
d Heckklappe *f*
f trappe *f* arrière
n achterklep
p klapa tylna

4021 rear fog lights
d Nebelrücklichter *npl*; Nebelschlussleuchten *fpl*
f feux *mpl* de brouillard arrière
n mistachterlichten
p reflektory przeciwmgłowe tylne

4022 rear hose
d Hintenschlauch *m*
f tuyau *m* arrière
n achterslang
p wąż tylny

4023 rear main bearing cover
d Schutzdeckel *m* des Kurbelwellenlagers
f couvre-palier *m* arrière
n achterste krukaslagerdeksel
p osłona tylnego łożyska

4024 rear overhang
d hintere Überhanglänge *f*
f porte-à-faux *m* arrière
n achteroverbouw
p zwis tylny

4025 rear panel
d Heckwand *f*

f panneau *m* arrière; jupe *f* arrière
n achterpaneel; achterwand
p ściana tylna

4026 rear panel window pane
d Rückwandscheibe *f*
f panneau *m* de glace de paroi arrière
n achterpaneel
p szyba tylna

4027 rear pillar shield
d Hintersäulenschild *n*
f revêtement *m* de montant arrière
n bekleding van de achterste stijl
p nakładka słupka tylnego

4028 rear seat
d Hintersitz *m*
f banquette *f* arrière
n achterbank
p siedzenie tylne

4029 rear silencer
d hinterer Schaltdämpfer *m*
f silencieux *m* arrière
n achterste uitlaatdemper
p tłumik tylny

4030 rear spring
d Hinterfeder *f*
f ressort *m* d'arrière
n achterveer
p resor tylny

4031 rear spring bracket
d Hinterfederbock *m*
f support *m* de ressort d'arrière
n achterveerdrager
p kozioł resoru tylnego

4032 rear stabilizer bar
d Hinterachsschubstange *f*
f bielle *f* stabilisatrice arrière
n achterste stabilisatorstang
p łącznik przesuwny tylnej osi

4033 rear support bracket
d hintere Tragpratze *f*
f support *m* arrière
n achterste draagsteun
p tylna łapa zawieszenia silnika

4034 rear suspension
d Hinterradaufhängung *f*
f suspension *f* arrière
n achterwielophanging
p zawieszenie kół tylnych

4035 rear wheel
d Hinterrad *n*
f roue *f* arrière
n achterwiel
p koło tylne

4036 rear wheel brake
d Hinterradbremse *f*
f frein *m* de roue arrière
n achterwielrem
p hamulec koła tylnego

4037 rear wheel drive; rear drive
d Hinterachsantrieb *m*
f propulsion *f* arrière
n achterwielaandrijving
p napęd tylny

4038 rear wheel hub
d Hinterradnabe *f*
f moyeu *m* de roue arrière
n achterwielnaaf
p piasta koła tylnego

4039 rear window; back window
d Rückfenster *n*
f lunette *f* arrière
n achterruit
p szyba tylna

4040 rear window defroster switch
d Rückfensterfrostschutzscheibenschalter *m*
f interrupteur *m* de dégivreur de lunette arrière
n schakelaar voor achterruitverwarming
p wyłącznik odmrażacza szyby tylnej

4041 rear window heating control light
d Beschlagentfernungskontrolleuchte *f*
f témoin *m* du chauffage électrique de vitre arrière
n controlelicht van achterruitverwarming
p lampka kontrolna ogrzewania tylnej szyby

4042 rear window shelf
d Ablageplatte *f*
f tablette *f* arrière
n hoedenplank
p półka okna tylnego

4043 rear window wiper
d Heckscheibenwischanlage *f*
f essuie-vitre *f* arrière
n achterruitenwisser
p wycieraczka szyby tylnej

* **rear wing** → 3391

4044 rebound valve
 d Zugstufendrosselventil *n*
 f clapet *m* de rebond
 n opveersmoorklep
 p zawór odciążania

4045 rechargeable battery
 d aufladbare Batterie *f*
 f batterie *f* rechargeable
 n oplaadbare batterij
 p baterie doładowywane

4046 reciprocating piston engine
 d Hubkolbenmotor *m*
 f moteur *m* à piston à mouvement alternatif
 n krukdrijfstangzuigermotor
 p silnik tłokowy suwowy

4047 reclining seat
 d verstellbarer Sitz *m*
 f strapontin *m*
 n verstelbare stoel
 p fotel przestawny

4048 recommended rim
 d Normfelge *f*
 f jante *f* standard
 n standaardvelg
 p felga standartowa

4049 reconditioning
 d Wiederinstandsetzung *f*
 f rectification *f*
 n herstelling
 p regeneracja

4050 recording instrument
 d registrierendes Instrument *n*
 f appareil *m* enregistreur
 n registrerende meter
 p przyrząd rejestrujący

4051 rectifier
 d Gleichrichter *m*
 f redresseur *m*
 n gelijkrichter
 p prostownik

4052 rectilinear motion
 d geradlinige Bewegung *f*
 f mouvement *m* rectiligne
 n rechtlijnige beweging
 p ruch prostoliniowy

 * **red brass** → 4053

4053 red bronze; red brass
 d Rotguss *m*

 f bronze *m* rouge
 n rood messing
 p mosiądz czerwony

4054 redress couple
 d Rückstellmoment *n*
 f couple *m* d'autoalignement
 n herstelkoppel
 p moment odwodzący; moment cofający

4055 reduce *v*
 d mindern; verringern
 f réduire; diminuer
 n beperken; reduceren
 p ograniczyć; zredukować

 * **reducer** → 4060

4056 reduce *v* **speed**
 d Geschwindigkeit reduzieren
 f diminuer la vitesse
 n snelheid verminderen; snelheid beperken
 p zmniejszyć szybkość

4057 reduction gear
 d Reduktionsgetriebe *n*
 f engrenage *m* de réduction
 n reductietandwiel
 p przekładnia redukcyjna

4058 reduction gearbox; auxiliary gearbox
 d Nachschaltgetriebe *n*
 f réducteur *m*
 n naschakelbak
 p terenowa skrzynka biegów

4059 reduction gearbox housing
 d Nachschaltgetriebegehäuse *n*
 f carter *m* de réducteur
 n reductiekast
 p obudowa reduktora

4060 reduction socket; reducer
 d Zwischenstück *n*
 f manchon *m* de réduction
 n verloopstuk
 p króciec redukcyjny; element pośredniczący

4061 reference gauge; master gauge
 d Prüflehre *f*; Kontrollehre *f*
 f rapporteur *m*
 n standaardkaliber
 p przeciwsprawdzian

4062 reference sensor
 d Bezugsmarkensensor *m*
 f détecteur *m* de repères

n referentievoeler
p czujnik wielkości odniesienia

4063 reflector
d Reflektor *m*
f réflecteur *m*
n reflector
p reflektor

4064 reforming dolly
d Neugestalterhandfaust *f*
f tas *m* à reformer
n uitdeuktas bij omvorming
p kowadełko do wgnieceń

4065 refrigerating machine
d Kältemaschine *f*
f machine *f* frigorifique
n koelmachine
p chłodziarka

4066 refuse collector
d Müllwagen *m*
f benne *f* à ordures
n vuilniswagen
p śmieciarka

4067 registration card
d Fahrzeugzulassungsschein *m*
f carte *f* grise
n kentekenbewijs
p dowód rejestracyjny

4068 registration number
d Registriernummer *f*
f numéro *m* d'enregistrement
n kentekennummer
p numer rejestracyjny

4069 regular dolly
d Universalhandfaust *f*
f tas *m* américain
n Amerikaanse tas
p kowadełko uniwersalne

4070 regulating resistor
d Reglerwiderstand *m*
f résistance *f* de réglage
n regelweerstand; reostaat
p rezystor; opornik

4071 regulating spindle
d Einstellspindel *f*
f axe *m* de réglage
n spindelregelaar
p wrzeciono nastawcze

4072 regulating valve
d Regulierventil *n*

f soupape *f* de réglage
n regelklep
p zawór regulacyjny

* **regulator → 2505**

4073 regulator core
d Reglerkern *m*
f noyau *m* de régulateur
n regelkern
p rdzeń regulatora

4074 regulator drive gear
d Antriebsrad *n* des Reglers
f commande *f* de régulateur
n regulateuraandrijving
p napęd regulatora

4075 reinforcement
d Verstärkung *f*
f renforcement *m*
n versteviging
p wzmocnienie

4076 reinforcing rib
d Versteifungsrippe *f*
f nervure *f* de raidissement
n verstevigingsrib
p żebro usztywniające

4077 relative motion
d Relativbewegung *f*
f mouvement *m* relatif
n relatieve beweging
p ruch względny

4078 relay
d Relais *n*
f relais *m*
n relais
p przekaźnik

4079 relay control
d Relaissteuerung *f*
f commande *f* par relais
n relaiscontrole
p sterowanie przekaźnikowe

4080 relay emergency valve
d Relaisnotbremsventil *n*
f valve *f* relais d'urgence
n relaisnoodremklep
p zawór przekaźnikowo-awaryjny

4081 relay lever
d Übertragungshebel *m*
f levier *m* de renvoi

n overbrengingshefboom
p dźwignia przekaźnikowa

4082 relay valve
d Relaisventil *n*
f soupape *f* relais
n relaisklep
p zawór przekaźnikowy

4083 release bowl
d Ausrückschale *f*
f cloche *f* de débrayage
n ontkoppelingsplaat
p czasza wyciskowa

4084 release lever
d Auskupplungshebel *m*
f levier *m* de débrayage
n drukvinger van koppeling
p dźwignia zwalniająca

4085 release lever adjusting screw
d Ausrückhebeleinstellschraube *f*
f écrou *m* réglable de levier de débrayage
n stelschroef van loshefboom
p śruba regulacyjna dźwigni wyłączającej

4086 release lever support
d Ausrückhebelstütze *f*
f support *m* de levier de débrayage
n steun voor drukvinger
p wspornik dźwigni wyłączającej

4087 release rod; operating rod
d Ausrückstange *f*
f tige *f* de débrayage
n drukstang
p drążek odwodzący

4088 release rod adjusting screw
d Zugstangeneinstellschraube *f*
f écrou *m* réglable de tige
n drukstanginstelschroef
p śruba regulacyjna drążka

4089 release sleeve guide
d Führungsbuchse *f*
f guide *m* de centrage de butée
n geleidbus
p tuleja prowadząca

4090 release thrust bearing
d Ausrücklager *n*
f butée *f* de débrayage
n koppelingsdruklager
p łożysko wyciskowe

4091 release tool
(tool to track rod ends)
d Schlüssel *m* zum Entsperren
f outil *m* de déblocage
n deblokkeersleutel
p przyrząd odblokowujący

4092 relief valve
d Überströmventil *n*
f soupape *f* de décharge
n ontlastklep
p zawór upustowy

4093 remote control
d Fernschaltgerät *n*
f télécommande *f*
n afstandsbediening
p urządzenie sterowania zdalnego

4094 remote control gear
d Fernsteuerung *f*; Fernschaltung *f*
f commande *f* à distance
n afstandsbesturing
p sterowanie zdalne

4095 removal tool
(tool to remove parking brake mechanism)
d Demontagewerkzeuge *npl*
f outil *m* de dépose
n demontagegereedschap
p narzędzie do demontażu

* **removal van** → 2434

* **rental car** → 2681

4096 repaint *v*
d nachlackieren
f repeindre
n opnieuw lakken
p pomalować na nowo

4097 repair
d Reparatur *f*
f réparation *f*
n reparatie; herstelling
p reperacja

4098 repair kit
d Reparatursatz *m*
f complet *m* outillage
n reparatieset
p zestaw naprawczy

4099 repair order
d Reparaturauftrag *m*
f ordre *m* de réparation

n reparatieopdracht
p zlecenie naprawy

4100 repair shop; workshop
d Reparaturwerkstatt *f*
f atelier *m* de réparation
n reparatiewerkplaats; werkplaats
p warsztat naprawczy

4101 repair work
d Montagearbeit *f*
f travail *m* montage
n montagewerk
p praca montażowa

4102 requirement characteristic
d Bedarfskennung *f*
f caractéristique *f* de la demande
n kenmerk van het benodigde
p wymagana charakterystyka

4103 requirement graph
d Bedarfskennfeld *n*
f graphique *m* des demandes
n benodigdegrafiek
p wymagany wykres

4104 reservoir
d Behälter *m*
f réservoir *m*
n reservoir; tank
p zbiornik

4105 reservoir fastener
d Behälterklemmhalter *m*
f attache *f* de réservoir
n haak voor reservoir
p zaczep zbiornika

4106 residual imbalance
d Restunwucht *f*
f balourd *m* résiduel
n overgebleven onbalans
p niewyważenie resztkowe

4107 residual spring deflection
d Restfederweg *m*
f course *f* résiduelle
n overgebleven veerdoorbuiging
p szczątkowe wgięcie sprężyny

4108 resilient material
d Prallstoff *m*
f matière *f* amortissante
n elastisch materiaal
p materiał elastyczny

4109 resilient pin
d Federstift *m*

f goupille *f* élastique
n elastische pen
p kołek sprężynujący

4110 resistance plate
d Druckteller *m*
f disque *m* de pression
n drukplaat
p talerzyk oporowy

4111 resistance welding
d Widerstandschweissen *n*
f soudage *m* par résistance
n weerstand lassen
p zgrzewanie oporowe

4112 resonance
d Widerklang *m*
f résonance *f*
n resonantie
p rezonans

4113 retainer bush
d Fixierhülse *f*
f douille *f* de retenue
n borgbus
p tulejka ustalająca

4114 retainer pin
d Verschlussstift *m*
f épingle *f* de fixation
n borgpen
p kołek ustalający

4115 retaining spring
d Formfeder *f*
f ressort *m* façonné
n profielveer
p sprężyna kształtowa

4116 retaining wire
d Drahtsprengring *m*
f jonc *m*
n draadspringring
p druciany pierścień sprężynowy

4117 retarded closing
d Nachschliessung *f*
f fermeture *f* retardée
n nasluiting
p zamknięcie dodatkowe

4118 retarder
d Retarder *m*
f ralentisseur *m*
n retarder
p zwalniacz

4119 retracted axle
d Liftachse *f*
f essieu *m* relevable
n sleepas; lift-as
p oś wleczona

4120 return pipe
d Rückleitung *f*
f conduit *m* de retour
n retourslang; retourleiding
p przewód nadmiarowy

4121 return spring
d Rückfeder *f*
f ressort *m* de rappel
n terugtrekveer
p sprężyna powrotna

4122 reverse *v*
d umsteuern
f renverser
n omkeren
p nawracać

4123 reverse clutch
d Kupplung *f* des Rückwärtsganges
f embrayage *m* de marche arrière
n koppeling voor achteruit
p sprzęgło biegu wstecznego

4124 reverse driven gear
d angetriebenes Rückwärtsgangzahnrad *n*
f pignon *m* mené de marche arrière
n aangedreven tandwiel voor achteruit
p napędzające koło biegu wstecznego

4125 reverse gear
d Rückwärtsgang *m*
f marche *f* arrière
n achteruitversnelling
p przekładnia biegu wstecznego

4126 reverse gear fork rod
d Rückwärtsganggabelwelle *f*
f axe *m* de fourchette de marche arrière
n gaffelas voor achteruitversnelling
p wałek widełek biegu wstecznego

4127 reverse gear stop
d Rückwärtsganganschlag *m*
f butée *f* de marche arrière
n achteruitaanslag
p występ uniemożliwiający błędne włączenie
  biegu wstecz

4128 reverse gear stop spring
d Rückwärtsganganschlagfeder *f*

f butée *f* à ressort de marche arrière
n aanslagveer van achteruitversnelling
p sprężyna biegu wstecznego

4129 reverse idler gear spindle
d Rückwärtsgangzwischenrad *n*
f pignon *m* intermédiaire de marche arrière
n tussentandwiel voor achteruit
p wałek biegu wstecznego

4130 reverse idler shaft
d Rücklaufachse *f*
f arbre *m* de marche arrière
n achteruit-stationaire-as
p oś biegu wstecznego

4131 reverse pinion
d Rücklaufrad *n*
f pignon *m* de marche arrière
n omkeertandwiel
p koło biegu wstecznego

4132 reverse speed fork
d Schaltgabel *f* für Rücklauf
f fourchette *f* de marche arrière
n achteruitversnellingvork
p widełki biegu wstecznego

4133 reverse straight jaws
d umgekehrte gerade Backen *mpl*
f becs *mpl* droits inversés
n omgekeerde rechte bekken
p szczęki proste odwrotne

4134 reversible long blades
d umkehrbare lange Klingen *fpl*
f lames *fpl* longues réversibles
n omkeerbare lange klingen
p odwracalne długie ostrza

4135 reversible ratchet
d Umschaltknarre *f*
f cliquet *m* réversible
n omschakelbare ratel
p pokrętło zapadkowe przełączania

4136 reversible ratchet adapter
d Ansteckumschaltknarre *f*
f cliquet *m* adaptateur réversible
n adapter van omschakelbare ratel
p przystawka pokrętła zapadkowego

4137 reversing light
d Rückfahrtscheinwerfer *m*
f projecteur *m* de recul
n achteruitrijlicht
p światła cofania

4138 reversing shaft
d Umsteuerungswelle *f*
f arbre *m* de renversement de marche
n as voor achteruitgangbeweging
p wałek zwrotny

4139 reversing valve
d Umsteuerungsventil *n*
f soupape *f* de renversement de marche
n achteruitklep
p zawór zwrotny

4140 revolution counter
d Tourenzähler *m*
f compte-tours *m*
n toerenteller
p licznik obrotów; obrotomierz

4141 ribbed cap
d gerändelter Pressattel *m*
f tête *f* striée
n gekarteld perszadel
p kołpak karbowany

4142 ribbed individual cylinder
d gerippter Einzylinder *m*
f cylindre *m* à ailettes
n cilinder met koelribben
p pojedynczy cylinder użebrowany

4143 rich mixture
d reiches Gemisch *n*; fettes Gemisch *n*
f mélange *m* riche
n rijk mengsel; vet mengsel
p mieszanka bogata

4144 right door
d rechte Tür *f*
f porte *f* droite
n rechter portier
p drzwi prawe

4145 right hand drive
d rechts angeordnete Lenkung *f*
f direction *f* à droite
n rechtse besturing
p układ kierowniczy prawostronny

4146 right of way
d Vorfahrt *f*
f priorité *f* de passage
n voorrang
p pierwszeństwo przejazdu

* **rigid → 4811**

4147 rigid axle shaft
d starre Achswelle *f*

f arbre *m* de roue rigide
n drijfashuis
p półoś sztywna

4148 rigid coupling
d starre Kupplung *f*
f accouplement *m* rigide
n vaste koppeling
p sprzęgło sztywne

4149 rigidity
d Starrheit *f*
f rigidité *f*
n stijfheid
p sztywność

4150 rigid pipe
d starre Leitung *f*
f tuyau *m* rigide
n starre leiding
p przewód sztywny

4151 rig test
d Prüfstandversuch *m*
f essai *m* sur banc
n uitrustingstest
p próba na hamowni

* **rim base → 5445**

4152 rim flange
d Felgenhorn *m*; Felgenrand *m*
f bord *m* de jante
n velgrand
p obrzeże obręczy koła

4153 rim type
d Felgenart *f*
f type *m* de jante
n velgtype
p typ obręczy koła

4154 rim width
d Felgenmaulweite *f*; Felgenbreite *f*
f largeur *f* de jante
n velgbreedte
p szerokość obręczy

4155 ring
d Ring *m*
f anneau *m*
n ring
p pierścień

4156 ring end crowfoot
d Radschraubenschlüssel *m*
f clé *f* coudée

n kopboutsleutel
p klucz do głowicy cylindra

4157 ring gauge
d Ringkaliber *m*
f calibre *m* annulaire
n ringkaliber
p sprawdzian pierścieniowy

4158 ring gear; crown wheel
d Tellerrad *n*
f couronne *f* dentée
n kroonwiel
p koło talerzowe

4159 ring gear hub
d Tellerradnabe *f*
f moyeu *m* de couronne
n kroonwielnaaf
p piasta koła talerzowego

4160 ring gear overload plug
d Tellerradstützschraube *f*
f vis *f* de frotteur de couronne
n overbelastingplug voor het kroonwiel
p śruba oporowa koła talerzowego

* **ring groove** → 3779

4161 ring nut
d Ringmutter *f*
f bague *f* écrou
n ringschroef
p nakrętka oczkowa

4162 ring road
d Ringstrasse *f*
f périphérique *m*
n ringweg
p obwodnica

4163 ring spanner
d Ringschlüssel *m*
f clé *f* à douille; clé *f* fermée
n ringsleutel
p klucz oczkowy

4164 ring spring
d Ringfeder *f*
f ressort *m* en anneau
n ringveer
p sprężyna pierścieniowa

4165 rinsing
d Spühlung *f*
f rinçage *m*
n doorspoeling
p płukanie

4166 rivet
d Niet *m*
f rivet *m*
n klinknagel
p nit

4167 riveted joint
d Zickzacknietung *f*
f rivure *f* en quinconce
n zigzagnaad
p szew nitowy

4168 rivet head
d Nietkopf *m*
f tête *f* de rivet
n nagelkop
p główka nitu

4169 rivet hole
d Nietloch *n*
f trou *m* de rivet
n klinknagelgat
p otwór nitowy

4170 riveting
d Nieten *n*
f rivetage *m*
n klinken
p nitowanie

4171 riveting hammer
d Schlosserhammer *m*
f marteau *m* rivoir
n bankhamer
p młotek do nitowania

4172 rivet shank
d Nietschaft *f*
f tige *f* du rivet
n klinknagelsteel
p trzon nitu; szyjka nitu

* **road ability** → 4175

4173 road crossing
d Strassenkreuzung *f*
f carrefour *m*; croisement *m* des routes
n kruispunt; wegkruising
p skrzyżowanie dróg

4174 road flare
d Warndreieck *n*
f triangle *m* de signalisation
n gevarendriehoek
p trójkąt ostrzegawczy

4175 road holding; road ability
d Strassenlage *f*

f tenue *f* de route
n wegligging; wegvastheid
p trzymanie sie drogi

4176 road service
d Störungsdienst *m*; Pannenhilfe *f*
f service *m* de dépannage
n wegenwacht
p pomoc drogowa

4177 road traffic
d Strassenverkehr *m*
f trafic *m*
n wegverkeer
p ruch drogowy

4178 road user
d Verkehrsteilnehmer *m*
f participant *m* de la circulation
n weggebruiker
p użytkownik drogi

4179 road works ahead
d Strassenarbeiten *fpl*
f travaux *mpl* routiers
n werk in uitvoering
p roboty drogowe

4180 rocker
d Kipphebel *m*; Schwinghebel *m*
f culbuteur *m*
n tuimelaar
p dźwigienka zaworowa

4181 rocker shaft
d Kipphebelachse *f*
f axe *m* de culbuteur
n tuimelaaras
p oś dźwigienek zaworowych

4182 rocker shaft bracket
d Lagerbock *m* für Kipphebelachse;
   Kipphebelachsenbock *m*
f support *m* d'axe de culbuteur
n support van de tuimelaaras
p wspornik osi dźwigni zaworów

* rod → 397

4183 roller
d Rolle *f*
f rouleau *m*
n rol
p rolka

4184 roller bearing
d Rollenlager *n*
f roulement *m* à rouleaux

n rollenlager
p łożysko wałeczkowe

4185 roller bearing guideway
d Rollenführung *f*
f glissière *f* sur rouleaux
n rollengeleiding
p prowadnica rolkowa

4186 roller bow
d Rollenspriegel *m*
f arceau *m* à galets
n rolbeugel
p pałąk rolkowy

4187 roller chain
d Rollenkette *f*
f chaîne *f* à rouleaux
n rollenketting
p łańcuch sworzniowy tulejkowy

4188 roller plate
d Wälzplatte *f*
f plaque *f* à rouleaux
n rolplaat
p płyta walcowana

4189 roller shutter
d Rolladen *m*
f volet *m* roulant
n rolluik
p żaluzja

4190 roller tappet
d Rollenstössel *m*
f poussoir *m* à galet
n stoter met rol
p popychacz rolkowy

4191 roller tray
d zerlegbarer Werkstattwagen *m*
f servante *f* démontable d'atelier
n gereedschapskabinet
p wózek narzędziowy rolkowy

4192 roller type stud drivers
d Stehbolzenausdreher *m* mit Walzen
f dégoujonneur *m* à rouleaux
n tapeinddemontagedop met rollen
p rolkowy mechanizm zaciskowy przyrządu do
   usuwania kołków

4193 rolling chest
d Werkzeugwagen *m* mit Sitzbank
f rouleuse *f*
n gereedschapswagen met zitbank
p wózek narzędziowy z siedzeniem

4194 rolling diaphragm air spring
d Rollbalgluftfeder *f*
f diaphragme *m* pneumatique
n rolbalgluchtveer
p przeponowy resor pneumatyczny

4195 rolling element
d Wälzkörper *m*
f roulement *m* de palier
n wentellichaam
p część toczna

4196 rolling resistance
d Rollwiderstand *f*
f résistance *f* au roulement
n rolweerstand
p opór toczenia

4197 roll over bar
d Überrollschutzbogen *m*
f arceau *m* de sécurité
n rolbeugel
p pałąk ochronny

4198 roof bow
d Dachgurt *m*
f courbe *f* de parillon
n dakoverspanning; daktoog
p poprzeczna belka dachu

4199 roof clamp
d Dachklammer *f*
f agrafe *f* de toit
n dakklamp
p zacisk dachowy

4200 roof front top rail
d vorderer Dachbogen *m*
f arceau *m* frontal de pavillon
n voorste daktoog
p łuk czołowy dachu

* **roof lamp** → 934

4201 roof lamp manual switch
d Deckenlampehandschalter *m*
f commutateur *m* à main de plafonnier
n schakelaar voor plafonnière
p ręczny wyłącznik lampy sufitowej

4202 roof reinforcing crosspiece
d Dachverstärkungstraverse *f*
f traverse *f* de renfort de toit
n daktoog
p poprzeczka usztywniająca dachu

4203 roof vent
d Dachentlüftung *f*

f reniflard *m* de toit
n ontluchting via het dak
p wywietrznik dachowy

4204 roominess
d Geräumigkeit *f*
f largeur *f*
n ruimte
p przestronność wnętrza

4205 rotary burs
d Rotorfräser *mpl*
f fraises *fpl* rotatives
n rotatiefrezen; stiftfrezen
p frezy trzpieniowe

4206 rotary coil
d Drehspule *f*
f bobine *f* rotative
n draaispoel
p cewka obrotowa

4207 rotary piston
d Drehkolben *m*
f piston *m* rotatif
n draaizuiger
p tłok obrotowy

4208 rotary piston engine
d Kreiskolbenmotor *m*
f moteur *m* à piston rotatif
n rotatiemotor
p silnik o tłoku obrotowym

4209 rotary piston pump
d Kreiskolbenpumpe *f*
f pompe *f* rotative
n rotatiepomp
p pompa o wirujących tłokach

* **rotary snow plough** → 4562

4210 rotary valve
d Drehscheibeventil *n*
f tiroir *m* de rotation
n roterende klep; draaischuif
p zawór obrotowy

4211 rotating axle
d drehende Achse *f*
f essieu *m* tournant
n rotatieas
p oś obrotu

4212 rotation
d Drehung *f*
f rotation *f*

n rotatie
p rotacja; ruch obrotowy

4213 **rotational speed**
d Drehzahl *f*
f vitesse *f* régime
n toerental
p prędkość obrotowa

4214 **rotational speed sensor**
d Drehzahlregler *m*
f capteur *m* de régime
n rijsnelheidssensor
p czujnik zakresu prędkości

4215 **rotative saw**
(tool used for all circular cuttings into wood
through partitions, into leather pouch)
d Sägekreisel *m*
f scie *f* rotative
n roterende zaag
p piła tarczowa

4216 **rotodynamic pump**
d Strömungsarbeitmaschine *f*
f pompe *f* dynamique
n stromingswerkpomp
p pompa wirowa; pompa rotodynamiczna

4217 **rotoflex coupling**
d elastische Kupplung *f*
f joint *m* de type flector
n flexibele koppeling
p pierścieniowy przegub elastyczny

4218 **rotor**
d Läufer *m*
f rotor *m*
n rotor m
p rotor

4219 **rotor core claw**
d Klauenpolläufer *m*
f pièce *f* polaire d'inducteur; rotor *m* à griffes
n klauwpoolrotor
p kief rdzenia twornika

4220 **rotor disc**
d Rotorscheibe *f*
f disque *m* de rotor
n rotorschijf
p tarcza wirnika

4221 **rotor drum**
d Trommelläufer *m*
f tambour *m*
n turbinetrommel
p wirnik bębnowy

4222 **rotor shaft**
d Rotorwelle *f*
f arbre *m* d'inducteur
n rotoras
p wał twornika

4223 **rotor winding**
d Läuferwicklung *f*
f bobinage *m* de rotor; bobinage *m* d'inducteur
n rotorwikkeling
p uzwojenie twornika

4224 **rough file; coarse file**
d Grobfeile *f*; Armfeile *f*
f lime *f* grosse; lime *f* rude
n grove vijl; armvijl
p pilnik zdzierak

4225 **roundabout**
d Kreisplatz *m*; Wendeplatz *m*
f point *m* rond
n verkeersplein; rotonde
p rondo

4226 **round belt**
d Rundriemen *m*
f courroie *f* ronde
n ronde snaar
p pas okrągły

4227 **round file**
d Rundfeile *f*
f lime *f* ronde
n ronde vijl; rattenstaart
p pilnik okrągły

* **round head bolt** → 1022

4228 **rounding cutter**
d Stabfräser *m*
f fraise *f* pour barreaux
n staaffrees
p frez zaokrąglony

4229 **round nut with drilled holes inside**
d Kreuzlochmutter *f*; Stirnlochmutter *f*
f écrou *m* cylindrique à trois latéraux
n ronde moer met kruisgaten; pensleutelmoer
p nakrętka okrągła; nakrętka otworowa

4230 **round plier**
d Rundzange *f*
f pince *f* ronde
n rondbektang; platbektang
p szczype okrągłe

4231 **rubber**
d Gummi *n*

f caoutchouc *m*
n rubber
p guma

4232 rubber annulus coupling
d Gummimetallkupplung *f*
f accouplement *m* caoutchouteux métallique
n koppeling met rubberen schijf
p sprzęgło gumowo-metalowe

4233 rubber bellows coupling
d Gummiwulstkupplung *f*
f accouplement *m* par manchon en caoutchouc
n rubberbalgkoppeling
p sprzęgło gumowe oponowe

4234 rubber block compression spring
d Gummidruckfeder *f*
f ressort *m* à compression en caoutchouc
n rubberen drukveer
p gumowa sprężyna dociskowa

4235 rubber bracket
d Gummistütze *f*
f support *m* en caoutchouc
n rubberen doorvoertule
p wspornik gumowy

4236 rubber buffer
d Gummipuffer *m*
f tampon *m* en caoutchouc
n stootrubber
p zderzak gumowy

4237 rubber bushing
d Gummifederbuchse *f*
f manchon *m* de caoutchouc
n silentbloc
p tuleja elastyczna

4238 rubber cap
d Gummischutzbalg *m*
f protection *f* en caoutchouc
n rubberen kap
p osłona gumowa

4239 rubber coupling
d Gummigelenk *n*
f joint *m* en caoutchouc
n rubberen scharnier
p przegub gumowy

4240 rubber cushioned spring hanger
d Gummifederlager *n*
f articulation *f* sur caoutchouc
n ophangrubber
p gumowy element elastyczny

4241 rubber grommet
d Gummitülle *f*
f embout *m* en caoutchouc
n rubberen doorvoertule
p przelotka gumowa

4242 rubber hose
d Gummischlauch *m*
f tuyau *m* en caoutchouc
n rubberen slang
p przewód gumowy; wąż gumowy

4243 rubber insert; elastic insert
d elastische Einlage *f*; Gummieinsatz *m*
f segment *m* caoutchouc; pièce *f* rapportée en caoutchouc
n rubberen inzetstuk
p kształtka gumowa; kształtka elastyczna

4244 rubberization
d Gummierung *f*
f gommage *m*
n behandelen met rubber
p gumowanie

4245 rubber mat
d Gummimatte *f*
f tapis *m* en caoutchouc
n rubberen mat
p dywanik gumowy

4246 rubber metal shackle
d Metallgummilasche *f*
f œil *m* avec bague élastique
n rubber-metaallager
p łącznik metalowo-elastyczny

4247 rubber pipe
d Gummirohr *n*
f tuyau *m* en caoutchouc
n rubberen pijp
p rurka gumowa

4248 rubber ring
d Gummiring *m*
f anneau *m* en caoutchouc
n rubberen ring
p pierścień gumowy

4249 rubber seal
d Gummidichtung *f*
f rondelle *f* en caoutchouc
n afdichtingsrubber
p uszczelka gumowa

4250 rubber shoe valve
d Gummifussventil *n*

f valve *f* à pied gommé
n rubberen schoenklep
p zawór z gumowa podstawą

4251 rubber solution
d Gummilösung *f*
f solution *f* de caoutchouc
n rubbersolutie
p klej kauczukowy

4252 rubber stop
d Gummianschlag *m*
f butée *f* en caoutchouc
n aanslagrubber
p ogranicznik gumowy

4253 rubber stopper
d Gummipfropfen *m*
f butoir *m* en caoutchouc
n rubberen stop
p korek gumowy

4254 rubber universal joint
d Gummikreuzgelenk *n*
f joint *m* universel en caoutchouc
n rubberen kruiskoppeling
p gumowy przegub uniwersalny

4255 rubber valve
d Gummiventil *n*
f valve *f* en caoutchouc
n rubberen klep
p zawór gumowy

4256 ruler supporting the dial gauge
d Lineal *n* als Halterung für Messuhr
f règle *f* support de comparateur
n liniaal tot steun van wijzerplaat
p liniał jako zamocowanie czujnika zegarowego

4257 run *v* in
d einlaufen
f roder
n inlopen
p docierać

4258 runner stick
d Eingussstock *m*
f broche *f* de coulée
n ingietstok
p wlew formierski

4259 running plate
d Trittplatte *f*
f plateau *m* de marchepied
n drempel; treeplank
p stopka pedału

4260 running repair
d Kleininstandsetzung *f*
f réparation *f* courante
n lopend onderhoud
p naprawa bieżąca

4261 rush hours
d Hauptverkehrsstunden *fpl*
f heures *fpl* de pointe
n spitsuren
p godziny szczytu

4262 rust
d Rost *m*
f rouille *f*
n roest
p rdza

4263 rustproofing agent
d Rostschutzmittel *n*
f produit *m* anticorrosif
n antiroestmiddel
p środek przeciwkorozyjny; środek rdzochronny

4264 Rzeppa joint
d Rzeppagelenk *n*
f joint *m* homocinétique
n Rzeppa-kruiskoppeling
p przegub Rzeppa

S

4265 saddle key
d Nasenhohlkeil *m*
f clavette *f* évidée
n holle spie
p klin wzdłużny wklęsły

4266 safety
d Sicherheit *f*
f sécurité *f*
n veiligheid
p bezpieczeństwo

4267 safety belt; seat belt
d Sicherheitsgurt *m*
f ceinture *f* de sécurité
n veiligheidsgordel
p pas bezpieczeństwa

4268 safety brake light; additional brake light
d Zusatzbremsleuchte *f*; Sicherheitsbremsleuchte
   f
f feu *m* stop sécurité
n derde remlicht
p dodatkowe światło stopu

4269 safety catch
d Sicherungshaken *m*
f crochet *m* de sécurité
n veiligheidsvergrendeling
p zapadka zabezpieczająca

4270 safety clutch
d Sicherheitskupplung *f*
f accouplement *m* de sécurité
n veiligheidskoppeling
p sprzęgło przeciążeniowe

4271 safety glass
d Sicherheitsglas *n*
f glace *f* de sécurité
n veiligheidsglas
p szkło bezpieczne

4272 safety goggles
d Schutzbrille *f*
f lunettes *fpl*
n slijpbril
p okulary ochronne

4273 safety lock
d Verriegelung *f* für Achse
f verrou *m* d'axe
n asvergrendeling
p blokada osi

4274 safety pin
d Sicherheitsnadel *f*
f épingle *f* de nourrice
n veiligheidsspeld
p agrafka

4275 safety plug
d Sicherheitsschraube *f*
f bouchon *m* de sûreté
n veiligheidsplug
p korek bezpieczeństwa

4276 safety spark gap
d Sicherheitsfunkstrecke *f*
f parafoudre *m*
n beveiligingsvonkbrug
p iskiernik zabezpieczający

4277 safety switch
d Sicherheitsschalter *m*
f interrupteur *m* de sécurité
n veiligheidsschakelaar
p wyłącznik bezpieczeństwa

4278 safety valve
d Sicherheitsventil *n*
f soupape *f* de limiteur de pression
n veiligheidsklep
p zawór bezpieczeństwa

4279 salvage trailer
d Abschleppanhänger *m*
f remorque *f* dépanneuse
n aanhanger voor wrakkenvervoer
p przyczepa holownicza

4280 sandblast blower; grit sprayer
d Sandstrahlgebläse *n*; Sandstreuer *m*
f machine *f* à jet de sable; sableuse *f*
n zandstraalapparaat
p piaszczarka

4281 sandblasting
d Sandstrahlreinigung *f*
f décapage *m* au jet de sable
n zandstraalreiniging
p piaskowanie

4282 sandblasting apparatus
d Gussputzmaschine *f*
f dessableuse *f*
n poetsmachine
p oczyszczarka

4283 sandblasting cabin
  d Sandstrahlkabine *f*
  f cabine *f* de sablage
  n zandstraalcabine
  p komora do piaskowania

4284 sanding scaler
  d Porenfüller *m*
  f enduit *m* bouche-micropores
  n vuller van microporiën
  p wypełniacz porów

4285 sanding vehicle
  d Sandstreuwagen *m*
  f véhicule *m* de distribution de sable
  n strooiwagen
  p piaskarka

4286 sandpaper
  d Sandpapier *n*
  f papier *m* de sable
  n schuurpapier
  p papier ścierny piaskowy

4287 saponify *v*
  d verseifen
  f saponifier
  n verzepen
  p zmydlać

4288 sash cramps
  d Schnellschraubzwingen *fpl*
  f serre-joints *mpl*
  n kozijnklemmen
  p ściski szybkodziałające

4289 saw
  d Säge *f*
  f scie *f*
  n zaag
  p piła

4290 sawdust
  d Sägemehl *n*
  f sciure *f*
  n zaagmeel; zaagsel
  p mączka drzewna

4291 sawing
  d Sägeführung *f*
  f guide *m* de sciage
  n zaaggeleiding
  p prowadnica piły

4292 saw planing anvil
  d Sägenamboss *m*
  f tas *m* à planer les scies

  n aanbeeld om zagen te richten; aambeeld om zagen te richten
  p kowadło piły

4293 scale interval
  d Skalateilung *f*
  f division *f* d'échelle
  n schaalwijzer
  p podziałka

4294 scavenge pump
  d Spülpumpe *f*
  f pompe *f* de balayage
  n spoelpomp
  p pompa przepłukująca

4295 scavenging air
  d Spülluft *f*
  f air *m* de balayage
  n spoellucht
  p powietrze przepłukujące

4296 scavenging duct
  d Spülleitung *f*
  f tubulure *f* d'évacuation
  n valpijp
  p przewód płuczkowy

4297 scavenging pressure
  d Spüldruck *m*
  f pression *f* de balayage
  n spoeldruk
  p ciśnienie płukania

  * **scissors jack** → 269

4298 scouring brush
  d Scheuerbürste *f*
  f brosse *f* à écurer
  n schuurborstel
  p pędzel do oczyszczania

4299 scrap
  d Schrott *m*
  f grenaille *f*
  n schroot
  p złom

4300 scrape *v*
  d verschrotten
  f casser; démolir
  n slopen
  p złomować

4301 scraper
  d Kratzer *m*
  f racloir *m*

n schraper
p skrobak

4302 scraper ring
d Abstreifring *m*
f segment *m* refouleur
n olieschrapveer
p pierścień tłokowy zgarniający

4303 scrap yard
d Schrottplatz *m*; Autofriedhof *m*
f cimetière *m* automobile
n sloperij
p złomowisko; składowisko złomu

* **screened wire** → **4436**

4304 screw
d Schraube *f*
f vis *f*
n schroef
p śruba; wkręt

4305 screw bolt
d Schraubenbolzen *m*
f boulon *m*
n bout zonder moer
p śruba z dwustronnym gwintem

4306 screw brake
d Spindelbremse *f*
f frein *m* à vis
n schroefrem
p hamulec śrubowy

4307 screw cutting latch
d Zugspindeldrehbank *f*
f tour *m* à fileter
n draadsnijbank
p tokarka pociągowa

4308 screw jack
d Schraubenwinde *n*
f cric *m* à vis
n schaarkrik
p dźwignik śrubowy

4309 screw plate
d Gewindeschneidbacke *f*
f filière *f*
n draadsnijplaat
p gwintownica ramkowa

4310 screw puller
d Schraubabzieher *m*
f extracteur *m*
n trekker; trekgereedschap
p ściągacz śrub

4311 screw pump
d Schraubenpumpe *f*
f pompe-hélice *f*
n schroefpomp
p pompa śrubowa

4312 screw with conical head
d Schraube *f* mit konischem Kopf
f vis *f* à tête conique
n bout met conische kop
p śruba z łbem stożkowym

4313 scriber
d Reissnadel *f*
f pointe *f* à tracer
n kraspen
p rysik

4314 scuffing rib
d Scheuerleiste *f*
f pare-trottoir *m*
n schuurstrip
p pasek ochronny

4315 seal *v*
d abdichten; dichten
f étancher
n afdichten
p uszczelniać

4316 sealed beam
d Sealed-Beam-Scheinwerfer *m*
f projecteur *m* sealed beam
n sealed beam
p wkład reflektora

4317 sealed cooling system
d Überdruckkühlflüssigkeitsumlauf *m*
f système *m* de refroidissement scellé
n gesloten koelsysteem
p zamknięty układ chłodzenia cieczą

4318 sealing bush
d Dichthülse *f*
f bague *f* d'étanchéité
n afdichtingshuls
p tujejka uszczelniająca

4319 sealing cement
d Dichtungskitt *m*
f mastic *m*
n afdichtkit
p kit uszczelniający

4320 sealing compound
d Dichtungsmasse *f*
f pâte *f* de joint

n afdichtpasta
p masa uszczelniająca; kit uszczelniający

4321 sealing lip
  d Dichtlippe *f*
  f lèvre *f* de joint
  n afdichtingslip
  p krawędź uszczelniająca

4322 sealing ridge
  d Dichtungsgrille *f*
  f nervure *f* d'étanchéité
  n dichtingskam
  p rowek uszczelniający

4323 sealing ring; packing ring
  d Dichtring *m*
  f segment *m* d'étanchéité
  n afdichtingsring
  p pierścień uszczelniający

4324 sealing strip
  d Abdichtband *n*
  f bande *f* d'étanchéité
  n rubberen afdichtstrip
  p gumowa taśma uszczelniająca

4325 sealing surface
  d Dichtungsfläche *f*
  f plan *m* de joint; surface *f* de joint
  n pasvlak
  p powierzchnia uszczelniająca

4326 seamless tube
  d nahtloses Rohr *n*
  f tube *m* sans soudure
  n niet-gelaste pijp
  p rura bez szwu

4327 seam welding
  d Nahtschweissen *n*
  f soudage *m* par ligne
  n naaldlassen
  p zgrzewanie liniowe

4328 seasonal service
  d Saisondurchsicht *f*
  f entretien *m* saisonnier
  n seizoenonderhoudsbeurt
  p obsługa sezonowa

4329 seat
  d Sitz *m*
  f siège *m*
  n zitplaats
  p siedzenie

4330 seat adjusting lever
  d Sitzverstellhebel *m*
  f levier *m* de déplacement de siège
  n verstelhefboom van de zitting
  p dźwignia przesuwu siedzenia

4331 seat back
  d Rückenlehne *f*
  f inclinaison *f* du dossier; dossier *m* de siège
  n rugleuning
  p oparcie

4332 seat back adjusting handle
  d Rückenlehneverstellgriff *m*
  f poignée *f* de réglage de dossier de siège
  n stelknop voor de rugleuning
  p pokrętło regulacji oparcia siedzenia

4333 seat back adjusting rod
  d Rückenlehneeinstellstange *f*
  f manchon *m* de réglage de dossier de siège
  n stelstang van rugleuning
  p drążek regulacji oparcia siedzenia

4334 seat back return spring
  d Rückenlehnerückholfeder *f*
  f ressort *m* de rappel de dossier de siège
  n terugtrekveer voor rugleuning
  p sprężyna powrotna oparcia siedzenia

* **seat belt** → **4267**

4335 seat belt anchorage
  d Sicherheitsgurtverankerung *f*
  f ancrage *m* de ceinture de sécurité
  n verankering van veiligheidsgordel
  p zakotwiczenie pasa bezpieczeństwa

4336 seat bottom panel
  d Sitzunterhaube *f*
  f panneau *m* inférieur de siège
  n onderkap van de zitting
  p pokrywa dolna siedzenia

4337 seat center runner
  d Mittelsitzschiene *f*
  f glissière *f* centrale de siège
  n middenrail van de zitting
  p środkowa prowadnica siedzenia

4338 seat cover
  d Sitzschonbezug *m*
  f revêtement *m* de siège
  n bekleding van de zitting
  p pokrycie siedzenia

4339 seat cushion
  d Sitzkissen *n*

f coussin *m* de siège
n zitkussen
p poduszka siedzenia

4340 seat frame
d Sitzgestell *m*
f carcasse *f* de siège
n stoelframe
p szkielet siedzenia

4341 seating capacity
d Platzzahl *f*
f nombre *m* de places
n aantal zitplaatsen
p liczba miejsc siedzących

4342 seating position
d Sitzposition *f*
f position *f* assise
n zitpositie
p pozycja siedzenia

4343 seating width
d Sitzbreite *f*
f largeur *f* des sièges
n breedte van zitkussen
p szerokość siedzenia

4344 seat side rail
d Seitensitzschiene *f*
f glissière *f* latérale de siège
n zijrail van de zitting
p boczna prowadnica siedzenia

4345 seat support
d Sitzstütze *f*
f support *m* de fauteuil
n steun van de zitplaats
p wspornik fotela

* secondary brake shoe → 5159

4346 secondary circuit
d Sekundärstromkreis *m*
f circuit *m* secondaire
n secundaire stroomkring
p obwód wtórny

4347 secondary coil
d Sekundärspule *f*
f bobine *f* secondaire
n secundaire spoel
p cewka wtórna

4348 secondary cup
d Abstreifmanschette *f*
f coupelle *f* secondaire

n schraapmanchet
p kołnierz zgarniający

4349 secondary winding; high tension winding
d Sekundärwicklung *f*
f bobinage *m* secondaire; enroulement *m* secondaire
n secundaire wikkeling
p uzwojenie wtórne; uzwojenie wysokonapięciowe

4350 secondhand car; used car
d Gebrauchtwagen *m*
f voiture *m* d'occasion
n tweedehands auto
p samochód używany

4351 second speed
d zweiter Gang *m*
f seconde vitesse *f*
n tweede versneling
p drugi bieg

4352 second stage choke tube
d Lufttrichter *m* der zweiten Stufe
f diffuseur *m* de deuxième flux
n tweedetraps chokebuis
p gardziel drugiego przelotu

4353 secret dovetail joint
d verdecker Schwalbenschwanz *m*
f assemblage *m* à queue perdue
n verdekte zwaluwstaartverbinding
p połączenie na wczepy kryte

4354 sector
d Sektor *m*
f secteur *m*
n sector
p wycinek

4355 sector gear
d Lenksegment *n*
f secteur *m* denté de direction
n wormwielsector
p wycinek ślimacznicy

4356 securing clip
d Haltelasche *f*
f patte *f* de maintien
n bevestigingsklem
p zacisk zabezpieczający

4357 securing plate
d Befestigungsplakette *f*
f plaquette *f* de fixation
n bevestigingsplaatje
p płytka zabezpieczająca

4358 sediment chamber
d Schlammraum *m*
f chambre *f* de sédimentation
n bezinkselruimte
p komora osadowa

* **sediment element** → 1226

* **seeger clip** → 4681

4359 selecting pin
d Fühlnadel *f*
f tige *f* de contrôle
n vulnaald
p igła dozująca

4360 selective oil
d Öl *n* aus Selektivraffination
f huile *f* de raffinage sélectif
n selectief geraffineerde olie
p olej rafinacji selektywnej

4361 selector
d Gangwähler *m*
f doigt *m* de commande et de sélection;
sélecteur *m*
n selector; keuzehendel
p wybierak suwakowy zmiany biegów

4362 selector rod
d Gabelwelle *f*; Schaltwelle *f*
f levier *m* de commande de changement de
vitesse
n schakelas
p wałek widełek

4363 selector shaft
d Gangwählerwelle *f*
f arbre *m* de sélecteur
n selectoras
p wałek wybieraka

4364 selector slider
d Gangwählerschuber *m*
f coulisseau *m* de sélecteur
n glijstukje van schakelstang
p suwak wybieraka

4365 selector spring
d Gangwählerfeder *f*
f ressort *m* de sélecteur
n veer van de selector
p sprężyna selektora

4366 selector spring seat
d Gangwählerfederteller *m*
f cuvette *f* de ressort de sélecteur
n veerschotel van de selector
p miseczka sprężyn selektora

4367 self-adhesive tape
d Selbstklebeband *n*
f ruban *m* autocollant
n plakband; kleefband
p taśma samoklejąca

4368 self-adjusting nut
d selbsteinstellbare Mutter *f*
f écrou *m* autofreiné
n zelfafstellende moer
p nakrętka samonastawna

4369 self-adjusting variable pitch fan
d selbsttätig regulierbarer Verstellflügellüfter *m*
f ventilateur *m* à pales autoréglables
n ventilator met zelfinstelbare schoepen
p wentylator o łopatkach samonastawnych

4370 self-aligning roller bearing
d Pendelrollenlager *n*
f roulement *m* à rouleaux autocentreurs
n zelfstellend rollager
p łożysko wałeczkowe samonastawne

4371 self-checking
d selbstkontrollierend
f autocontrôlant
n zichzelf controlerend
p samokontrolujący się

4372 self-cleaning
d Selbstreinigung *f*
f autonettoyage *m*
n zelfreiniging
p samooczyszczanie

4373 self-discharge
d Selbstentladung *f*
f décharge *f* spontanée
n zelfontlading
p samowyładowywanie

4374 self-energizing brake
d selbststärkende Bremse *f*
f frein *m* à renforcement automatique
n zelfbekrachtigende rem
p hamulec samowzmacniający

4375 self-gripping clamp
d selbstschliessende Klemme *f*
f pince *f* autoserreuse
n zelfsluitende klem
p zacisk samozaciskający się

4376 self-gripping puller
d Aussenabzieher *m*
f extracteur *m* autoserreur

n buitentrekker
p ściągacz do łożysk

4377 self-gripping swivel clamp
d selbstschliessende Drehklemme *f*
f pince *f* autoserreuse orientable
n zelfsluitende draaibare klem
p zacisk obrotowy samozaciskający się

4378 self-holding taper
d selbsthemmender Kegel *m*
f cône *m* d'autofreinage
n zelfklemmende conische pen
p stożek samonastawny

4379 self-ignition
d Selbstzündung *f*
f autoallumage *m*
n zelfontbranding
p zapłon samoczynny

4380 self-levelling headlamp device
d automatische Scheinwerferbetätigung *f*
f phare *m* à commande automatique de niveau
n meesturende koplampen
p urządzenie automatycznego wyrównywania reflektorów

4381 self-loading truck
d Selbstladewagen *m*
f camion-autochargeur *m*
n zelflader
p samochód samozaładowczy

4382 self-lubricating bearing
d selbstschmierendes Lager *n*
f palier *m* autograisseur
n zelfsmerend lager
p łożysko samosmarujące

4383 self-protecting screw
d Schraube *f* für Selbstsicherung
f vis *f* autofreinée
n borgmoer
p śruba samozabezpieczająca

* self-tapping screw → 3673

4384 semiconductor
d Halbleiter *m*
f semi-conducteur *m*
n halfgeleider
p półprzewodnik

4385 semiconductor device
d Halbleiterbauelement *n*
f dispositif *m* à semi-conducteurs

n halfgeleiderapparaat
p przyrząd półprzewodnikowy

4386 semidry friction
d Grenzreibung *f*
f frottement *m* mixte
n slipkoppeling
p tarcie półsuche

4387 semielliptic flat spring
d halbelliptische Blattfeder *f*
f ressort *m* demi-pincette
n halfelliptische bladveer
p resor piórowy półepileptyczny

4388 semigloss
d Seidenglanz *m*
f satinée *f*
n zijdeglans
p połysk jedwabisty

4389 semiheavy oil
d Halbschwieröl *n*
f huile *f* semi-fluide
n halfzware olie
p olej półciężki

4390 semitrailer
d Sattelschlepperanhänger *m*
f semi-remorque *f*
n oplegger
p naczepa

4391 semitrailer van
d Kastenauflieger *m*
f fourgon *m* semi-remorque
n gesloten oplegger
p naczepa furgon

4392 semitrailing arm
d Schräglenker *m*
f bras *m* diagonal
n dwarse geleiderarm
p wahacz skośny

4393 sensor
d Messgeber *m*
f témoin *m*
n sensor
p czujnik

4394 sensor plate switch
d Luftmesserkontakt *m*
f contacteur *m* de plateau-sonde
n schakelaar van voelerplaat
p styk przysłony powietrza

4395 separate *v*
d trennen

f séparer
n scheiden
p oddzielać

4396 separator ring
  d Stützring *m*
  f anneau *m* de renfort
  n versterkingsring
  p pierścień usztywniający

4397 serial equipment
  d Serienausstattung *f*
  f équipement *m* en série
  n serie-uitrusting
  p wyposażenie seryjne

4398 serial number
  d Seriennummer *f*
  f numéro *m* de la série
  n serienummer
  p numer seryjny

4399 series connection
  d Reihenschaltung *f*
  f montage *m* en série
  n serieschakeling
  p połączenie szeregowe

4400 serpentine
  d Serpentinenstrasse *f*
  f lacet *f*
  n bochtige weg
  p serpentyna

4401 serrated
  d gekerbt
  f cannelé
  n getand; zaagvormig
  p wielorowkowy; wielokarbowy

4402 serrated joint
  d Zahnwellenverbindung *f* mit Kerbzahnprofil
  f cannelures *fpl* à flancs triangulaires
  n getande spieverbinding
  p połączenie wielokarbowe

4403 serrated plate
  d gekerbte Platte *f*
  f plateau *m* cannelé
  n gegroefde plaat
  p płyta karbowana; płyta wielorowkowa

4404 service brake
  d Betriebsbremse *f*
  f frein *m* à pied
  n bedrijfsrem
  p hamulec główny

4405 service instruction; instruction book
  d Bedienungsanweisung *f*; Betriebsanleitung *f*
  f notice *f* d'emploi; mode *m* d'emploi; notice *f* technique
  n instructieboekje
  p instrukcja obsługi

4406 service life
  d Lebensdauer *f*
  f durée *f* de vie
  n gebruiksduur
  p trwałość; okres użytkowania

4407 service station
  d Servicestation *f*
  f station-service *f*
  n benzinestation
  p stacja benzynowa

4408 servo brake
  d Servobremse *f*
  f servofrein *m*
  n servorem
  p serwohamulec; hamulec ze wspomaganiem

4409 servo cover
  d Verstärkerdeckel *m*
  f carter *m* postérieur de servo
  n achterste gedeelte van servohuis
  p pokrywa siłownika

4410 servo shell
  d Verstärkergehäuse *n*
  f carter *m* de servo
  n servohuis
  p obudowa siłownika

4411 servo steering; power steering
  d Servolenkung *f*
  f direction *f* assistée
  n stuurbekrachtiging
  p wspomagany układ kierowniczy

  * **set** → 280

  * **set of blades** → 526

4412 set screw
  d Stellschraube *f*
  f vis *f* de réglage
  n stelschroef
  p wkręt dociskowy; śruba regulująca

4413 shackle joint
  d Laschengelenk *n*
  f accouplement *m* à plaquettes
  n schalmverbinding
  p połączenie sworzniowe

4414 shaft
d Welle *f*
f arbre *m*
n as
p wał

4415 shaft arm
d Betriebsbremseumstellhebel *m*
f levier *m* de renvoi
n asarm
p ramię wału

4416 shaft bearing; journal bearing
d Wellenlager *n*
f palier *m* d'arbre
n taplager
p łożysko główne wału

4417 shaft butt
d Wellenstumpf *m*
f bout *m* d'arbre
n asstomp
p czop końcowy wału

4418 shaft collar
d Wellenbund *m*
f collet *m* d'arbre
n askraag; asborst
p wieniec oporowy wału

4419 shaft coupling
d Wellengelenk *n*
f joint *m* d'arbre
n askoppeling
p przegub wału

4420 shaft yoke
d Endstück *n* der Gelenkwelle
f embout *m* d'arbre
n eindstuk van de cardanas
p końcówka wału

4421 shake *v*
d zittern
f brouter
n trillen
p trząść

4422 shape of combustion chamber
d Brennraumform *f*
f forme *f* de la chambre de combustion
n vorm van verbrandingsruimte
p kształt komory spalania

4423 shape of gear tooth
d Zahnprofil *n*
f profil *m* de dent
n tandprofiel
p kształt uzębienia

4424 shaping
d Formgebung *f*
f profilage *m*
n vormgeving
p kształtowanie; formowanie

4425 shaping dolly
d Konturenhandfaust *f*
f tas *m* américain plat
n Amerikaanse tas plat
p płaskie klepadło do formowania

4426 sharp turn
d enge Kurve *f*; Steilkurve *f*
f virage *m* aigu
n scherpe bocht
p ostry zakręt

4427 shear force
d Scherkraft *f*
f force *f* de cisaillement
n dwarskracht
p siła ścinająca; siła tnąca

4428 shearing pin
d Abscherstift *m*
f goupille *f* de cisaillement
n breekpen
p kołek zabezpieczający ścinany

4429 shear pin
d Anschlagstift *m*; Sperrstift *m*
f goupille *f*
n borgpen
p klin ścinany; kołek ścinany

4430 shear strength
d Scherfestigkeit *f*
f résistance *f* au cisaillement
n afschuifsterkte
p wytrzymałość na ścinanie

4431 sheeting
d Blechverkleidung *f*
f blindage *m*
n bekleding met plaatstaal
p osłona blaszana

4432 sheet metal
d Blech *n*
f tôle *f*
n staalplaat
p blacha

4433 shelf
d Ablageplatte *f*

f tablette *f*
n plank
p półka

4434 shell bit; gouge bit
  d Löffelbohrer *m*
  f mèche *f* à cuiller
  n lepelboor
  p świder łyżkowy

4435 shield
  d Abdeckplatte *f*
  f plaque *f* de couverture
  n afdekplaat
  p płyta ochronna

4436 shielded wire; screened wire
  d Abschirmleitung *f*
  f fil *m* blindé
  n afgeschermde draad
  p drut osłonięty

  * **shift bar** → **4440**

4437 shift bar end
  d Schaltstangenendstück *n*
  f embout *m* de tige
  n eind van schakelstang
  p końcówka cięgna

4438 shifting
  d Umschalten *n*
  f changement *m*
  n overschakelen
  p przełączanie; przełączać

4439 shifting gate
  d Schaltkulisse *f*
  f secteur *m* du levier de changement de vitesse
  n schakelsjabloon ter "geleiding" van versnellingspook
  p jarzmo przełączenia przekładni

  * **shifting spanner** → **70**

4440 shift rod; shift bar
  d Schaltstange *f*
  f tige *f* fixe
  n schakelstang
  p cięgno sztywne

4441 shift valve
  d Schaltventil *n*
  f vanne *f* de passage
  n schakelklep
  p zawór przełączania

4442 shimming pliers
  d Zange *f* für Ventilkopf
  f pince *f* pour déposer les pastilles des poussoirs
  n tang voor klepschotels
  p szczypce do grzybka zaworu

4443 shock
  d Stoss *m*
  f choc *m*
  n schok
  p wstrząs

4444 shock absorber; damper
  d Stossdämpfer *m*
  f amortisseur *m* de choc
  n schokdemper
  p amortyzator

4445 shock absorber bracket
  d Stossdämpferbock *m*
  f support *m* d'amortisseur
  n schokdempersteun
  p wspornik amortyzatora

4446 shock absorber casing
  d Stossdämpfergehäuse *n*
  f boîtier *m* d'amortisseur
  n schokdemperhuis
  p obudowa amortyzatora

4447 shock absorber lever
  d Stossdämpferhebel *m*
  f levier *m* d'amortisseur
  n schokdemperhefboom
  p ramię amortyzatora

4448 shock absorber piston
  d Stossdämpferkolben *m*
  f piston *m* d'amortisseur
  n schokdemperzuiger
  p tłok amortyzatora

4449 shock absorber suspension
  d Stossdämpferaufhängung *f*
  f suspension *f* d'amortisseur
  n schokdemperophanging
  p zawieszenie amortyzatora

4450 shock absorber tester
  d Stossdämpfertester *m*
  f banc *m* d'essai pour amortisseurs
  n schokdempertestbank
  p urządzenie do kontroli amortyzatorów

4451 shock absorption; damping
  d Federung *f*
  f amortissement *m*

n vering
p amortyzacja

4452 shock damper fluid
d Stossdämpferöl *n*
f huile *f* d'amortisseur
n schokdemperolie
p olej amortyzatorowy

4453 shock power screwdriver
d Schlagschrauber *m*
f tournevis *m* à frapper
n slagmoersleutel
p wkrętak udarowy

4454 shoe brake hydromechanical expander
d hydromechanische Bremsspannvorrichtung *f*
f expandeur *m* hydromécanique de frein
n hydromechanische remschoen
p rozpieracz hamulcowy hydromechaniczny

4455 shoe tappet
d Druckbolzen *m*
f poussoir *m* de mâchoire
n nok van remschoen
p popychacz szczęki hamulcowej

* **shoptruck** → **3361**

4456 short circuit
d Kurzschluss *m*
f court-circuit *m*
n kortsluiting
p zwarcie; spięcie

4457 short stroke engine
d Kurzhubmotor *m*
f moteur *m* supercarré
n korteslagmotor
p silnik krótkoskokowy

4458 shoulder
d Schulter *f*
f épaulement *m*
n borst
p występ

4459 shoulder belt; diagonal seat belt
d Schulterschräggurt *m*
f ceinture *f* boudrier
n schoudergordel; diagonaalgordel
p przekątny pas ramieniowy

4460 shoulder nipple
d Doppelnippel *m*
f raccord *m* à double filetage
n dubbelnippel
p złączka podwójna

4461 shrink *v*
d schwinden
f contracter
n krimpen
p kurczyć się

4462 shrinkage
d Schrumpfung *f*
f retrait *m*
n krimp
p kurczenie się

4463 shrinking bit
d Mundstück *n* für Schrimpfen
f embout *m* de retreint
n mondstuk voor krimpen
p nasadka do kurczenia się

4464 shrinking hammer
d Spann- und Schlichthammer *m*
f marteau *m* à rétreindre
n krimphamer
p młotek do kurczenia

4465 shrink ring
d Schrumpfring *m*
f anneau *m* de frettage
n krimpring
p pierścień skurczowy

4466 shunt dynamo
d Nebenschlussdynamo *n*
f dynamodérivation *f*
n shunt-dynamo
p prądnica bocznikowa

4467 shunt off valve
d Absperrventil *n*
f obturateur *m*
n afsluitklep
p zawór zaporowy

4468 shunt wound motor
d Nebenschlussmotor *m*
f moteur *m* excité en dérivation
n shunt-motor
p silnik bocznikowy

4469 side and end cutting chisel
d Trennmeissel *m*
f burin *m* à coupe latérale et en bout
n beitel met zijlemmet
p przecinak o ostrzu niesymetrycznym

4470 side bearing fit
d Flankenzentrierung *f*
f centrage *m* sur flancs

n flankcentrering
p pasowanie prowadnicy

4471 side board
d Bordwand *f*
f ridelle *f*
n zijschot
p boczna ściana skrzyni

4472 side cutter; disc milling cutter
d Scheibenfräser *m*
f fraise *f* latérale
n schijffrees
p frez tarczowy; frez krążkowy

4473 side cutting pliers
d Seitenschneider *m*
f pince *f* coupante de côté
n zijkniptang
p szczypce z bocznymi nożami

* **side dump truck → 4484**

4474 side electrode; lateral electrode
d Seitenelektrode *f*
f électrode *f* latérale
n massaelektrode
p elektroda boczna

4475 side force
d Seitenzug *m*
f force *f* latérale
n zijkracht
p siła boczna

4476 side gear sliding ring
d Gleitring *m* des Antriebskegelrades
f anneau *m* coulissant de pignon planétaire
n glijring van neventandwiel
p pierścień ślizgowy koła koronowego

4477 side lamp control light
d Begrenzungsleuchtenanzeige *f*
f témoin *m* de feux de position
n controlelamp voor parkeerlicht
p lampa kontrolna świateł pozycyjnych

4478 side lamp weld
d Flankenkehlnaht *f*
f soudure *f* latérale
n flanklas
p spoina pachwinowa wzdłużna

4479 side marker light
d Peilstableuchte *f*
f feu *m* d'indicateur d'encombrement
n markeringslicht
p światła gabarytowe

4480 side outline
d Seitensicht *f*
f encombrement *m* en vue de côté
n zijaanzicht; aanzicht van opzij
p obrys podłużny; obrys pionowy

4481 side panel
d Seitendeckel *m*
f panneau *m* latéral
n zijpaneel
p pokrywa boczna; płat boczny

4482 side shield
d Fussraumseitenschutz *m*
f revêtement *m* latéral de repose-pied
n zijbekleding van de voetenruimte
p osłona boczna podnóżka

4483 side sill
d Trittbrett *n*
f seuil *m*
n drempel
p próg

4484 side tipper; side dump truck
d Seitenkipper *m*
f chariot *m* basculant latéralement
n zijkiepauto
p wywrotka boczna

4485 side valve
d seitlich angeordnetes Ventil *n*
f soupape *f* latérale
n zijklep
p zawór stojący; zawór dolny

4486 side valve engine
d untergesteuerter Motor *m*
f moteur *m* à soupapes latérales
n zijklepmotor
p silnik dolnozaworowy

4487 side whip
d Axialschlag *m*
f voile *m*
n zijwaartse schokbelasting
p bicie toczne

4488 side window
d Seitenfenster *n*
f baie *f* latérale
n zijruit
p okno boczne

4489 sieve
d Sieb *n*
f tamis *m*

n zeef
p sito

4490 signpost
d Wegweiser *m*
f poteau *m* indicateur
n wegwijzer
p drogowskaz

4491 silencer
d Geräuschdämpfer *m*
f amortisseur *m* de bruit
n luchtfilterdemper
p wkład tłumika szmerów ssania

4492 silencer flexible mounting
d elastische Schalldämpferaufhängung *f*
f suspension *f* élastique de silencieux
n elastische ophangstrip voor knaldemper
p elastyczny wieszak tłumika

* **silencer tube** → 2923

4493 silica gel
d Kieselgel *n*
f gel *m* de silice
n silicagel
p żel krzemionkowy; silikażel

4494 silicone
d Silikon *n*
f silicone *f*
n silicone
p silikon

4495 silicone varnish
d Silikonlack *m*
f vernis *m* silicone
n siliconelak
p lakier silikonowy

4496 simplex brake
d Simplexbremse *f*
f frein *m* simplex
n simplex-rem
p zwykły hamulec szczękowy; hamulec Simplex

4497 single acting hydraulic cylinder
d einfach wirkender Hydrozylinder *m*
f cylindre *m* hydraulique à simple effet
n enkelwerkende hydraulische cilinder
p cylinder hydrauliczny jednostronnego działania

4498 single arm lever
d einarmiger Hebel *m*
f levier *m* à un seul bras

n eenarmige hefboom
p dźwignia jednoramienna

4499 single bevel weld
d halbe V-Naht *f*
f soudure *f* à V asymétrique
n halve V-naad
p spoina czołowa na V

4500 single caliper disc brake
d Sattelscheibenbremse *f*
f frein *m* à disque à étrier fixe
n schijfrem met vast remzadel
p jednozaciskowy hamulec tarczowy

4501 single core cable
d einadriges Kabel *n*
f câble *m* unipolaire
n eenaderige kabel
p kabel jednożyłowy

4502 single cylinder engine
d Einzylindermotor *m*
f moteur *m* monocylindre
n eencilindermotor
p silnik jednocylindrowy

4503 single direction ball thrust bearing
d einseitig wirkendes Axialrillenkugellager *n*
f roulement *m* de butée à une rangée de billes
n enkelwerkend drukkogellager
p łożysko kulkowe jednostronnie obciążalne

4504 single hole nozzle
d Einlochdüse *f*
f injecteur *m* simple dissymétrique
n eengatsverstuiver; enkelgatsverstuiver
p rozpylacz jednootworkowy; rozpylacz niesymetryczny

4505 single piston brake cylinder
d einfach wirkender Bremszylinder *m*
f cylindre *m* hydraulique à simple effet
n eenplunjerremcilinder
p rozpieracz hamulcowy jednostronny

4506 single plate clutch
d Einscheibenkupplung *f*
f embrayage *m* monodisque
n enkelschijfswrijvingskoppeling
p sprzęgło jednotarczowe

4507 single riveted joint
d einreihige Nietverbindung *f*
f rivure *f* à une rangée
n enkel klinknagelverband
p szew nitowy jednorzędowy

4508 single row bearing
  d einreihiges Wälzlager *n*
  f roulement *m* à une rangée
  n eenrijig wentellager
  p łożysko jednorzędowe

4509 single spring clutch; central spring clutch
  d Zentralfederkupplung *f*;
   Hauptfederkupplung *f*
  f embrayage *m* à ressort central
  n centrale veerkoppeling
  p sprzęgło o sprężynie centralnej

4510 single start thread
  d eingängiges Gewinde *n*
  f filet *m* au pas unique
  n enkelvoudige schroefdraad
  p gwint pojedynczy; gwint jednozwojny

4511 single step
  d einstufig
  f simple
  n eentraps
  p jednostopniowy

4512 single throw crankshaft
  d einfach gekröpfte Kurbelwelle *f*
  f arbre *m* coudé
  n krukas met één kruk
  p wał jednokorbowy

4513 single U weld
  d U-Naht *f*
  f soudure *f* à U
  n kelknaad
  p spoina czołowa na U; spoina jednostronna

4514 single V weld
  d V-Naht *f*
  f soudure *f* à V
  n open V-naad
  p spoina czołowa na V

4515 sintered steel
  d Sinterstahl *m*
  f acier *m* fritté
  n sinterstaal
  p stal spiekana

4516 skeleton type body
  d Skelettaufbau *m*
  f carrosserie *f* à ossature
  n kooiconstructie
  p karoseria samonośna

4517 skew spanner
  d Schlüssel *m* mit schrägem Maul

  f clé *f* coudée
  n steeksleutel met scheve bek
  p klucz skośny

4518 skid mark
  d Bremsspur *f*
  f trace *f* de freinage
  n remspoor
  p ślad hamowania

4519 skid pad
  d Schleuderpiste *f*
  f piste *f* de dérapage
  n slipbaan
  p bieżnia ślizgowa

4520 skid row
  d Gleiten *n*
  f dérapage *m*
  n slipgevaar
  p poślizg

4521 ski rack
  d Skiträger *m*
  f porte-skis *m*
  n skidrager; ski-imperiaal
  p belka na narty

4522 slave cylinder piston
  d Kolben *m* des Ausrückzylinders
  f piston *m* de cylindre récepteur
  n zuiger van de ontkoppelingscilinder
  p tłok siłownika wyprzęgającego

4523 slave cylinder push rod
  d Kolbenstange *f* des Ausrückzylinders
  f poussoir *m* de piston de cylindre récepteur
  n zuigerstang van de ontkoppelingscilinder
  p popychacz siłownika wyprzęgającego

4524 slave cylinder return spring
  d Rückzugfeder *f* des Ausrückzylinders
  f ressort *m* de rappel de cylindre récepteur
  n trekveer van de ontkoppelingscilinder
  p sprężyna siłownika wyprzęgającego

4525 sleeve; socket
  d Muffe *f*; Tülle *f*
  f manchon *m*; douille *f*
  n huls; mof
  p tuleja; mufa

4526 sleeve clutch
  d Hülsenkupplung *f*
  f accouplement *m* à manchon
  n koppeling met koppelbus
  p sprzęgło tulejowe

4527 sleeve insert
 d Hülseneinlage *f*
 f fourrure *f* de douille
 n inzetstuk voor huls
 p wkładka tulei

4528 sleeve nut
 d Hülsenmutter *f*
 f écrou *m* à douille
 n hulsmoer
 p nakrętka tulejowa

4529 sleeve valve engine
 d Schiebermotor *m*
 f moteur *m* à distribution par chemises
 n schuivenmotor
 p silnik suwakowy

4530 sleeve yoke; splined end
 d Vielkeilgelenkgabel *f*
 f chape *f* coulissante; manchon *m* coulissant
 n schuifstuk
 p rozwidlone złącze wielowypustowe

4531 slide face
 d Gleitfläche *f*
 f surface *f* de glissement
 n glijvlak
 p płaszczyzna ślizgowa

4532 slide hammer
 d Gleithammer *m*
 f masse *f* à inertie
 n slagtrekker
 p młotek przesuwny

4533 slide hammer hub puller
 d Schlagnabenabzieher *m*
 f extracteur *m* de moyeux à inertie
 n slagnaaftreker
 p młotek do ściągaczy piast kół

4534 slide hammer valve lifter
 d Ventilfederzange *f* mit Schlagmechanismus
 f lève-soupape *m* à inertie
 n klepveertang met slagmechanisme
 p szczypce do sprężyny zaworu z mechanizmem uderzeniowym

4535 slider
 d Schieber *m*
 f coulisseau *m*
 n schuif
 p suwak

4536 slider coupling
 d Kreuzscheibenkupplung *f*
 f accouplement *m* à glissières
 n kruisschijfkoppeling
 p sprzęgło odsuwne

4537 slide valve
 d Schieberventil *n*
 f soupape *f* coulissante
 n schuifklep
 p zawór suwakowy

4538 sliding air piston carburetter
 d Steuerkolbenvergaser *m*
 f carburateur *m* à piston de commande
 n doseercarburateur
 p gaźnik suwakowy

4539 sliding bearing; plain bearing
 d Gleitlager *n*
 f palier *m* lisse; palier *m* à glissement
 n glijlager
 p łożysko ślizgowe

4540 sliding caliper
 d Schublehre *f*
 f pied *m* à coulisse
 n schuifmaat
 p suwmiarka

4541 sliding caliper housing
 d Schwimmbremssattelkörper *m*
 f corps *m* d'étrier flottant
 n zwevend remzadelhuis
 p korpus siodła

 * **sliding conical disc** → **3600**

 * **sliding coupling** → **4545**

4542 sliding door
 d Schiebetür *f*
 f porte *f* à glissière; portière *f* glissante; portière *f* à glissière
 n schuifdeur
 p drzwi przesuwne

4543 sliding fit
 d Gleitsitz *m*
 f ajustement *m* glissant
 n schuifpassing
 p pasowanie suwliwe

4544 sliding guide
 d Gleitführung *f*
 f glissière *f* de guidage
 n glijdende geleiding
 p prowadnica ślizgowa

4545 sliding joint; sliding coupling
  d Schiebegelenk *n*
  f joint *m* coulissant
  n schuifkoppeling
  p połączenie przesuwne

4546 sliding plate
  d Gleitplatte *f*
  f plaque *f* de glissement
  n slijtplaat
  p płyta przesuwna

4547 sliding roof; sunshine roof; sunshade roof
  d Sonnendach *n*; Schiebedach *n*
  f toit *m* ouvrant
  n schuifdak; open dak
  p dach rozsuwany

4548 sliding sleeve
  d Schiebehülse *f*
  f douille *f* de coulissement
  n schakelmof
  p przesuwka

4549 sliding window
  d Schiebefenster *n*
  f fenêtre *f* coulissante
  n schuifraam
  p okno przesuwne

4550 slipper oil pump
  d Gleitsteinölpumpe *f*
  f pompe *f* à huile à segments
  n slipperoliepomp
  p ślizgaczowa pompa oleju

4551 slippery road
  d glatte Fahrbahn *f*
  f chaussée *f* glissante
  n gladde weg
  p śliska jezdnia

4552 slip ring
  d Schleifring *m*; Gleitring *m*
  f anneau *m* de coulissement
  n sleepring
  p pierścień ślizgowy

4553 slip universal joint
  d Gleitkreuzgelenk *n*
  f cardan *m* à joint coulissant
  n slipkruiskoppeling
  p uniwersalne połączenie przesuwne

4554 slip velocity
  d Schlupfgeschwindigkeit *f*
  f vitesse *f* de glissement

  n slipsnelheid
  p prędkość poślizgu; prędkość ślizgania się

4555 slit skirt piston
  d Schlitzmantelkolben *m*
  f piston *m* à jupe fendue
  n zuiger met gespleten mantel
  p tłok z rozciętym płaszczem

4556 slot cutter; groove milling tool
  d Nutenfräser *m*
  f fraise *f* à rainer
  n sleuffrees
  p frez tarczowy do rowków

4557 slotted cheese head screw
  d Zylinderkopfschraube *f*
  f vis *f* à tête fendue cylindrique
  n vlakkopschroef
  p wkręt o płaskim łbie walcowym

  * slotted inner joint → 2119

4558 slotted round head screw
  d Halbrundkopfschraube *f*
  f vis *f* à tête fendue ronde
  n bolkopschroef
  p wkręt o łbie kulistym

4559 slotted round nut
  d Nutmutter *f*
  f écrou *m* cylindrique à fente
  n cilindrische sleufmoer
  p nakrętka okrągła rowkowa

  * slow running jet → 2780

4560 slow running stop screw
  d Leerlaufbegrenzungsschraube *f*
  f vis *f* de butée de ralenti
  n aanslagschroef; stopschroef
  p wkręt oporowy

4561 smooth v
  d glätten
  f polir
  n polijsten
  p gładzić

  * smooth finishing → 1745

  * snap gauge → 842

  * snap ring → 1049

  * snorkel tube → 2923

4562 snow blower; rotary snow plough
  d Schneeschleuderwagen *m*
  f chasse-neige *m* rotatif
  n sneeuwblazer
  p odśnieżarka wirnikowa

4563 snug bolt
  d Nasenbolzen *m*
  f boulon *m* à ergot
  n pasbout
  p sworzeń dociskowy

4564 snug washer
  d Nasenscheibe *f*
  f rondelle *f* à ergot
  n pasvulring
  p podkładka nosowa

  * socket → 4525

4565 socket dispenser
  d Ausstellungselement *n* für Steckschlüssel
  f présentoirs *mpl* douilles
  n hulsapparaat
  p końcówka do założenia klucza nasadowego

4566 sodium base grease
  d sodahaltiges Schmierfett *n*
  f graisse *f* consistante au sodium
  n natriumhoudend consistentvet
  p smar sodowy

4567 softener
  d Weichmacher *m*
  f plastifiant *m*
  n weekmaker
  p zmiękczacz; plastyfikator

4568 softness of a spring
  d Federweichheit *f*
  f mollesse *f* de ressort
  n zachtheid van een veer
  p miękkość sprężyny

  * soft plug → 2720

4569 soft rubber
  d Weichgummi *m*
  f gomme *f* molle
  n zacht rubber
  p guma miękka

4570 solder *v*
  d löten
  f souder
  n solderen
  p lutować

4571 soldering iron
  d Lötkolben *m*
  f fer *m* à souder
  n soldeerbout
  p lutownica

4572 solenoid type starter
  d Magnetschalteranlasser *m*
  f démarreur *m* à solénoïde
  n startmotor met elektromagnetische schakelaar
  p rozrusznik włączony przekaźnikiem

4573 solid fuel
  d Festkraftstoff *m*
  f combustible *m* solide
  n vaste brandstof
  p paliwo stałe

4574 solvent
  d Lösungsmittel *n*
  f solvant *m*
  n oplosmiddel
  p rozpuszczalnik

4575 soot
  d Russ *m*
  f suie *f*
  n roet
  p sadza

4576 sort of oil
  d Ölsorte *f*
  f qualité *f* d'huile
  n oliesoort
  p rodzaj oleju

4577 sound insulating material
  d Dämmstoff *m*
  f isolant *m* phonique
  n isolatiemateriaal
  p materiał izolacji dźwięków

4578 sound level
  d Schallpegel *m*
  f niveau *m* sonore
  n geluidsniveau
  p poziom głośności

4579 sound level meter
  d Schallpegelmesser *m*
  f sonomètre *m*
  n geluidsniveaumeter
  p sonometr; miernik poziomu głośności

4580 souped up engine
  d frisierter Motor *m*
  f moteur *m* gonflé

   n opgefokte motor
   p silnik zatarty

4581 source of electric power
   d Stromquelle *f*
   f source *f* de courant
   n stroombron
   p źródło prądu

4582 space between twin wheels
   d Felgenabstand *m*
   f écartement *m* des jantes
   n hartafstand
   p odstęp między obręczami kół

4583 space preservation
   d Hohlraumkonservierung *f*
   f conservation *f* des parties creuses
   n conservering van holle ruimtes
   p konserwacja przestrzeni zamkniętych

4584 spacer
   d Abstandsstück *n*
   f entretoise *f*
   n vulstuk
   p element dystansowy

4585 spacer ring
   d Abstandring *m*
   f anneau *m* entretoise
   n afstandsring
   p pierścień rozstawczy

4586 span saw; frame saw
   d Spannsäge *f*
   f scie *f* à châssis
   n spanzaag
   p piła ramowa

4587 spare battery
   d Ersatzbatterie *f*
   f batterie *f* de réserve
   n reservebatterij
   p baterie zapasowe

4588 spare bulb
   d Ersatzbirne *f*
   f lampe *f* de rechange
   n reservelamp
   p żarówka zapasowa

4589 spare part
   d Ersatzteil *n*
   f pièce *f* de rechange
   n reserveonderdeel
   p część zamienna; część zapasowa

4590 spare parts number
   d Ersatzteilnummer *f*

   f numéro *m* de pièces de rechange
   n reserveonderdeelnummer
   p numer części zamiennej; numer części zapasowej

4591 spare wheel
   d Ersatzrad *n*
   f roue *f* de secours
   n reservewiel
   p koło zapasowe

4592 spare wheel carrier
   d Ersatzradhalter *m*
   f support *m* de roue de secours
   n reservewielsteun
   p podpora koła zapasowego

4593 spark
   d Funke *f*
   f étincelle *f*
   n vonk
   p iskra

4594 spark advance selector
   d Zündpunktversteller *m*
   f correcteur *m* d'avance à l'allumage
   n ontstekingsversteller
   p nastawnik wyprzedzenia zapłonu

4595 spark arrester
   d Funkenfänger *m*
   f grille *f* de flammèches
   n vonkvanger
   p chwytacz iskier

4596 spark formation
   d Funkenbildung *f*
   f formation *f* d'étincelles
   n vonkvorming
   p iskrzenie

   * **spark gap** → 1908

4597 spark ignition
   d Funkenzündung *f*
   f allumage *m* par bougie
   n vonkontsteking
   p zapłon iskrowy

4598 sparking plug wrench
   d Zündkerzenschlüssel *m*
   f clé *f* de bougie
   n bougiesleutel
   p klucz nasadowy do świec

4599 sparking voltage
   d Überschlagspannung *f*

f tension *f* d'étincelles
n overslagspanning
p napięcie przeskoku iskry

4600 spark patch
d Funkenstrecke *f*
f voie *f* d'étincelle
n vonktraject
p tor iskry

4601 spark plug
d Zündkerze *f*
f bougie *f* d'allumage
n bougie
p świeca zapłonowa

4602 spark plug brush
(brush to clean electrodes and threads)
d Zündkerzenbürste *f*
f brosse *f* à bougie
n bougieborstel
p szczotka do czyszczenia świec zapłonowych

4603 spark plug cable terminal
d Kerzenkabelanschluss *m*
f prise *f* de câble de bougie
n contactbus voor bougiekabel
p gniazdo wysokiego napięcia

4604 spark plug engine
d Vergasermotor *m*
f moteur *m* à carburateur
n motor met carburateur
p silnik z zapłonem iskrowym

4605 spark plug heat range
d Wärmewert *m* der Zündkerze
f degré *m* thermique de bougie
n warmtegraad van bougie
p wartość cieplna świecy

4606 spark plug hole
d Zündkerzensitz *m*
f puits *m* de bougie d'allumage
n bougiegat
p gniazdo świecy zapłonowej

4607 spark plug insulator
d Isolierkörper *m*
f isolant *m* de bougie
n isolator van bougie
p izolator świecy zapłonowej

4608 spark plug shell
d Zündkerzengehäuse *n*
f corps *m* de bougie
n bougiehuis
p kadłub świecy

4609 spark plug socket spanner
d Zündkerzensteckschlüssel *m*
f douille *f* pour bougie
n bougiesleutel
p klucz do świecy zapłonowej

4610 spark plug terminal pin
d Zündbolzen *m*
f tige *f* centrale de bougie
n eindstuk van bougie
p rdzeń świecy

4611 spark plug tester
d Zündkerzenprüfer *m*
f contrôleur *m* de bougies
n bougiespanningzoeker
p przyrząd do pomiaru napięcia świecy
       zapłonowej

4612 spark plug tube
d Zündkerzenschacht *f*
f tube *m* de bougie
n bougiehuls
p tuleja świecy zapłonowej

4613 spark test
d Funkenprobe *f*
f épreuve *f* aux étincelles
n vonktest
p próba iskrowa

4614 special equipment
d Sonderzubehör *n*
f équipement *m* spécial
n speciale uitrusting
p wyposażenie specjalne

4615 special tool
d Spezialwerkzeug *n*
f outil *m* spécial
n speciaal gereedschap
p narzędzie specjalne

4616 special transport vehicle
d Sondernutzfahrzeug *n*
f véhicule *m* adapté
n wagen voor speciaal gebruik
p samochód specjalizowany

4617 special type
d Sonderausstattung *f*
f construction *f* spéciale
n bijzondere uitvoering; niet-standaard
       uitvoering
p typ specjalny

4618 specific fuel consumption
d spezifischer Kraftstoffverbrauch *m*

f consommation *f* spécifique de combustible
n specifiek brandstofverbruik
p jednostkowe zużycie paliwa

4619 specific heat
d spezifische Wärme *f*
f chaleur *f* spécifique
n soortelijke warmte
p ciepło właściwe

4620 specific weight
d spezifisches Gewicht *n*
f poids *m* spécifique
n soortelijk gewicht
p ciężar jednostkowy

4621 speed
d Geschwindigkeit *f*
f vitesse *f*
n snelheid
p prędkość

4622 speed change lever knob
d Schalthebelknopf *m*
f pommeau *m* de levier de vitesses
n schakelpookhendel
p gałka dźwigni zmiany biegów

4623 speed governor
d Drehzahlregler *m*
f régulateur *m* de vitesse
n snelheidsregelaar
p regulator prędkości

* **speed indicator** → 4625

* **speed limit** → 3271

4624 speed of revolution
d Umdrehungsgeschwindigkeit *f*
f vitesse *f* de rotation
n omwentelingssnelheid
p prędkość obrotowa

4625 speedometer; speed indicator; tachometer
d Geschwindigkeitsmesser *m*
f compteur *m* de vitesse
n snelheidsmeter
p prędkościomierz; szybkościomierz

4626 speedometer cable
d Geschwindigkeitsmesserantriebskabel *n*
f câble *m* de compteur
n snelheidsmeterkabel
p linka prędkościomierza; linka szybkościomierza

4627 speedometer drive
d Antrieb *m* des Geschwindigkeitsmessers
f prise *f* de commande d'indicateur de vitesse
n aandrijving voor kilometerteller
p napęd szybkościomierza

4628 speedometer reading
d Geschwindigkeitsanzeige *f*
f indication *f* de vitesse
n snelheidsmeter
p znak zalecanej prędkości jazdy

4629 spherical ball joint
d Kugelgelenk *n*
f rotule *f* sphérique
n kogelgewricht
p przegub kulowy

4630 spherical ball pin
d Kugelzapfen *m*
f queue *f* de rotule sphérique
n kogeltap
p sworzeń kulowy

4631 spherical cap
d kugelförmiges Ansatzstück *n*
f flasque *m* cuvette
n balvormig verlengstuk
p nasadka kulista

4632 spherical pressured gas chamber; spring chamber
d sphärische Druckgaskammer *f*
f chambre *f* sphérique à gaz comprimé
n bolvormige persgaskamer
p kulista przestrzeń sprężonego gazu

4633 spherical roller bearing
d Tonnenlager *n*
f roulement *m* sphérique à rouleaux
n tonrollager
p łożysko baryłkowe

4634 spheric pair
d Kugelpaar *n*
f charnière *f* universelle
n kogelscharnier
p para kulista

4635 spider
d Gelenkkreuz *n*; Kreuzstück *n*
f croisillon *m*
n kruisstuk
p krzyżak

4636 spigot inner joint
d Zentrierhalbgelenk *n*
f noix *f* à tenon

n   centreergewricht
p   ślizgacz środkujący

4637 spigot trunnion
d   Tragzapfen *m*
f   tourillon *m*
n   draagtap
p   czop nośny

*   **spigot valve → 1411**

4638 spill deflector
d   Drosselhülse *f*
f   manchon *m* de réflexion
n   centrale deflector
p   tuleja odbijająca

*   **spill gallery → 2111**

4639 spiral final drive; bevel type final drive
d   Spiralachsgetriebe *n*; Kegelradachsgetriebe *n*
f   couple *m* conique; couple *m* à taille spirale
n   schroefvormige eindaandrijving
p   stożkowa przekładnia główna; spiralna przekładnia główna

4640 spiral housing
d   Spiralgehäuse *n*
f   carter *m* en spirale
n   spiraalvormig huis
p   osłona spiralna; korpus spiralny

4641 spiral spring
d   Spiralfeder *f*
f   ressort *m* en spirale
n   spiraalveer
p   sprężyna spiralna

4642 spiral tooth coupling
d   Bogenzahnkupplung *f*
f   accouplement *m* à denture en spirale
n   koppeling met spirale vertanding
p   sprzęgło zębate spiralne

4643 splash apron
d   Schutzblech *n*
f   garde-boue *m*
n   beschermkap
p   osłona blaszana

4644 spline bits
d   mehrfachverzahnte Schraubendrehernüsse *fpl*
f   douilles *fpl* tournevis à denture multiple
n   schroevendraaierdoppen meervoudig vertand
p   nasadki na śrubokręt wielokrotnie uzębione

4645 splined bolt
d   Keilbolzen *m*
f   boulon *m* à clavette
n   keilbout; spiebout
p   czop stożkowy

*   **splined end → 4530**

4646 splined flange sleeve; splined sliding flange
d   Vielkeilbundbuchse *f*; Vielkeilflanschscheibe *f*
f   flasque *m* coulissant
n   overlangs gegroefde flensmanchet
p   kołnierzowe złącze wielowypustowe

*   **splined sliding flange → 4646**

4647 splines
d   Vielkeil *m*
f   cannelure *f*
n   spiebanen
p   wielowypust

4648 spline shaft
d   Keilwelle *f*; Nutenwelle *f*
f   arbre *m* cannelé
n   spieas; as met spiebanen
p   wał o wielopustach

4649 split axle housing
d   geteiltes Achsgehäuse *n*
f   corps *m* de pont à trompettes
n   gespleten achteras
p   dzielona obudowa mostu pędnego

4650 split bearing shell
d   Lagerschale *f*
f   demi-coussinet *m*
n   gedeelde lagerschaal
p   półpanew

4651 split bush
d   Spaltring *m*
f   bague *f* fendue
n   gesleten mof
p   panew dzielona

4652 split housing
d   geteiltes Gehäuse *n*
f   corps *m* en plusieurs pièces
n   gedeeld huis
p   obudowa dzielona

4653 split muff coupling
d   Schalenkupplung *f*
f   accouplement *m* à coquilles
n   deelbare koppelingsbus
p   sprzęgło łubkowe

4654 split pin; cotter pin
 d Splint *m*
 f goupille *f* fendue
 n splitpen
 p nawleczka rozginana

4655 split pin removal set
 d Werkzeugsatz *m* mit
   Scharnierachsenaustreibern
 f composition *f* d'extraction des goupilles
   fendues
 n set scharnierpendrevels
 p komplet narzędzi do usuwania zawleczek

4656 split skirt piston
 d Kolben *m* mit geschlitztem Körper
 f piston *m* à jupe fendue
 n spleetzuiger
 p tłok o przeciętej części prowadzącej

4657 splitting saw
 d Websäge *f*
 f scie *f* à refendre
 n splijtzaag
 p piła wzdłużna

4658 spoiler
 d Spoiler *m*
 f becquet *m*
 n spoiler
 p spoiler

4659 sponge rubber
 d Schaumgummi *n*
 f caoutchouc *m* mousse
 n schuimrubber
 p guma gąbczasta

4660 spontaneous ignition
 d Selbstzündung *f*
 f allumage *m* automatique
 n spontane ontsteking
 p samozapłon; zapłon samoczynny

4661 spoon dolly
 d Löffeleisenhandfaust *f*
 f tas *m* cuillère
 n lepeltas
 p kowadełko ręczne

4662 spoon tool
 d Polierlöffel *m*
 f lissoir *m* à cuiller
 n polijstlepel
 p łyżeczka formierska

4663 spot weld cutter
 d Schweisspunktbohrer *m*

 f fraise *f* pour points de soudure
 n puntlasboor
 p przecinak do rozłączania zgrzein
   punktowych

4664 spot welder
 d Punktschweissgerät *n*
 f pince *f* à souder par points
 n puntlastang
 p zgrzewarka punktowa

4665 spray *v*
 d spritzen
 f projeter
 n spuiten
 p natryskiwać

4666 spray cone
 d Zerstäubungskegel *m*
 f cône *m* de pulvérisation
 n bereik van straal uit verstuiver
 p stożek rozpylania

4667 spray gun; spraying pistol
 d Spritzpistole *f*
 f pistolet *m* de vernissage
 n spuitpistool
 p pistolet natryskowy

 * **spraying equipment** → **4670**

4668 spraying nozzle
 d Spritzdüse *f*
 f gicleur *m*
 n sproeier
 p rozpylacz

 * **spraying pistol** → **4667**

4669 spray painting
 d Spritzlackieren *n*
 f peinture *f* au pistolet
 n spuiten
 p malowanie natryskowe

**4670 spray painting equipment; spraying
   equipment**
 d Farbspritzanlage *f*
 f installation *f* de peinture au pistolet
 n spuitinrichting
 p wyposażenie do malowania natryskowego

4671 spray tip
 d Düseneinsatz *m*
 f buse *f* d'injecteur
 n verstuiverinzetstuk; verstuivermondstuk
 p końcówka wtryskiwacza

4672 **spreader ring**
  d Spreizring *m*
  f anneau *m* d'écartement
  n stelring; afstandsring
  p pierścień rozprężny

4673 **spread jack**
  d Streuungsgerät *n*
  f vérin *m* écarteur
  n spreider
  p rozpórka

4674 **spring**
  d Feder *f*
  f ressort *m*
  n veer
  p sprężyna

4675 **spring assembly bolt**
  d Federverbindungbolzen *m*
  f boulon *m* d'assemblage de ressort
  n torenbout
  p sworznie łączenia sprężynowego

4676 **spring bolt**
  d Federbolzen *m*
  f boulon *m* étoquiau
  n veerbout
  p śruba resoru

4677 **spring bracket**
  d Federhand *f*
  f main *f* du châssis
  n veerhand
  p przedni wspornik resoru

4678 **spring bushing**
  d Federbuchse *f*
  f douille *f* de ressort
  n veerbus
  p tuleja sprężysta

4679 **spring button knob**
  d Federknopf *m*
  f bouton *m* de ressort
  n spanveerzitting
  p kołek sprężyny

4680 **spring catch**
  d Federhaken *m*
  f attache *f* de ressort
  n veerpal
  p zaczep sprężyny

  * **spring chamber** → **4632**

4681 **spring clamp; spring clip; seeger clip**
  d Federklammer *m*

  f bride *f* de ressort
  n veerstrop
  p opaska resoru

  * **spring clip** → **4681**

4682 **spring coil**
  d Federwicklung *f*
  f spire *f* de ressort
  n veerwinding
  p zwój sprężyny

4683 **spring compressor and removal tool body**
  d Federspanner *m* und Zerlegungswerkzeug *n*
  f corps *m* compresseur et démonte *m* ressort
  n veerspanner en demontagegereedschap
  p naciągacz sprężyny z przyrządem demontującym

4684 **spring compressor body**
  d Federspanner *m*
  f corps *m* compresseur de ressort
  n veerspanner
  p naciągacz sprężyny

4685 **spring drive**
  d Federantrieb *m*
  f commande *f* à ressort
  n veeraandrijving
  p napęd na wpust

4686 **spring eye**
  d Federauge *f*
  f œil *m* du rouleau de ressort
  n veeroog
  p ucho resoru

4687 **spring fork**
  d Federgabel *f*
  f fourche *f* à ressort
  n veervork
  p widełki resoru

4688 **spring guide**
  d Gleitfeder *f*
  f ressort *m* de glissement
  n schuifspie; inlegspie
  p prowadzenie sprężyny

  * **spring leg** → **4850**

4689 **spring loaded hydraulic accumulator**
  d Federspeicher *m*
  f accumulateur *m* hydraulique à ressort
  n met drukveer op druk gebrachte hydraulische accumulator
  p akumulator hydrauliczno-sprężynowy

4690 spring lock washer
d Federring *m*
f rondelle *f* à ressort
n verende onderlegring
p nakładka sprężynująca

4691 spring manometer
d Federmanometer *n*
f manomètre *m* de ressort
n veermanometer
p manometr sprężynowy

4692 spring of round wire
d Feder *f* mit Kreisquerschnitt
f ressort *m* à fil rond
n veer uit verenstaaldraad
p sprężyna z drutu okrągłego

4693 spring pad
d Federunterlage *f*
f semelle *f* de ressort
n veerring
p pierścień sprężysty

* **spring perch** → 3047

4694 spring pin
d Federhaken *m*
f crochet *m* à ressort
n spanstift
p kołek sprężynujący

4695 spring plate
d Plakette *f* für Feder
f plaquette *f* de ressort
n veerplaat
p płytka sprężysta

4696 spring plate mechanism
d Federplattenmechanismus *n*
f mécanisme *m* à diaphragme
n veerbladmechaniek
p mechanizm pióra resoru

4697 spring rebound clip
d Federanschlag *m*
f butoir *m* de ressort
n aanslagrubber
p ogranicznik gumowy

4698 spring retaining ring
d Sperring *m*
f anneau *m* de fixation; anneau *m* de
verrouillage
n blokkeerring
p pierścień oporowy

4699 spring ring
d Anlauffederring *m*
f anneau *m* élastique de décollage
n veerring
p pierścień sprężysty

4700 spring seat
d Federteller *m*
f cuvette *f* de ressort
n veerschotel
p miseczka sprężyny

* **spring shackle** → 1722

4701 spring sliding plate
d Blattfedergleitplatte *f*
f plaque *f* de glissement de ressort
n glijplaat van bladveer
p ślizgacz resoru

4702 spring steel
d Federstahl *m*
f acier *m* pour ressorts
n verenstaal
p stal sprężynowa

* **spring strut** → 4850

4703 spring tension
d Federspannung *f*
f tension *f* de ressort
n veerspanning
p napięcie sprężyny

4704 spring tension plate
d Federspannplatte *f*
f patin *m* d'étrier de ressort
n veerspanplaat
p płyta naciągu sprężyny

* **sprocket wheel** → 988

4705 sprung weight
d abgefedertes Gewicht *n*
f poids *m* suspendu
n afgeveerd gewicht
p masa resorowana

4706 spur geared pinion
d Stirnritzel *n*
f pignon *m* cylindrique
n koptandrondsel
p zębnik walcowy

4707 spur gear final drive
d Stirnachsgetriebe *n*
f couple *m* à engrenages parallèles

n eindaandrijving van recht tandwiel
p walcowa przekładnia główna; czołowa
przekładnia główna

4708 spur gear wheel
d Stirnrad *n*
f roue *f* à denture extérieure
n koptandwiel
p koło o uzębieniu zewnętrznym

4709 square anvil
d Kubikamboss *m*
f tas *m* cubique
n vierkant aambeeld
p kowadło sześcienne

4710 square collar head bolt
d Kranzkopfverbindungsschraube *f*
f vis *f* à tête carrée à embase
n bout met vierkante kraagkop
p śruba złączna o łbie wieńcowym

4711 squared end turn; closed end turn
d angelegte Endwindung *f*
f spire *f* fermée de ressort
n vlakgewikkeld veereinde
p koniec skrętu sprężyny przyłożonej

4712 square end set bolt
d Vierkantzapfenkopfschraube *f*
f vis *f* de pression à tête carrée
n bout met vierkante tapkop
p śruba o czopikowym łbie czworokątnym

4713 square face weld
d I-Naht *f*
f soudure *f* sans chanfrein
n I-naad
p spoina czołowa na I

4714 square head bolt
d Vierkantkopfverbindungsschraube *f*
f vis *f* à tête carrée
n bout met vierkante kop
p śruba złączna o łbie czworokątnym

4715 square nut
d Vierkantmutter *f*
f écrou *m* carré
n vierkante moer
p nakrętka czworokątna

4716 squeak *v*
d knirschen
f grincer
n knarsen
p piszczeć

4717 squirrel cage induction motor
d Käfigankermotor *m*; Kurzschlussläufermotor
m
f électromoteur *m* à cage
n kooiankermotor
p silnik klatkowy

4718 stability
d Stabilität *f*
f stabilité *f*
n stabiliteit
p stabilność

4719 stabilizer
d Stabilisator *m*
f stabilisateur *m*
n stabilisator
p stabilizator

4720 stabilizer plies
d Gürtel *m*
f ceinture *f* de pneu; berceau *m* de pneu
n gordel van autoband
p podkład bieżnika

4721 stainless steel
d nichtrostender Stahl *m*
f acier *m* inoxydable
n roestvrij staal
p stal nierdzewna

4722 stain remover
d Fleckentferner *m*
f détachant *m*
n vlekkenverwijderingsmiddel
p środek usuwający plamy; usuwacz plam

4723 stair step iron
d Stufeneisen *n*
f fer *m* à marche
n traptrededeel
p żelazo stopniowane

4724 stamp
d Stempelung *f*
f marquage *m*
n markering
p oznaczenie

4725 standard size
d Nennmass *n*
f cote *f* nominale
n standaardmaat
p norma wymiarowa

4726 standby
d Fahrbereitschaft *f*

f état *m* de marche
n standby
p stan gotowości do jazdy

4727 standing room
d Stehplatz *m*
f place *f* debout
n staanplaats
p miejsce stojące

4728 standing vice
d Schlosserschraubstock *m*
f étau *m* à pied
n vertikale bankschroef
p imadło ślusarskie; imadło zawiasowe

4729 St Andrew's cross
d Andreaskreuz *n*
f croix *f* de Saint André
n Andreaskruis
p krzyż świętego Andrzeja

4730 start clutch
d Andrehklaue *f*
f dent *f* de loup de la manivelle de lancement
n startkoppeling
p zazębiacz korby rozruchowej

4731 start code of antitheft device
d Anlasskode *f* von Diebstahlwarnlage
f antidémarrage *m* codé
n antistartcode van antidiefstalinrichting
p kod startowy urządzenia zabezpieczającego
  przed kradzieżą

4732 starter
d Anlasser *m*; Starter *m*
f démarreur *m*
n startmotor
p rozrusznik

4733 starter armature
d Anker *m*
f induit *m* de démarreur
n anker van startmotor
p wirnik rozrusznika

4734 starter battery; starting battery
d Starterbatterie *f*; Anlassbatterie *f*
f batterie *f* de démarrage
n startaccu
p akumulator rozruchowy

4735 starter button
d Anlassdruckknopf *m*
f bouton *m* de commande de démarreur
n startknop

p przycisk rozrusznika

* **starter cable → 422**

4736 starter drive head
d Polgehäusekopf *m*
f nez *m* de démarreur
n kop van starteraandrijving
p głowica stojana

4737 starter generator
d Lichtanlasser *m*; Startergenerator *m*
f démarreur-dynamo *m*
n startgenerator
p prądnica-rozrusznik

4738 starter inertia mass
d Anlasserschwungrad *n*
f volant *m* d'inertie de lanceur
n startmotorvliegwiel
p koło zamachowe rozrusznika

4739 starter motor control lead
d Anlasserzug *m*
f cordon *m* de démarrage à distance
n kabel voor afstandsbediening van starten
p kontrolka rozrusznika

4740 starter pinion
d Anlasserritzel *n*
f pignon *m* de démarreur
n starterrondsel
p zębnik

4741 starter ring
d Anlasserzahnkranz *m*
f couronne *f* dentée de volant
n starterkrans
p wieniec zębaty koła zamachowego

4742 starter switch
d Anlasserschalter *m*
f contacteur *m* de démarreur
n startschakelaar
p wyłącznik rozrusznika

4743 starting
d Anlassen *n*
f démarrage *m*
n aanzetten
p uruchamianie; rozruch

4744 starting acceleration
d Anfahrtbeschleunigung *f*
f accélération *f* au démarrage
n acceleratie vanuit staande start
p przyśpieszenie rozruchu

* starting battery → 4734

4745 starting clutch
  d Anlaufkupplung f
  f embrayage m de démarrage
  n aanloopkoppeling
  p sprzęgło rozruchowe

* starting crank → 4750

4746 starting crankshaft
  d Andrehkurbelwelle f
  f arbre m de manivelle de lancement
  n startkrukas
  p zazębiacz korby na wale korbowym

4747 starting device jet
  d Starterkraftstoffdüse f
  f gicleur m de démarrage
  n startsproeier
  p dysza urządzenia rozruchowego

4748 starting ease
  d Lenkbarkeit f
  f facilité f de braquage
  n bestuurbaarheid
  p kierowalność; sterowalność

4749 starting flag; checkered flag
  d Startflagge f
  f drapeau m de départ
  n startvlag
  p flaga startowa

4750 starting handle; starting crank
  d Andrehkurbel f
  f manivelle f de démarrage
  n aanzetkruk; aanzetslinger
  p korba rozruchowa

4751 starting rod
  d Startstange f
  f starter m
  n starterstang
  p cięgno rozruchu

4752 starting swivel bush
  d Achsschenkelbuchse f
  f douille f de pivot de fusée
  n startzwenkbus
  p tuleja zwrotnicy

4753 starting time
  d Anfahrzeit f
  f temps m de démarrage
  n aanlooptijd
  p czas rozruchu

4754 starting torque
  d Anlassdrehmoment n
  f couple m de démarrage
  n startkoppel
  p moment rozruchowy

4755 starting valve piston
  d Kolben m der Startereinrichtung
  f piston m de soupape de démarrage
  n startzuiger
  p tłoczek zaworu rozruchowego

4756 start valve
  d Startventil n; Elektrostartventil n
  f injecteur m de départ à froid; injecteur m de démarrage à froid
  n elektromagnetische startklep
  p elektromagnetyczny wtryskiwacz rozruchowy

4757 state
  d Stand m
  f état m
  n stand
  p stan

4758 static screen
  d Entstörkappe f
  f capuchon m antiparasite
  n ontstoringskap
  p kapa przeciwzakłóceniowa

4759 stationary engine
  d Industriemotor m
  f moteur m stationnaire
  n stationaire motor
  p silnik stały

4760 stator
  d Ständer m
  f induit m; stator m
  n stator
  p stojan

4761 stator vane
  d Leitschaufel f
  f pale f de directrice
  n statorschoep; schoep van schoepenwiel
  p łopatka kierownicy

4762 stator winding
  d Ständerwicklung f
  f bobinage m d'induit
  n statorwikkeling
  p uzwojenie stojana

4763 stauffer grease
  d Staufferfett n

f graisse *f* Stauffer
n consistentvet
p smar stały Stauffera

4764 steady spring tool
d Bremsfederwerkzeug *n*
f tournevis *m* pour le ressort de placage
n remveergereedschap
p przyrząd do ustalania sprężystych pierścieni

4765 steam roller
d Dampfwalze *f*
f rouleau *m* compresseur
n wals voor wegenbouw
p walc drogowy

4766 steel
d Stahl *m*
f acier *m*
n staal
p stal

4767 steel diaphragm
d Stahlmembrane *f*
f membrane *f* d'acier
n stalen diafragma
p membrana stalowa

4768 steel ring
d Stahlring *m*
f anneau *m* en acier
n stalen ring
p pierścień stalowy

4769 steel rule screw
d biegsamer Stahlmassstab *m*
f réglet *m* en acier inoxydable
n stalen lineaal
p przymiar stalowy

4770 steel sleeve
d stählerne Sprenghülse *f*
f douille *f* en acier
n stalen bus
p tulejka stalowa

4771 steel tape
d Stahlbandmasse *f*
f mètre *m* roulant
n rolmaat
p taśma stalowa

4772 steel wheels
d Stahlfelgen *fpl*
f jantes *fpl* métalliques
n metalen velgen
p felgi metalowe

4773 steel wire
d Stahldraht *m*
f fil *m* d'acier
n staaldraad
p drut stalowy

4774 steep hill
d steile Auffahrt *f*
f pente *f* raide
n steile helling
p stromy podjazd

4775 steering bogie
d Lenkachsaggregat *n*
f train *m* roulant orientable
n stuurdraaistel
p wózek prowadzący

* **steering box → 4789**

4776 steering box cover
d Lenkgehäusedeckel *m*
f couvercle *m* du boîtier de direction
n stuurhuisdeksel
p pokrywa obudowy przekładni kierowniczej

4777 steering box flange
d Lenkgehäuseflansch *m*
f bride *f* de fixation du boîtier de direction
n stuurhuisflens
p kołnierz obudowy przekładni kierownicy

4778 steering cam
d Lenknocken *m*
f came *f* de direction
n stuurnok
p ślimak

4779 steering column
d Lenksäule *f*
f colonne *f* de direction
n stuurkolom
p kolumna kierownicy

4780 steering column adjustment
d Lenkrohrverstellung *f*
f réglage *m* du volant
n verstelling in stuurkolom
p przestawienie kolumny kierownicy

4781 steering column bracket
d Lenksäulenstütze *f*; Lenksäulenhalter *m*
f support *m* de colonne de direction
n ophangbeugel van de stuurkolom
p wspornik kolumny kierownicy

4782 steering column lock; steering lock
d Lenkradschloss *n*

f antivol *m* sur la direction; serrure *f* de blocage
de la direction
n stuurslot
p zamek kierownicy

4783 steering column protecting cover
d Lenksäulenabdeckung *f*
f protection *f* de colonne de direction
n kast van de stuurkolom
p osłona kolumny kierownicy

4784 steering cross rod
d Spurstange *f*
f barre *f* directrice transversale
n spoorstang
p drążek kierowniczy poprzeczny

4785 steering damper
d Lenkungsdämpfer *m*
f amortisseur *m* de direction
n stuurdemper
p amortyzator układu kierowniczego

* **steering finger** → 4792

4786 steering gear
d Lenkvorrichtung *f*
f dispositif *m* de direction
n stuurinrichting
p mechanizm kierowniczy

4787 steering gear bracket
d Lenkgetriebelagerbock *m*
f support *m* de boîtier de direction
n stuurhuisconsole
p wspornik przekładni kierowniczej

* **steering gear case** → 4789

4788 steering gear ratio
d Übersetzung *f* des Lenkgetriebes
f rapport *m* de couple de direction
n stuurreductie
p przełożenie przekładni kierowniczej

**4789 steering housing; steering gear case;
steering box**
d Lenkgehäuse *n*
f boîtier *m* de direction
n stuurhuis
p obudowa przekładni kierowniczej

* **steering knuckle pivot** → 4928

4790 steering lever
d Lenkstockhebel *m*
f levier *m* de commande de direction

n stuurarm; pitman-arm
p ramię kierownicze

4791 steering linkage
d Lenkgestänge *f*
f barres *fpl* de direction
n stuurstangenstelsel
p mechanizm zwrotnicy

* **steering lock** → 4782

4792 steering peg; steering finger
d Lenkfinger *m*
f doigt *m* de direction
n stuurvinger
p trzpień korby; zabierak korby

4793 steering play
d Lenkradspiel *n*
f jeu *m* de volant
n stuurspeling
p luz kierownicy

4794 steering rack
d Lenkzahnstange *f*
f crémaillère *f* coulissante
n tandheugel
p zębatka przekładni kierowniczej

4795 steering roller
d Lenkrolle *f*
f galet *m* de direction
n rol van worm-en-rol-stuurinrichting
p rolka kierownicza

4796 steering screw
d Lenkschraube *f*
f vis *f* de direction
n schroefspil
p śruba przekładni kierowniczej

4797 steering shaft
d Lenkspindel *f*
f arbre *m* de direction
n stuuras
p wał kierowniczy

4798 steering shaft gasket
d Lenkspindeldichtung *f*
f joint *m* d'étanchéité d'arbre de direction
n stuuraspakking
p osłona wału kierowniczego

4799 steering shaft jacket
d Lenkrohr *n*
f jupe *f* de colonne
n stuurbuis
p obudowa wału kierowniczego

4800 steering stop
d Lenkanschlag *m*
f butée *f* de direction
n stuuraanslag
p zderzak kierujący

4801 steering stop pin arm
d Achsschenkelarm *m*
f levier *m* de commande de fusée
n fuseearm
p ramię zwrotnicy

4802 steering swivel arm
d Spurstanghebel *m*
f levier *m* de fusée
n spoorstangarm
p ramię zwrotnicy; dźwignia zwrotnicy

4803 steering wheel
d Lenkrad *n*
f volant *m* de direction
n stuurwiel
p koło kierownicze

4804 steering wheel lock
d Lenkradsperre *f*
f blocage *m* du volant
n stuurslot
p blokada kierownicy

4805 steering wheel play
d Lenkradspiel *n*
f jeu *m* de volant de direction
n stuurspeling
p luz koła kierownicy

4806 steering wheel puller
d Lenkradabzieher *m*
f arrache-volant *m*
n stuurwieltrekker
p ściągacz do kół kierowniczych

4807 step bearing
d Spurlager *n*
f crapaudine *f*
n taatslager
p łożysko stopowe

4808 stepped shaft
d Absatzwelle *f*
f arbre *m* à gradins
n faconas
p wałek stopniowany; wałek schodkowy

4809 stepped shoulder
d Rundschulter *f*
f épaulement *m* arrondi

n getrapte schouder
p odsadzenie okrągłe

* stepping motor → 3929

* sticker → 3016

4810 sticky valves
d klebende Ventile *npl*
f soupapes *fpl* gommées
n klevende kleppen
p zawór klejony

4811 stiff; rigid
d steif; starr
f raide; rigide
n stijf
p sztywny

4812 stiffener
d Versteifung *f*; Versteifungsstück *n*
f renfort *m*
n verstevigingsplaat
p usztywniacz

* stiff spring → 3273

4813 stock
d Vorrat *m*; Reserve *f*
f réserve *f*
n voorraad
p zapas

4814 stock car
d Serienwagen *m*
f voiture *f* de série
n serieauto
p samochód seryjny

4815 stock service cabinet
d Verkaufständer *m*
f composition *f* meubles stock service
n verkoopstandaard
p stojak do sprzedaży

4816 stock shears
d Stockschere *f* mit Hebelübersetzung
f cisailles *fpl* à banc à levier
n bankschaar
p nożyce pniakowe do blach

4817 stone chisel
d Breiteisen *n*
f ciseau *m* de tailleur
n slagbeitel
p dłuto szerokie; dłuto kamieniarskie

4818 stone ejector
 d Steinabweiser *m*
 f éjecteur *m* de pierres
 n steenuitwerper
 p odrzutnik kamieni

4819 stop cock
 d Absperrhahn *m*
 f robinet *m* d'arrêt
 n afsluitklep
 p zawór odcinający

4820 stop lever
 d Abstellhebel *m*
 f levier *m* d'arrêt
 n stophendel
 p dźwignia stop

4821 stop nut; stop pin
 d Anschlagbolzen *m*
 f tige *f* de butée
 n aanslagbout
 p sworzeń zderzakowy

4822 stop order
 d Haltgebot *n*
 f stop *m*
 n stopgebod
 p nakaz zatrzymania

 * **stop pin → 4821**

4823 stopping
 d Anhalten *n*
 f arrêt *m*
 n stoppen
 p zatrzymywać się

4824 stopping radiator
 d schräger Kühler *m*
 f radiateur *m* incliné
 n hoekradiateur
 p chłodnica pochyła

4825 stop valve
 d Absperrventil *n*
 f soupape *f* d'arrêt
 n afsluitventiel
 p zawór zamykający

4826 storage case
 d Metalldose *f*
 f coffret *m* tôle
 n stalen doos
 p pudełko metalowe

4827 storage tank
 d Lagerbehälter *m*

 f réservoir *m* de manutention
 n opslagtank
 p zbiornik zasobnikowy

4828 storage tray
 d Abstellüste *f*
 f plateau *m* de rangement
 n opbergplaat
 p paleta magazynowa

4829 storeroom
 d Vorratsraum *m*
 f magasin *m*
 n magazijn
 p magazyn

4830 stowaway for board documentation
 d Ablagelach *n* für Borddokumentation
 f rangement *m* documents de bord
 n opbergvak boorddocumentatie
 p kieszeń na dokumenty pokładowe

4831 straightener
 d Richtvorrichtung *f*
 f appareil *m* à redresser
 n richtbank
 p urządzenie do prostowania

4832 straightening
 d Richten *n*
 f redressage *m*
 n richten
 p prostowanie

4833 straightening iron
 d Biegeeisen *n*
 f pince *f* à dégauchir
 n buigijzer
 p żelazo zginane

 * **straightening plate → 4901**

4834 straightening tool
 d Richtwerkzeug *n*
 f outil *m* à dresser
 n richtgereedschap
 p narzędzie do prostowania

4835 straight extension
 d gerade Verlängerung *f*
 f rallonge *f* droite
 n recht verlengstuk
 p wydłużenie proste

4836 straight pincers
 d Flachzange *f*
 f pince *f* plate

n platbektang
p szczypce płaskie; kleszcze płaskie

4837 straight spanner
d Schlüssel *m* mit geradem Mäul
f clé *f* droite
n rechte steeksleutel
p klucz maszynowy

4838 strainer
d Siebfilter *m*
f crépine *f*
n zeeffilter
p filtr sitowy

4839 strap
d Gurt *m*
f sangle *f*
n draagriem
p pas

* **strap belt** → 3025

4840 strap for glovebox
d Band *n* für Handschuhkasten
f sangle *f* de boîte à gants
n handschoenenkaststeun
p podpora wzmacniająca schowka podręcznego

4841 strap wrench
d Gurtrohrzange *f*
f clé *f* à sangle
n bandsleutel
p klucz taśmowy zaciskowy

4842 street sprayer
d Sprengwagen *m*
f arroseuse *f* automobile
n sproeiwagen
p polewarka

4843 strickling
d Schablonieren *n*
f troussage *m*
n sjabloneren
p formowanie wzornikiem; wzornikowanie

4844 striking face box wrench
d Ringschlagschlüssel *m*
f clé *f* polygonale à frapper
n pijpsleutel met slagkant
p klucz oczkowy uderzeniowy

* **striking fork** → 995

4845 stripped chassis
d nacktes Chassis *n*

f châssis *m* nu
n chassis zonder carrosserie
p podwozie bez karoserii

4846 stripping agent
d Trennmittel *n*
f produit *m* antiadhésif
n afbijtmiddel
p środek antyadhezyjny; środek zapobiegający przyleganiu

4847 stroboscope
d Stroboskop *n*
f stroboscope *m*
n stroboscoop
p stroboskop

4848 stroke bore ratio
d Hubverhältnis *n*
f rapport *m* de course
n slagverhouding
p stosunek skoku tłoka do średnicy cylindra

* **structural steel** → 3328

4849 structure
d Struktur *f*
f structure *f*
n structuur
p struktura

4850 strut; spring leg; spring strut; strut damper
d Federbein *n*; Aufhängungsbein *n*; Radaufhängungsbein *n*
f jambe *f* de suspension; montant *m* de suspension
n veerbeen; veerpoot
p goleń resorująca

* **strut damper** → 4850

4851 stub axle spindle
d Achsschenkelhubzapfen *m*
f pivot *m* de fusée
n gestuurde fuseetap
p czop łożyskowy zwrotnicy

4852 stub tenon
d Fusszapfen *m*
f tenon *m* invisible
n verdekte pen
p czop krótki

4853 stud bolt
d Bolzenschraube *f*
f goujon *m* prisonnier
n tapeind
p kołek gwintowany

4854 studded tyre
d Spikesreifen *m*
f pneu *m* à clous; pneu *m* à crampons
n band met spikes
p opona okolcowana

4855 stud extractor
d Schraubenausdreher *m*
f extracteur *m* de goujons cassés
n trekkers voor gebroken tapeinden
p przyrząd do usuwania złamanych kołków

4856 stud puller
d Bolzenausdreher *m*
f extracteur *m* de goujon
n tapeindentrekker
p ściągacz sworzni

4857 stuffing box
d Stopfbüchse *f*
f boîte *f* à bourage
n pakkingbus
p dławica

4858 styling
d Styling *n*
f styling *m*
n styling; vormgeving
p stylizacja

4859 stylist
d Formgestalter *m*
f styliste *m*
n vormgever
p stylista

4860 subframe side member
d Hilfsrahmenlängsträger *m*
f longeron *m* de soubassement auxiliaire
n langsligger van subframe
p podłużnica ramy dodatkowej

4861 submaster key
d Nebenschlüssel *m*
f clé *f* secondaire
n reservesleutel
p klucz pomocniczy

4862 submerged fuel pump
d Tauchpumpe *f*
f pompe *f* immersible
n dompelpomp
p pompa głębinowa

4863 substance
d Werkstoff *m*
f matière *f*
n grondstof
p tworzywo; materiał

4864 substitute
d Stellvertreter *m*
f remplacement *m*
n vervanging
p zamiana

4865 suction disc
d Unterdruckscheibe *f*
f disque *m* à dépression
n aanzuigschijf
p podciśnieniowa tarcza suwaka

4866 suction hole
d Unterdruckkanal *m*
f canal *m* de dépression
n onderdrukkanaal
p kanał podciśnieniowy

4867 suction of intake pressure
d Ansaugunterdruck *m*
f dépression *f* de l'admission
n onderdruk door aanzuiging
p podciśnienie zasysania; próżnia zasysania

4868 suction pad
d Gummisaugeinsatz *m*
f ventouse *f*
n zuignap
p przyssawka

4869 suction pipe
d Saugleitung *f*
f conduit *m* d'aspiration
n zuigleiding
p przewód ssawny

4870 suction pressure; intake pressure
d Ansaugspannung *f*
f pression *f* d'aspiration
n aanzuigdruk; inlaatdruk
p ciśnienie ssania

4871 suction side
d Ansaugseite *f*
f côte *m* de l'admission
n inlaatzijde
p strona wlotowa

4872 suction stroke; intake stroke
d Saughub *m*
f course *f* d'aspiration
n aanzuigslag; inlaatslag
p suw dolotu; suw ssania

4873 **suction valve**
  d Saugventil *n*
  f clapet *m* d'aspiration
  n zuigklep
  p zawór ssący

4874 **suction valve closing period delay**
  d Verzögerung *f* bei der Saugventilsperrung
  f retard *m* de fermeture de la soupape
    d'admission
  n vertraging van het sluiten van inlaatklep
  p opóźnienie zamknięcia zaworu ssącego

4875 **sulphur dioxide**
  d Schwefeldioxid *n*
  f bioxyde *m* de soufre
  n zwaveldioxide
  p dwutlenek siarki; bezwodnik siarkawy

4876 **sump filter**
  d Ansaugsieb *n*
  f crépine *f* de pompe à huile
  n oliezeef
  p smok pompy oleju

4877 **sump filter bottom**
  d Grobfilterboden *m*
  f fond *m* de crépine d'aspiration
  n groffilterbodem
  p denko smoka (pompy oleju)

* **sun screen** → 4878

* **sunshade** → 4878

* **sunshade roof** → 4547

* **sun shield** → 4878

* **sunshine roof** → 4547

4878 **sun visor; sun shield; sun screen; sunshade;
    antidazzle screen**
  d Sonnenblende *f*
  f pare-soleil *m*
  n zonneklep
  p osłona przeciwsłoneczna

4879 **sun wheel**
  d Mittenrad *n*
  f pignon *m* central
  n zonnewiel
  p koło słoneczne; koło centralne

4880 **supercharged engine**
  d Motor *m* mit Vorverdichtung
  f moteur *m* suralimenté
  n motor met drukvulling
  p silnik z doładowywaniem

4881 **supercharger**
  d Auflader *m*
  f compresseur *m* à suralimentation
  n lader
  p sprężarka doładowująca

4882 **supercharger clutch**
  d Kompressorkupplung *f*
  f embrayage *m* de compresseur
  n compressorkoppeling
  p sprzęgło sprężarki

4883 **supercharger impeller**
  d Kompressorrotor *m*
  f rotor *m* du compresseur
  n compressormeenemer
  p wirnik sprężarki

4884 **supercharging ratio**
  d Aufladegrad *m*
  f taux *m* de suralimentation
  n drukvullingsfactor
  p stopień doładowywania

4885 **super petrol**
  d Superbenzin *n*
  f essence *f* super; supercarburant *m*
  n superbenzine
  p benzyna super

* **supplementary air** → 108

4886 **supplementary pump**
  d Zusatzpumpe *f*
  f pompe *f* auxiliaire
  n hulppomp
  p pompa wspomagająca

* **supplementary spring** → 3635

* **supply of fuel** → 2399

* **supply pump** → 2395

4887 **supply system**
  d Kraftstoffanlage *f*
  f système *m* d'alimentation
  n toevoersysteem
  p układ zasilania

4888 **support bracket**
  d Stützgerüst *m*

f charpente *f* de support
n draagsteun
p wspornik

4889 support frame
d Tragrahmen *m*
f traverse *f* de suspension
n draagframe
p rama nośna

* **supporting axle** → 445

4890 supporting flange
d Tragflansch *m*
f flasque *m* support
n draagflens; steunflens
p kołnierz oporowy

4891 supporting member
d Abstützung *f*
f support *m*
n stutwerk
p podparcie

4892 supporting rod
d Achstragzapfen *m*
f barre *f* support d'essieu
n steunstang
p drążek nośny mostu

4893 support leg
d Stützbein *n*
f jambe *f* de béquille
n steunvoet
p noga oporowa

4894 support plate
d Plakette *f*
f plaquette *f* de support
n steunplaat; draagplaat
p płyta nośna

4895 support spring
d Druckfeder *f*
f ressort *m* de pression
n drukveer
p sprężyna dociskowa

4896 suppressor sleeve
d Entstörmuffe *f*
f gaîne *f* antiparasite
n demphuls
p mufa przeciwzakłóceniowa

4897 surface
d Oberfläche *f*
f surface *f*

n oppervlakte
p powierzchnia

4898 surface cutter
d Planfräser *m*; Walzenfräser *m*
f fraise *f* cylindrique
n cilinderfrees
p frez walcowy

4899 surface grinding machine
d Planschleifmaschine *f*
f rectifieuse *f* plane
n vlakslijpmachine
p szlifierka do płaszczyzn

4900 surface ignition
d Oberflächenzündung *f*
f allumage *m* de surface
n gloei-ontsteking
p zapłon powierzchniowy

4901 surface plate; straightening plate
d Richtplatte *f*
f marbre *m* à dresser
n vlakplaat
p płyta do prostowania

4902 surfacer
d Spachtelmasse *f*
f mastic *m*
n afdichtkit
p masa szpachlowa

4903 surface radiator
d Oberflächenkühler *m*
f radiateur *m* à grande surface
n oppervlaktekoeler
p chłodnica powierzchniowa

4904 surge tank
d Ausgleichbehälter *m*
f réservoir *m* de compensation
n buffertank
p zbiornik wyrównawczy

4905 surveyor's chain
d Messkette *f*
f chaîne *f* d'arpenteur
n meetketting
p łańcuch mierniczy

4906 suspension
d Aufhängung *f*
f suspension *f*
n ophanging
p zawieszenie

4907 suspension adjuster
 d Aufhängungnachsteller *m*
 f correcteur *m* de suspension
 n ophangingsregelaar
 p korektor zawieszenia

4908 suspension bridge
 d Hängebrücke *f*
 f pont *m* suspendu
 n hangbrug
 p most wiszący

4909 suspension coil spring
 d Schraubentragfeder *f*
 f ressort *m* hélicoïdal de suspension
 n schroefdraagveer
 p sprężyna resorowa śrubowa

4910 suspension levelling control
 d Höhenverstellung *f*
 f réglage *m* de hauteur; réglage *m* d'assiette de voiture
 n hoogteregeling; niveauregeling
 p regulacja wzniosu nadwozia

4911 suspension unit
 d Federungselement *n*
 f élément *m* pneumatique
 n veerelement
 p element pneumatyczny; element zawieszenia

4912 swan neck dolly
 d Schwanenhalshandfaust *f*
 f tas *m* col-de-cygne
 n zwanenhalstafel
 p kowadło w postaci szyi łabędzia

4913 swan neck spoon
 d Schwanenhalslöffel *m*
 f palette *f* "col-de-cygne"
 n zwanenhalslepel
 p wygięta łyżka blacharska

4914 swash plate
 d Taumelscheibe *f*
 f plateau *m* incliné
 n tuimelschijf
 p krzywka tarczowa skośna

4915 swash plate engine
 d Taumelscheibenmotor *m*
 f moteur *m* à disc en nutation
 n tuimelschijfmotor
 p silnik hydrauliczny wielotłokowy osiowy ze skośną tarczą

4916 swing axle carrier
 d Pendelachsgehäuse *n*

 f carter *m* de pont oscillant
 n pendelashuis
 p obudowa mostu przegubowego

4917 swing axle vertical support
 d Gelenkzapfen *m* des Achsgehäuses
 f biellette *f* verticale de carter de pont
 n gewrichttap van het achterashuis
 p pionowy wspornik obudowy mostu

4918 swinging axle shaft tube
 d Pendelachswellentragrohr *n*
 f carter *m* de demi-arbre oscillant
 n draagbuis voor "swing-as"
 p pochwa półosi wahliwej

4919 swinging caliper housing
 d Pendelbremssattelkörper *m*
 f corps *m* d'étrier oscillant
 n kantelremzadel
 p stożkowy korpus siodła

4920 switch
 d Schalter *m*
 f interrupteur *m*
 n schakelaar
 p wyłącznik; przełącznik

4921 switch box
 d Schaltkasten *m*
 f tableau *m* de distribution
 n schakelpaneel
 p tablica rozdzielcza

4922 switch knob
 d Schalterdruckknopf *m*
 f poussoir *m* de contacteur
 n schakelaarknop
 p gałka przełącznika

4923 switch lever
 d Schalthebel *m*
 f levier *m* de renvoi
 n schakelhefboom
 p dźwignia przełącznikowa

4924 switch *v* off
 d ausschalten
 f interrompre
 n uitschakelen
 p wyłączać

4925 swivel axle pin bearing; king pin bush
 d Achsschenkelbolzenlager *n*
 f roulement *m* d'axe de fusée
 n lager voor de fuseepen
 p łożysko sworznia zwrotnicy

4926 swivel base
 d Drehsolen *n*
 f base *f* tournante
 n zwenkvoet
 p podstawa z osadzeniem wahliwym

4927 swivel joint
 d Drehgelenk *n*
 f articulation *f* pivotante
 n scharnierende koppeling
 p przegub płaski; węzeł obrotowy

4928 swivel pin; steering knuckle pivot
 d Achsschenkelbolzen *m*
 f pivot *m* de fusée d'essieu
 n fuseepen
 p sworzeń zwrotnicy

4929 swivel pin inclination
 d Spreizung *f*
 f inclinaison *f* des pivots de fusée
 n fuseepenhoek
 p pochylenie sworznia zwrotnicy

4930 synchromesh cone
 d Gleichlaufkonus *m*
 f cône *m* de synchronisation
 n synchromeshconus
 p stożek synchronizatora

4931 synchromesh hub
 d Synchronnabe *f*
 f moyeu *m* de synchroniseur
 n synchronisatienaaf
 p piasta synchronizatora

4932 synchronized speed
 d synchronisierter Gang *m*
 f vitesse *f* synchronisée
 n gesynchroniseerde versnelling
 p przekładnia synchronizowana

4933 synchronizing ball
 d Synchronkugel *f*
 f bille *f* de synchroniseur
 n synchronisatiekogel
 p kulka synchronizująca

4934 synchronizing cone
 d Synchronkegel *m*
 f cône *m* dispositif de synchronisation
 n synchronisatiekegel
 p stożek synchronizujący

4935 synchronizing lock
 d Synchronriegel *m*
 f verrou *m* de synchroniseur
 n synchronisatieslot
 p rygiel synchronizujący

4936 synchronizing slide collar
 d Synchronschiebehülse *f*
 f manchon *m* coulissant de synchronisation
 n synchronisatieglijstuk
 p tuleja przesuwna synchronizująca

4937 synchronizing spring
 d Synchronfeder *f*
 f ressort *m* de synchroniseur
 n veer van synchronisatie-inrichting
 p sprężyna synchronizująca

4938 synchronous speed
 d Synchrodrehzahl *f*
 f vitesse *f* synchrone
 n synchroon toerental
 p prędkość obrotowa synchronicznna

4939 synchro ring
 d Zahnring *m* für Synchronisation
 f bague *f* de synchro
 n synchromeshring
 p pierścień synchronizacyjny

4940 synthetic oil
 d synthetisches Öl *n*
 f huile *f* de synthèse
 n synthetische olie
 p olej syntetyczny

T

4941 tab washer
d Sicherungsblech *n*
f rondelle *f* à aileron
n borgonderlegring
p podkładka odginana

4942 tachograph
d Tachograph *m*
f tachographe *m*
n tachograaf; snelheidsmeter
p tachograf

* **tachometer** → **4625**

4943 tandem axle
d Tandemhinterachse *f*
f essieu *m* en tandem
n tandemas
p układ dwóch osi posobnych

4944 tangential key
d Tangentkeil *m*
f clavette *f* tangente
n tangentiaalspie
p klin kwadratowy

4945 tank capacity
d Behälterinhalt *m*
f contenance *f* de réservoir
n tankinhoud; tankcapaciteit
p pojemność zbiornika

4946 tanker semitrailer
d Tankanhänger *m*
f citerne *f* semi-remorque
n tankoplegger
p naczepa-cysterna

4947 tank filler cap
d Behälterdeckel *m*
f bouchon *m* de réservoir
n tankvuldop
p pokrywa wlewu

* **tank mounting bracket** → **4948**

4948 tank mounting element; tank mounting bracket
d Behälteraufhängungselement *n*
f sangle *f* de fixation de réservoir

n tankbevestigingsbeugel
p element zawieszenia zbiornika

4949 tap
d Zapfhahn *m*
f robinet *m* de soutirage
n kraan
p kurek czerpalny; zawór kurkowy czerpalny

* **tapered needle** → **3320**

4950 tapered roller; conical roller
d Kegelrolle *f*
f galet *m* conique
n tapse rol
p wałeczek stożkowy

4951 tapered roller thrust bearing
d Axialkegelrollenlager *n*
f butée *f* à galets coniques
n conische rollenlager
p wzdłużne łożysko stożkowe

4952 taper gauge
d Kegellehre *f*
f calibre *m* de conicité
n tapsheidmeter
p sprawdzian stożkowy

4953 taper journal
d Kegelzapfen *m*
f tourillon *m* conique
n kegelvormige tap
p czop główny stożkowy

4954 taper pin
d konischer Stift *m*; Kegelstift *m*
f goupille *f* conique
n conische pen; tapse pen
p kołek stożkowy

4955 taper roller bearing; conical bearing
d Kegelrollenlager *n*
f roulement *m* à rouleaux coniques
n rollager met kegelrollen
p łożysko stożkowe

4956 taper sunk key
d Einlegekeil *m*
f clavette *f* inclinée encastrée
n tapse inlegspie; verzonken tapse spie
p klin opuszczany

* **taper tap** → **2013**

4957 taper thread
d kegeliges Gewinde *n*

f filet *m* conique
n tapse schroefdraad; conische schroefdraad
p gwint stożkowy

4958 **tappet**
d Stössel *m*
f poussoir *m*
n klepstoter
p popychacz

4959 **tappet adjusting screw**
d Stösseleinstellschraube *f*
f vis *f* de réglage de poussoir
n stelschroef van klepstoter
p śruba regulacyjna popychacza

4960 **tappet depressor**
(tool to shim overhead camshaft engines)
d Ventilfederheber *m*
f levier *m* de compression des poussoirs
n klepveerlichter
p klucz do wyciskania sprężyn zaworów

4961 **tappet guide**
d Stösselführung *f*
f guide *m* de poussoir de soupape
n geleider van klepstoter
p prowadnica popychacza

4962 **tappet roller**
d Stösselrolle *f*
f galet *m* de poussoir
n stoterrol
p rolka popychacza

4963 **tare weight; dead load**
d Eigengewicht *n*
f poids *m* à vide
n eigengewicht; leeggewicht
p ciężar własny

4964 **tarp material**
d Planenstoff *m*
f tissu *m* pour bâche
n dekkleedmateriaal
p brezent impregnowany

4965 **tarp supporting structure**
d Planengestell *n*
f ossature *f* de bâchage
n frame voor huif
p rama do plandeki

4966 **tar remover**
d Teerentferner *m*
f produit *m* pour enlever le goudron
n teerverwijderaar
p usuwacz smoły

4967 **taxable horsepower**
d Steuerleistung *f*
f puissance *f* fiscale
n fiscaal vermogen
p moc podatkowa

4968 **taximeter**
d Fahrpreisanzeiger *m*
f taximètre *m*
n taximeter
p taksometr

4969 **taxi rank**
d Taxistand *m*
f station *f* de taxis
n taxistandplaats
p postój taksówek

4970 **technical backup facilities**
d technische Basis *f*
f service *m* technique
n technische basis
p zaplecze techniczne

4971 **technical service vehicle**
d technisches Einsatzfahrzeug *n*
f véhicule *m* de secours technique
n voertuig voor technische diensten
p pojazd służby technicznej

4972 **technical term**
d technischer Ausdruck *m*
f terme *m* technique
n vakterm
p termin fachowy

4973 **technician**
d Techniker *m*
f technicien *m*
n technicus
p technik

4974 **technological test**
d technologische Probe *f*
f essai *m* technologique
n technologische test
p próba technologiczna

4975 **technology**
d Technologie *f*
f technologie *f*
n technologie
p technologia

4976 **tee fitting**
d T-Stück *n*
f raccord *m* à trois voies

n T-stuk
p trójnik rurowy

4977 tee wrench for hinge screws
d T-Schlüssel *m* zum Scharnierbolzen
f clé *f* abéquille pour vis de charnières
n T-sleutel voor scharnierbouten
p klucz do śrub zawiasów

4978 telescopic piston
d Teleskopkolben *m*
f piston *m* télescopant
n telescopische plunjer
p tłok teleskopowy

4979 telescopic shaft
d Teleskopwelle *f*
f arbre *m* télescopique
n telescopische as
p wałek teleskopowy

4980 telescopic shock absorber
d Teleskopstossdämpfer *m*
f amortisseur *m* télescopique
n telescopisch werkende schokdemper
p amortyzator teleskopowy

4981 telescopic suspension
d Hülsenfederführung *f*
f suspension *f* télescopique
n telescopische ophanging
p zawieszenie teleskopowe

4982 temperature
d Temperatur *f*
f température *f*
n temperatuur
p temperatura

4983 temperature control
d Temperaturregelung *f*
f régulation *f* de température
n temperatuurregeling
p regulacja temperatury

4984 temperature control valve
d Regulierventil *n* der Temperatur
f volet *m* de commande de la température
n temperatuurregelklep
p zawór regulacji temperatury

4985 temperature of air
d Lufttemperatur *f*
f température *f* de l'air
n luchttemperatuur
p temperatura powietrza

4986 temperature selector button
d Temperaturwahlknopf *m*

f commande *f* de sélection de la température
n temperatuurkeuzeknop
p guzik wybierakowy temperatury

4987 temperature sensor
d Temperaturfühler *m*
f sonde *f* de température
n temperatuursensor
p czujnik temperatury

4988 tempered glass
d gehärtetes Glas *n*
f verre *m* trempé
n enkellaagsveiligheidsglas
p szkło hartowane

4989 template
d Schablone *f*; Modell *n*
f gabarit *m*; échantillon *m*
n sjabloon
p wzornik; model

4990 temporary repair
d Behelfsreparatur *f*
f réparation *f* provisoire
n noodreparatie
p naprawa prowizoryczna

4991 tensile force
d Zugkraft *f*
f force *f* de traction
n trekkracht
p siła rozciągająca

4992 tensile ring
d Zerreissring *m*
f anneau *m* de rupture
n trekring
p pierścień rozpychający

4993 tension; voltage
d Spannung *f*
f tension *f*
n spanning
p napięcie

4994 tensioner for antiroll bar removal
d Spanner *m* für Stabilisatorstange
f tendeur *m* pour la dépose de la barre antiroulis
n stabilisatorstangspanner voor het verwijderen
   van de stabilisatorstang
p napinacz drążka poprzecznego

4995 tensioner roller
d Spannerrolle *f*
f galet *m* de tendeur
n spanrol
p rolka napinana

4996 tension spring
  d Spannfeder *f*
  f ressort *m* tendeur
  n spanveer
  p sprężyna napinajaca

4997 terminal
  d Anschlussklemme *f*
  f borne *f*
  n aansluitklem
  p zacisk

4998 terminal board
  d Klemmbrett *n*
  f tablette *f* des contacts
  n klemmenbord
  p płytka zaciskowa

4999 terminal corrosion
  d Polkorrosion *f*
  f corrosion *f* des bornes
  n corrosie van accupolen
  p korozja końcówck

5000 terminal marking
  d Klemmenbezeichnung *f*
  f marquage *m* des plots
  n markering van aansluitklem
  p oznakowanie końcówek zacisku

5001 terminal nut
  d Klemmutter *f*
  f écrou *m* de serre-fil
  n klemmoer
  p nakrętka zaciskowa

5002 terminal voltage
  d Klemmenspannung *f*
  f tension *f* aux bornes
  n klemspanning
  p napięcie na zaciskach

5003 test
  d Probe *f*
  f recherche *f*; épreuve *f*
  n proef; test
  p próba; test

5004 test certificate
  d Prüfbescheinigung *f*
  f certificat *m* de vérification
  n testbewijs
  p świadectwo próby

5005 testing stand
  d Prüfstand *m*
  f banc *m* d'essai

  n testbank
  p stół probierczy

5006 test sample
  d Prüfkörper *m*
  f éprouvette *f*
  n testexemplaar
  p próbka

5007 tetraethyl lead
  d Tetraethylblei *n*
  f tétraéthyle *m* de plomb
  n tetra-ethyl lood
  p czteroetylek ołowiu

5008 textile cleaner
  d Textilreinigungsmittel *n*
  f nettoyeur *m* textile
  n textielreiniger
  p środek czyszczący do materiałów
     włókienniczych

 * **thermal conductivity** → **2628**

5009 thermal efficiency
  d thermischer Wirkungsgrad *m*
  f rendement *m* thermique
  n thermisch rendement
  p sprawność cieplna

5010 thermal relay
  d thermisches Relais *n*
  f relais *m* thermique
  n thermisch relais
  p przekaźnik cieplny

 * **thermal stability** → **2641**

5011 thermal time switch
  d Thermozeitschalter *m*
  f thermocontact *m* temporisé
  n thermotijdschakelaar
  p temperaturowy przekaźnik czasowy

5012 thermic wear
  d Wärmeverschleiss *m*
  f usure *f* thermique
  n thermische slijtage
  p zużycie cieplne

5013 thermistor
  d Thermistor *m*
  f thermistance *f*
  n thermistor
  p termistor

5014 thermocouple
  d Thermoelement *n*

f thermocouple *m*
n thermokoppel; thermo-element
p termoelement

5015 thermoelectric effect
d thermoelektrischer Effekt *m*
f effet *m* thermoélectrique
n thermo-elektrisch effect
p zjawisko termoelektryczne

5016 thermosiphon water circulation
d Thermosyphonkühlflüssigkeitsumlauf *m*
f circulation *f* d'eau par thermosiphon
n thermosyphon-circuit
p samoczynne krążenie cieczy chłodzącej

5017 thermostat
d Thermostat *m*
f thermostat *m*
n thermostaat
p termostat

5018 thermostat bypass valve
d Überströmventil *n* des Thermostats
f clapet *m* d'écoulement de thermostat
n nevenstroomklep
p dolny zawór termostatu

5019 thermostatic
d thermostatisch
f thermostatique
n thermostatisch
p termostatyczny

5020 thermostatic regulator
d Thermostatregler *m*
f régulateur *m* thermostatique
n thermostatische regelaar
p regulator termostatyczny

5021 thermostat switch
d Thermoschalter *m*
f interrupteur *m* thermique
n thermoschakelaar
p termowyłącznik; wyłącznik termostatyczny

5022 thermostat valve
d Thermostathahn *m*
f robinet *m* thermostatique
n thermostaatklep
p zawór termostatu

5023 thickener
d Viskositätsindexverbesserer *m*
f additif *m* améliorant de l'indice de viscosité
n viscositeitsverhogende dope
p dodatek podwyższający wskaźnik lepkości

5024 thin grooved dolly
d dünne Handfaust *f* mit Rinne
f tas *m* mince à rainure
n dunne gegroefde uitdeuktas
p cienkie kowadełko z rynną

5025 third and fourth speed synchronizer
d Synchronvorrichtung *f* für 3. und 4. Gang
f synchroniseur *m* de troisième et quatrième vitesse
n synchronisatie van derde en vierde versnelling
p synchronizator biegu trzeciego i czwartego

5026 thread
d Gewinde *n*
f filet *m*
n schroefdraad
p gwint

5027 thread cutter
d Schneidbacken *fpl*
f mâchoire *f* à peigne
n schroefdraadfrees
p frez do gwintów

5028 threaded adapter
(adapter for fitting accessories)
d kurzes Kupplungsteil *n* mit Aussengewinde
f raccord *m* fileté
n kort koppelstuk met buitendraad
p złączka gwintowana

5029 threaded bush
d Gewindebüchse *f*
f bague *f* taraudée
n draadsok
p tuleja gwintowana

5030 threaded fitting
(tool to extract rear suspension bars, in conjunction with slide hammer)
d Gewindekappe *f*
f douille *f* filetée
n dop met schroefdraad
p łącznik gwintowany

5031 threaded joint
d Schraubenverbindung *f*
f assemblage *m* par filet
n schroefverbinding
p połączenie gwintowe

5032 threaded male coupling
d Verbindungsrohr *n* für Schnellkupplung mit Aussengewinde
f raccord *m* fileté male
n verbindingspijp voor snelkoppeling met buitendraad
p gwintowany męski sprzęg

5033 threaded spindle
  d Gewindespindel f
  f broche f filetée
  n draadstang; schroefspil
  p wałek z gwintem

5034 thread gauge
  d Gewindelehre f
  f calibre m de filetage
  n schroefkaliber
  p sprawdzian gwintowy

5035 threading die
  d Gewindeschneideisen n
  f filière f à fileter
  n snijplaat
  p narzynka do gwintów

5036 thread profile in normal section
  d Zahnprofil m im Normalschnitt
  f profil m normal de spire
  n tandprofiel in normale doorsnede
  p zarys zęba w przekroju normalnym

5037 thread size
  d Gewindemass n
  f cote f de filetage
  n schroefdraadafmeting
  p wymiar gwintu

5038 three armed flange
  d Dreiarmflansch m
  f flasque m à trois bras
  n driearmige flens
  p kryza trójramienna

5039 three plate clutch
  d Dreiplattenkupplung f
  f embrayage m à trois plateaux
  n drieplaatkoppeling
  p sprzęgło trójtarczowe

5040 three point bearing
  d Dreipunktlager n
  f palier m à trois points portants
  n driepuntslager
  p łożysko trójpunktowe

5041 three point safety belt
  d Dreipunktgurt m
  f ceinture f à trois points; ceinture f combinée
  n driepuntsveiligheidsgordel
  p trójpunktowy pas bezpieczeństwa

5042 three square scraper
  d Dreikantschaber m
  f grattoir m triangulaire
  n driehoekkrabber
  p skrobak trójkątny

5043 three stage
  d dreistufig
  f à trois degrés
  n drietraps
  p trzystopniowy

5044 three way cock
  d Dreiwegehahn m
  f robinet m à trois voies
  n driewegkraan
  p kurek trójdrogowy

5045 three way tipper
  d Dreiseitenkipper m
  f basculeur m trilatéral; camion m tri-benne
  n driezijdige kiepauto
  p wywrotka trójstronna

5046 throttle control cable handle
  d Griff m des Drosselklappenbowdenzuges
  f tirette f d'accélérateur à main
  n bedieningsknop voor handgas
  p uchwyt cięgna przepustnicy

5047 throttle control lever
  d Drosselklappengestängehebel m
  f levier m de commande de papillon
  n hefboom van de smoorklep
  p dźwignia sterowania przepustnicy

5048 throttle housing
  d Drosselklappenstützen m; Drosselklappengehäuse n
  f corps m papillon; porte-papillon m; boîtier m papillon
  n gasklephuis
  p podpora przepustnicy

5049 throttle lever
  d Drosselklappenhebel m
  f levier m de papillon
  n smoorklephefboom
  p dźwignia przepustnicy

5050 throttle microswitch
  d Drosselklappenmikroschalter m
  f microcontact m de papillon
  n gasklepmicroschakelaar
  p mikrowyłącznik przepustnicy

5051 throttle ring
  d Drosselring m
  f anneau m d'étranglement
  n smoorring
  p pierścień dławiący

* throttle shaft → 5053

5052 throttle valve
d Drosselklappe f
f papillon m d'air
n smoorklep
p przepustnica

5053 throttle valve spindle; throttle shaft
d Drosselklappenwelle f
f axe m de papillon
n smoorklepas
p ośka przepustnicy

5054 throttle valve switch
d Drosselklappenschalter m
f interrupteur m sur axe de papillon
n smoorklepschakelaar
p wyłącznik przepustnicy

5055 throttling
d Drosselung f
f étranglement m
n smoren
p dławienie

5056 throttling governor
d Regler m mit Drosselbetätigung
f régulateur m de carburateur
n carburateurregelaar
p regulator gaźnika

5057 through hole
d Passloch n
f trou m débouchant
n door-en-door gat
p otwór przelotowy

5058 thrust bearing
d Drucklager n
f palier m de pression
n druklager
p łożysko oporowe

* thrust disc → 2335

5059 thrust flange
d Flansch m für Nockenwellenanschlag
f bride f de butée
n borgplaat
p kołnierz oporowy

* thrust plate → 3877

5060 thrust ring
d Stellring m
f arrêtoir m

n stelring
p pierścień osadczy

5061 thrust spring
d Kupplungsfeder f
f ressort m d'un embrayage
n drukveer van koppeling
p sprężyna dociskowa sprzęgła

5062 thrust washer
d Stossring m
f rondelle f de butée
n afstelring; stelring
p podkładka oporowa

5063 tightening lever
d Spannhebel m
f levier m de serrage
n spanhefboom
p dźwignia zaciskowa

* timer → 5064

5064 time switch; timer
d Zeitschalter m
f temporisateur m
n tijdschakelaar
p przekaźnik czasowy

5065 timing belt
d Zahnriemen m
f courroie f crantée
n getande riem
p pas zębaty

5066 timing chain
d Steuerungskette f
f chaîne f de distribution
n distributieketting
p łańcuch pędny

5067 timing fork
d Gelenkgabel f
f fourchette f
n verstelgaffel
p widełki przegubu

5068 timing gauge for injection pumps
d Messuhr f zum Bestimmung des
Einspritzzeitpunktes
f jauge f de mise au point pour pompe à
injection
n meter voor het bepalen van het inspuitmoment
p miernik do pomiaru punktu wtrysku

5069 timing gear cover
d Steuergehäusedeckel m

f couvercle *m* du carter de distribution
n distributiedeksel
p pokrywa napędu rozrządu

5070 timing gear system
d Steuerungssystem *n*
f système *m* de distribution
n distributiesysteem
p układ rozrządu

5071 timing gear wheel
d Steuerrad *n*
f roue *f* de distribution
n distributietandwiel
p koło zębate rozrządu

5072 timing lamp
d Zündpunktprüfungslampe *f*
f lampe *f* stroboscopique
n afstellamp
p lampa do sprawdzania punktu zapłonu

5073 timing relay
d Zeitrelais *n*
f relais *m* temporisé
n tijdrelais
p przekaźnik czasowy; przekaźnik zwłoczny

* **tinman → 556**

5074 tinsmit's hammer
d Tellerhammer *m*
f marteau *m* à vaisselle
n blikslagershamer
p klepak; młotek blacharski

5075 tinted glass
d getöntes Glas *n*
f glace *f* teintée
n getint glas
p szyba pzryciemniona

5076 tipper hinge
d Kipperscharnier *n*
f charnière *f* de basculateur
n kiepscharnier
p przegub urządzenia przechylnego

5077 tipper semitrailer
d Kippauflieger *m*
f semi-remorque *f* à benne basculante
n kipper-oplegger
p naczepa wywrotka

5078 tipping gear unit
d Kippvorrichtung *f*
f dispositif *m* de basculement

n kiepmechanisme
p urządzenie przechylne

5079 tipping trailer
d Kippanhänger *m*
f remorque *f* à benne
n kipper-aanhanger
p przyczepa-wywrotka

5080 tip relief
d Zahnkopfabrundung *f*
f chanfrein *m* de tête de dent
n ronding van tandkop
p zaokrąglenie głowy zęba

* **tire → 5237**

5081 t joint
d T-Stoss *m*
f assemblage *m* en T
n T-naad
p szew T

5082 toe dolly
d Handfaust *f* Zehenform
f tas *m* à courbure
n gewelfde tas
p kowadełko zagięte

5083 toe in
d Vorspur *f*
f pincement *m* des roues
n toespoor
p zbieżność kół

5084 toe out
d Nachspur *f*
f ouverture *f*
n uitspoor
p rozbieżność kół

5085 toggle brake
d Knebelbremse *f*
f frein *m* à machoires articulées
n sleutelrem
p hamulec z przetyczką

5086 toggle lever
d Kniehebel *m*
f barre *f* à genouillère
n kniehefboom
p dźwignia kolankowa

5087 toggle press
d Kniehebelpresse *f*
f presse *f* à genouillère
n kniehefboompers
p prasa kolanowa

5088 toggle switch
d Tumblerschalter *m*
f tumbler *m*
n tuimelschakelaar
p przełącznik migowy

5089 toggle system
d Gelenkhebelsystem *n*
f système *m* à leviers articulés
n kiepsysteem
p system dźwigni przegubowych

5090 tool
d Werkzeug *n*
f outil *m*
n gereedschap; werktuig
p narzędzie

5091 tool cabinet
d Werkzeugschränk *m*
f armoire *f* à outils
n gereedschapskast
p szafka narzędziowa

5092 tool kit
d Werkzeugkasten *m*
f caisse *f* à outils
n gereedschapskist
p skrzynka narzędzia

5093 tool set
d Werkzeugsatz *m*
f jeu *m* d'outils
n gereedschapsset
p zestaw narzędzi

5094 tool steel
d Werkzeugstahl *m*
f acier *m* pour outils
n gereedschapsstaal
p stal narzędziowa

5095 tooth
d Zahn *m*
f dent *f*
n tand
p ząb

5096 toothed belt
d Zahnriemen *m*
f courroie *f* dentée
n getande riem
p pas zębaty

5097 toothed chain
d Zahnkette *f*
f chaîne *f* dentée

n tandketting
p łańcuch zębaty

5098 toothed gear
d Zahnradgetriebe *n*
f engrenage *m* à crémaillère
n tandwieloverbrenging
p przekładnia zębata

5099 toothed ring
d Zahnkranz *m*
f courroie *f* dentée
n tandkrans
p wieniec zębaty

5100 toothed washer
d Zahnscheibe *f*
f rondelle *f* découpée
n getande onderlegring
p podkładka zębata

5101 tooth flank
d Zahnflanke *f*
f flanc *m* de dent
n tandflank
p powierzchnia boku zęba

5102 tooth gauge
d Zahnlehre *f*
f trusquin *m* à dents
n tandmaat
p sprawdzian do pomiaru zębów

5103 tooth pressure
d Zahndruck *m*
f pression *f* des dents
n druk op tandflank
p ciśnienie na ząb

5104 tooth sector
d Zahnbogen *m*
f secteur *m* denté
n tandsector
p sektor zęba

5105 tooth tip
d Zahnkrone *f*
f sommet *m* de dent
n tandkop
p wierzchołek zęba

* top cylinder lubricant → 5276

5106 top dead position; inner dead center
d oberer Totpunkt *m*
f point *m* mort haut
n bovenste dode punt
p górny punkt martwy

* **top gear** → 2662

5107 top ring
d Hauptring *m*
f segment *m* de feu
n bovenste zuigerveer
p pierścień główny

5108 torque adjustment
d Drehmomentregelung *f*
f réglage *m* de couple de rotation
n regulatie van het draaimoment
p regulacja momentu obrotowego

5109 torque ball
d Schubkugel *f*
f rotule *f* de poussée
n koppelkogel
p kulka sprzęgająca

5110 torque control rod
d Schubstange *f*
f barre *f* de poussée
n torsieopvangarm
p wzdłużny drążek stabilizacyjny

5111 torque convertor
d Drehmomentenwandler *m*
f convertisseur *m* de couple
n koppelomvormer
p przemiennik momentu obrotowego

5112 torque handle
d Drehmomentschlüssel *m* mit Momentanzeige
f poignée *f* dynamométrique
n handgreep met vast moment
p klucz dynamometryczny z momentem
   wskazywanym

5113 torque multiplier
d Kraftvervielfältiger *m*
f multiplicateur *m* de couple
n koppelvergroter
p wzmacniacz momentu obrotowego

5114 torque tube support
d Zentralgelenkbrücke *f*
f traverse *f* de tube de réaction
n ondersteuningsbrug van de torsiebuis
p wspornik rury reakcyjnej

5115 torque wrench
d Drehmomentschlüssel *m*
f clé *f* dynamométrique
n momentsleutel
p klucz dynamometryczny

5116 torsion
d Torsion *f*; Drehung *f*

f torsion *f*
n torsie
p skręcanie

5117 torsional strength
d Torsionsfestigkeit *f*; Drehfestigkeit *f*
f résistance *f* à la torsion
n torsiesterkte
p wytrzymałość na skręcanie

5118 torsional stress
d Drehbeanspruchung *f*
f effort *m* de torsion
n torsiespanning
p naprężenie skręcające

5119 torsion bar
d Drehstabfeder *f*
f barre *f* de torsion
n torsiestaaf
p drążek skrętny

5120 torsion bar locking seat
d Drehstabverankerung *f*
f ancrage *m* de barre de torsion
n verankering van de torsieveer
p gniazdo osadcze drążka skrętnego

5121 torsion bar suspension
d Drehstabfederung *f*
f suspension *f* à barres de torsion
n torsievering
p uresorowanie drążka skrętnego

5122 torsion spring
d Drehungsfeder *f*
f ressort *m* de torsion
n torsieveer
p sprężyna skręcana

5123 total brake lining area
d Gesamtbremsfläche *f*
f surface *f* totale de freinage
n totaal remoppervlak
p powierzchnia czynna okładzin ciernych

5124 total loss lubrication
d Durchlaufschmierung *f*
f graissage *m* à huile perdue
n doorstroomsmering
p smarowanie dopływowe

5125 total ratio
d Gesamtübersetzung *f*
f démultiplication *f* totale
n totale overbrengingsverhouding
p przełożenie całkowite

5126 toughness
d Zähigkeit f
f ténacité f
n taaiheid
p wiązkość

5127 touring bus
d Rundfahrtbus m
f car m "grand tourisme"
n touringcar
p autobus wycieczkowy

5128 tow v; haul v
d abschleppen
f remorquer
n slepen
p holować

5129 towing device
d Abschleppvorrichtung f
f dispositif m de remorquage
n sleepmechanisme
p mechanizm holowniczy

5130 towing hook
d Schleppseilhaken m
f crochet m d'avant
n trekhaak
p hak holowniczy

5131 towing pole
d Abschleppstange f
f barre f de remorquage
n trekstang
p drążek holowniczy

5132 tow ring
d Deichselzugöse f
f anneau m de triangle d'attelage
n trekoog van de dissel
p ogniwo dyszla

5133 tow rope
d Abschleppseil n
f câble m de remorque
n sleepkabel
p lina holownicza

5134 tow truck
d Schlepper m
f remorqeur m
n trekker
p holownik

5135 tow weight
d Anhanggewicht n
f poids m remorque

n aanhangwagengewicht
p ciężar holowany; masa ciągniona

5136 track front
d Spurweite f vorn
f voie f avant
n spoorbreedte voor
p rozstaw kół przednich

5137 track rear
d Spurweite f achter
f voie f arrière
n spoorbreedte achter
p rozstaw kół tylnych

5138 track rod center link
d Spurstangenmittenglied n
f barre f centrale
n middelste spoorstang
p środkowy człon drążka

5139 track rod guide lever
d Spurstangenführungshebel m
f levier m guide de barre
n spoorstanggeleidingshefboom
p dźwignia prowadząca drążka

5140 track rod left side link
d linkes Spurstangenglied n
f barre f gauche
n linker spoorstanglid
p lewy człon drążka

5141 track rod right side link
d rechtes Spurstangenglied n
f barre f droite
n rechter spoorstanglid
p prawy człon drążka

* **tracted tractor** → 1377

5142 tractive resistance
d Fahrwiderstand m
f résistance f à l'avancement
n rijweerstand
p opór czołowy

5143 tractor trailer unit
d Strassenzug m
f train m routier
n trekker met aanhangwagen
p zestaw przyczepowy

5144 traffic code
d Strassenverkehrsordnung f
f code m routier
n verkeersregels
p kodeks drogowy

5145 traffic jam
  d Verkehrsstau *f*
  f embouteillage *m*
  n verkeersknoop
  p korek uliczny

5146 traffic lights
  d Verkehrsampeln *f*
  f signalisation *f* lumineuse
  n verkeerslichten
  p sygnalizacja świetlna

5147 traffic signs
  d Verkehrszeichen *n*
  f signalisation *f* routière
  n verkeersborden
  p znaki drogowe

5148 trailer
  d Kraftwagenanhänger *m*
  f remorque *f*
  n aanhangwagen
  p przyczepa samochodowa

5149 trailer axle
  d Anhängerachse *f*
  f essieu *m* de remorque
  n aanhangwagenas
  p oś przyczepy

5150 trailer brake
  d Anhängerbremse *f*
  f frein *m* de la remorque
  n aanhangwagenrem
  p hamulec przyczepy

5151 trailer brake coupling
  d Anhängerbremskupplung *f*
  f connexion *f* pour frein de remorque
  n oplooprem van aanhangwagen
  p sprzęg hamulkowy przyczepy

5152 trailer brake valve
  d Anhängerbremsventil *n*
  f robinet *m* de commande
  n aanhangwagenremklep
  p zawór hamowania przyczepy

5153 trailer chassis
  d Anhängerchassis *n*
  f châssis *m* de remorque
  n chassis van aanhangwagen
  p podwozie przyczepy

5154 trailer coupling
  d Anhängerkupplung *f*
  f accouplement *m* pour remorques

  n aanhangwagenkoppeling
  p sprzęg przyczepowy

5155 trailer length
  d Anhängerlänge *f*
  f longueur *f* de remorque
  n aanhangwagenlengte
  p długość przyczepy

5156 trailer towing truck
  d Lastwagen *m* für Anhängerbetrieb
  f tracteur-porteur *m*
  n vrachtwagen met aanhangerfaciliteiten
  p samochód-ciągnik przyczep

5157 trailer van
  d Kastenanhänger *m*
  f remorque *f* fourgon
  n gesloten aanhanger
  p przyczepa furgon

5158 trailing arm
  d Längslenker *m*
  f bras *m* oscillant longitudinal; bras *m* de suspension
  n overlangse stabilisatiearm
  p wahacz wzdłużny

5159 trailing brake shoe; secondary brake shoe
  d gegenläufige Bremsbacke *f*; Sekundärbacke *f*
  f mâchoire *f* de frein étirée
  n aflopende remschoen; secundaire remschoen
  p szczęka hamulcowa przeciwbieżna

*   **transfer box → 5161**

5160 transfer canal
  d Überströmkanal *m*
  f canal *m* de transfert
  n overstroomkanaal
  p kanał przelotowy

5161 transfer case; transfer box
  d Verteilergetriebe *n*
  f boîte *f* de transfert
  n krachtoverbrengingsbak
  p rozdzielcza skrzynka biegów

*   **transfer chamber → 2262**

5162 transfer pump
  d Transferpumpe *f*
  f pompe *f* de transfert
  n doorvoerpomp
  p pompa przetłaczająca

5163 transformation
d Umwandlung *f*
f transformation *f*
n omzetting; transformatie
p przekształcenie

5164 transistor
d Transistor *m*
f transistor *m*
n transistor
p tranzystor

5165 transistorized ignition
d Transistorzündung *f*
f allumage *m* transistorisé
n transistorontsteking
p zapłon tranzystorowy

5166 transition fit
d Übergangspassung *f*
f ajustement *m* incertain
n overgangspassing
p pasowanie mieszane

* **translation screw** → 3850

* **transmission** → 3853

5167 transmission brake
d Getriebebremse *f*
f frein *m* de mécanisme
n transmissierem
p hamulec wału napędowego

5168 transmission case
d Getriebegehäuse *n*
f boîte *f* de vitesse
n versnellingsbakhuis
p obudowa skrzynki przekładniowej

* **transmission gear** → 2460

5169 transmission jack
d Getriebeheber *m*
f cric *m* pour transmission
n transmissiekrik
p podnośnik przekładniowy

5170 transmission tunnel
d Getriebetunnel *m*
f tunnel *m* de transmission
n cardanastunnel
p tunel mechanizmu przekładniowego

5171 transparent lacquer
d farbloser Lack *m*

f vernis *m* blanc
n transparante lak
p lakier bezbarwny

5172 transversal arm; wishbone
d Querlenker *m*
f bras *m* oscillant transversal
n dwarsgeplaatste wieldraagarm
p wahacz poprzeczny

5173 transversal scavenge
d Querspülung *f*
f balayage *m* transversal
n dwarsspoeling
p przepłukiwanie poprzeczne cylindra

5174 transverse compensating spring
d Ausgleichquerfeder *f*
f ressort *m* transversal de compensation
n dwarse compensatieveer
p sprężyna poprzeczna wyrównawcza

5175 transverse engine
d Quermotor *m*
f moteur *m* transversal
n dwarsmotor
p silnik usytuowany poprzecznie

5176 transverse engine rear support
d Quertraverse *f* der Motoraufhängung
f traverse *f* support arrière de moteur
n achterste motortraverse
p poprzeczka tylnego zawieszenia silnika

5177 transverse leaf spring
d Querblattfeder *f*
f ressort *m* transversal à lames
n dwarsgeplaatste bladveer
p poprzeczny resor piórowy

5178 transverse member carpet
d Querträgerbelag *m*
f tapis *m* de traverse
n dwarsbalkbekleding
p dywanik poprzeczki

5179 transverse pitch of rack
d Stirnteilung *f* der Zahnstange
f pas *m* transversal de crémaillère
n dwarssteek van een tandstang
p podziałka czołowa zębatki

5180 transverse stabilizer
d Querstabilisator *m*
f barre *f* stabilisatrice
n dwarsstabilisator
p stabilizator poprzeczny

5181 transverse support sandwich mounting
  d Gummikörper *m* der hinteren
     Motoraufhängung
  f tampon *m* élastique de support arrière de
     moteur
  n achterste silentbloc motorophanging
  p poduszka tylnego zawieszenia silnika

5182 trapezoidal thread
  d Trapezgewinde *n*
  f filet *m* trapézoïdal
  n trapeziumschroefdraad
  p gwint trapezowy

5183 travelling comfort
  d Fahrbequemlichkeit *f*
  f confort *m* dans un véhicule
  n reiscomfort
  p komfort jazdy

5184 tread groove
  d Profilrille *f*
  f rainure *f* de sculpture
  n profielgroef van autoband
  p rowek bieżnika opony

5185 tread shoulder
  d Reifenschulter *m*
  f épaulement *m* de bande de roulement
  n schouder van autoband
  p bark bieżnika

5186 tread wear
  d Laufflächenabnutzung *f*
  f usure *f* de bande de roulement
  n bandslijtage
  p ścieranie bieżnika opony

5187 trestle
  d Bock *m*
  f tréteau *m*
  n montageblok
  p wspornik

5188 trial run
  d Probefahrt *f*
  f essai *m* à la route
  n proefrit
  p jazda próbna

5189 triangle of forces
  d Kraftdreieck *n*
  f triangle *m* des forces
  n krachtendriehoek
  p trójkąt sił

5190 triangular
  d dreieckig

  f triangulaire
  n driehoekig
  p trójkątny

5191 triangular support frame
  d Fahrschemel *f*
  f traverse *f* triangulaire
  n driehoekig bevestigingsframe
  p trójkątna rama nośna

5192 trigger contacts
  d Auslösekontakte *mpl*
  f contacts *mpl* de déclenchement
  n timingcontacten in stroomverdeler
  p nadajnik impulsów

5193 trim removal tool
  d Werkzeug *n* für Demontage von Zierplatten
  f outil *m* pour déposer les garnitures
  n vork voor demontage voor sierpanelen
  p narzędzie do demontażu pokrycia ozdobnego

5194 trip computer
  d Fahrprozessrechner *m*
  f ordinateur *m* de bord
  n boordcomputer
  p komputer pokładowy; procesor pokładowy

5195 triple riveted joint
  d dreireihige Nietverbindung *f*
  f rivure *f* à triple rangée
  n drievoudig klinknagelverband
  p szew nitowy trójrzędowy

5196 trip mileage
  d Tageskilometerzähler *m*
  f compteur *m* kilométrique partiel
  n dagteller
  p dzienny licznik kilometrów

5197 trouble shooting
  d Störungssuche *f*
  f recherche *f* de la panne
  n opsporing van de storing
  p szukanie uszkodzenia

  * **truck mixer** → 1212

5198 true height
  d wirkliche Hohe *f*
  f hauteur *f* réelle
  n werkelijke hoogte
  p wysokość rzeczywista

  * **trunk lid** → 586

5199 trunk tool bag
  d Kofferraumtasche *f*

f trousse *f* d'outillage pour coffre de bagages
n kofferbaktas
p kieszeń bagażnika

5200 tube
d Rohr *n*
f tube *m*
n buis
p rura; tuba

5201 tubeless tyre
d schlauchloser Reifen *m*
f pneumatique *m* sans chambre à air
n binnenbandloze buitenband
p opona bezdętkowa

5202 tube protector
d Felgenband *n*
f protecteur *m* de chambre à air
n velglint
p ochraniacz dętki

5203 tube tyre
d Schlauchreifen *m*
f pneu *m* conventionnel
n autoband met binnenband
p opona dętkowa

5204 tubing cutter
d Rohrabschneider *m*
f coupe-tubes *m*
n pijpsnijder
p obcinak do rur; cęgi do cięcia rur

5205 tubular cross member
d Rohrquerträger *m*
f traverse *f* tubulaire
n buisdwarsdrager
p poprzeczka rurowa

5206 tubular frame
d Rohrrahmen *m*
f châssis *m* en tube
n buizenchassis; buizenframe
p rama rurowa

5207 tubular hexagonal offset box spanner
d gebogener Rohrsteckschlüssel *m*
f clé *f* en tube coudées
n verstekzeskantpijpsleutel
p wydłużony klucz nasadowy sześciokątny dwustronny

5208 tubular lamp
d Rohrenlampe *f*
f lampe *f* tubulaire
n buislamp
p żarówka rurowa

5209 tubular radiator
d Rohrenkühler *m*
f radiateur *m* tubulaire
n buiskoeler
p chłodnica wodnorurkowa

5210 tubular rear axle
d Rohrhinterachse *f*
f essieu *m* arrière tubulaire
n vaste holle achteras
p rurowa oś tylna

5211 tubular space frame
d Fachwerkrahmen *m*
f cadre *m* en treillis
n vakwerkchassis
p rama kratownicowa

5212 turbine bearing
d Turbinenradlager *n*
f roulement *m* de turbine
n turbinelager
p łożysko wału turbiny

5213 turbine sealing
d Turbinenraddichtung *f*
f joint *m* de turbine
n turbinewielafdichting
p uszczelnienie wirnika turbiny

5214 turbocompressor
d Turbokompressor *m*
f turbocompresseur *m*
n turbocompressor
p turbosprężarka

5215 turbulence chamber
d Wirbelkammer *f*
f chambre *f* de turbulence
n wervelkamer
p komora wirowa

5216 turbulent flow
d turbulente Strömung *f*
f courant *m* turbulent
n turbulente stroming
p przepływ turbulentny

5217 turn cone ring
d Dichtkegel *m*
f bague *f* d'étanchéité
n afdichtingsconus
p powierzchnia stożkowa uszczelnienia

5218 turning
d Drehung *f*
f tournage *m*

n   draaiing
p   przekręcenie

5219 turning point
d   Umkehrpunkt *m*
f   point *m* mort
n   draaipunt
p   punkt zwrotny

5220 turning radius
d   Kurvenradius *m*
f   rayon *m* de braquage
n   draaicirkel
p   promień skrętu

5221 turn of thread
d   Gewindegang *m*
f   spire *f* de filet
n   gang
p   zwój gwintu

5222 turn *v* on the lights
d   Lichter andrehen; Lichter einschalten
f   allumer les feux
n   lichten ontsteken
p   włącz swiatła

5223 turret platform
d   Drehscheibe *f*
f   plaque *f* tournante
n   draaischijf
p   tarcza obrotowa

5224 twin wheels
d   Zwillingsräder *npl*
f   roues *fpl* jumelles
n   tweelingwielen
p   koła bliźniacze

5225 twist bit; twist drill
d   Schlangenbohrer *m*
f   mèche *f* à hélice unique
n   slangenboor; spiraalboor
p   wiertło kręte; wiertło śrubowe

*   **twist drill → 5225**

5226 twisting moment
d   Drehmoment *n*
f   moment *m* de rotation
n   draaimoment
p   moment obrotowy

5227 twisting of the frame
d   Rahmenverwindung *f*
f   déformation *f* du châssis
n   vervorming van het chassis
p   zniekształcenie podwozia

5228 two component mixture
d   Zweistoffgemisch *n*
f   mélange *m* binaire
n   tweecomponentenmengsel
p   mieszanka dwuskładnikowa

5229 two convolution bellow
d   Doppelringbalg *m*
f   soufflet *m* double
n   dubbele ringbalg
p   miech dwufałdowy

5230 two direction road; dual carriage way
d   Zweibahnstrasse *f*
f   route *f* à deux sens
n   tweebaansweg
p   droga dwukierunkowa

5231 two fibre
d   zweifädig
f   bifilaire
n   tweevezel
p   dwuwłóknowy

5232 two pack paint
d   Zweikomponentenlack *m*
f   peinture *f* à deux composants
n   tweecomponentenlak
p   lakier dwuskładnikowy

5233 two shoe wheel brake
d   Zweibackenradbremse *f*
f   frein *m* de roue à deux mâchoires
n   trommelrem met twee remschoenen
p   hamulec dwuszczękowy koła

5234 two way ratchet gearing
d   zweiseitiges Klinkenschaltwerk *n*
f   mécanisme *m* bilatéral par rochet et cliquet
n   dubbelwerkend palwerk
p   dwustronny mechanizm zapadkowy

5235 type number
d   Typennummer *f*
f   numéro *m* dans la série du type
n   typenummer
p   numer wersji

5236 typical
d   Standard...
f   typique
n   standaard
p   typowy; standartowy

5237 tyre; tire
d   Reifen *m*
f   pneumatique *m*

n band
p opona

5238 tyre adhesion coefficient
d Kraftschlussbeiwert *m*
f coefficient *m* d'adhérence de pneu
n wrijvingscoëfficiënt
p współczynnik przyczepności opony

5239 tyre bead
d Deckenflansch *m*
f talon *m* de l'enveloppe
n hiel van autoband
p obrzeże opony

5240 tyre changing machine
d Reifenwechselmaschine *f*
f appareil *m* pour changer les pneus
n bandverwisselapparaat
p przyrząd do wymiany opon

5241 tyre drift angle
d Schwimmwinkel *m*
f angle *m* de dérive
n drifthoek van luchtband
p kąt znoszenia opony

5242 tyre engraving
d Reifenprofil *n*
f sculpture *f* d'un pneu
n bandprofiel
p profil opony

5243 tyre lever
d Reifenheber *m*
f démonte-pneu *m*
n bandenlichter
p łyżka do opon

5244 tyre pressure
d Luftdruck *m* im Reifen
f pression *f* du pneumatique
n bandspanning
p ciśnienie w ogumieniu

5245 tyre pressure gauge
d Reifendruckprüfer *m*
f vérificateur *m* de pression des pneumatiques
n bandspanningsmeter
p wskaźnik ciśnienia w ogumieniu

5246 tyre regeneration
d Reifengenerierung *f*
f rechappement *m* du pneu
n opwerking van banden
p regeneracja opony

5247 tyre rotation
d Austauschen *n* der Laufräder

f permutation *f* des roues
n bandrotatie
p przestawianie kół ogumionych

5248 tyre size
d Reifengrösse *f*
f dimensions *fpl* des pneumatiques
n bandmaat
p rozmiary ogumienia

5249 tyre valve
d Reifenventil *n*
f valve *f* de pneumatique
n velgventiel
p zawór powietrza obręczy opony

5250 tyre valve cap
d Luftventilschutzkappe *f*
f bouchon *m* de valve de pneumatique
n ventieldopje
p kaptur zaworu powietrza

5251 tyre wear
d Reifenverschleiss *m*
f usure *f* des pneus
n bandslijtage
p ścieranie opon

5252 tyre wear gauge
d Messuhr *f* für Reifenabnutzung
f contrôleur *m* d'usure des pneus
n meetklok voor bandslijtage
p miernik ścierania opon

* unit → 280

U

5253 U bolt
 d U-Bolzen *m*
 f tourillon *m* en U
 n klemstrop
 p śruba w ksztalcie U

5254 ultimate clearance
 d Grenzspiel *n*
 f jeu *m* limite
 n uiterste speling
 p luz graniczny

5255 unbalance
 d Unwicht *n*
 f balourd *m*
 n onbalans
 p niewyrównoważenie

5256 understeering
 d Untersteuerung *f*
 f sous-direction *f*
 n onderstuur
 p podsterowność

5257 unequal length wishbones
 d Trapezquerlenkeraufhängung *f*
 f suspension *f* à deux parallélogrammes
    déformables
 n trapeziumvormige wielophanging met
    dwarsarmen van ongelijke lengte
 p układ trapezowy dwóch wahaczy
    poprzecznych

5258 uniflow scavenge
 d Gleichstromspülung *f*
 f balayage *m* en équicourant
 n gelijkstroomspoeling
 p przepłukiwanie wzdłużne

5259 union fitting
 d Rohrverschraubung *f*
 f raccord *m* de séparation
 n pijpverbinding met wartel
 p złącze rurowe śrubowe; dwuzłączka
    rurowa

5260 union nipple
 d Stutzen *m*
 f raccord *m*
 n pijpverbindingsstuk
 p króciec rurowy

5261 united injector
 d Pumpedüseelement *n*
 f pompe-injecteur *m*
 n pomp- en verstuiverelement
 p pompowtryskiwacz

5262 universal
 d universal
 f universel
 n universeel
 p uniwersalny

5263 universal clutch aligner
 d Universalkupplungszentriergerät *n*
 f centreur *m* d'embrayage universel
 n universele hulpprise-as
 p trzpień centrujący tarczę sprzęgła

5264 universal hub stay
 d Nabensperrschlüssel *m*
 f bloque *m* de moyeux universel
 n universele naafblokkeersleutel
 p uniwersalny klucz blokujący piasty

5265 universal joint; cardan joint
 d Kardangelenk *n*
 f joint *m* de cardan
 n cardankoppeling
 p przegub Cardana; przegub uniwersalny

5266 universal knife set
 d Universalmessersatz *m* mit abbrechbaren
    Klingen
 f cutters *mpl* plastiques
 n messenset met intrekbare afbreekmesjes
 p noże uniwersalne z wymienialnymi
    ostrzami

5267 unleaded petrol; lead free gasoline
 d bleifreies Benzin *n*
 f essence *f* sans plomb
 n loodvrije benzine; ongelode benzine
 p benzyna bezołowiowa

5268 unlocking screwdriver
 d Schraubendreher *m* zum Lösen
 f tournevis *m* à débloquer
 n ontsluitschroevendraaier
 p śrubokręt odbezpieczony

5269 unlock security
 d Aufsperrsicherheit *f*
 f sécurité *f* d'ouvrir
 n ontsluitingsborging
 p zabezpieczenie otwierane kluczem

5270 unscrew *v*
d abschrauben
f dévisser
n losdraaien
p odkręcać

5271 unsprung weight
d ungefedertes Gewicht *n*
f poids *m* non-suspendu
n onafgeveerd gewicht
p ciężar nieuresorowany

5272 updraft carburetter
d Steigstromvergaser *m*
f carburateur *m* à courant vers le haut
n stijgstroomcarburateur
p gaźnik górnossący

5273 upholstery
d Polsterung *f*
f garnissage *m*
n interieurbekleding
p obicie tapicerskie wnętrza

5274 upholstery border
d Verkleidungsrand *m*
f rebord *m* de garniture intérieure
n bekledingslijst
p obrzeże tapicerskie

5275 upholstery moulding
d Polsterungabdeckleiste *f*
f moulure *f* de fixation d'embourrage
n bekledingslijst
p listwa mocująca obicie

5276 upper cylinder lubricant; top cylinder lubricant
d Oberschmieröl *n*;
   Oberschmiermittel *n*
f lubrifiant *m* de haut de cylindre
n bovensmeermiddel
p olej do górnego smarowania

5277 upper hinge
d oberes Scharnier *n*
f charnière *f* supérieure
n bovenste scharnier
p zawiasa górna

* **used car** → **4350**

5278 used oil
d Gebrauchsöl *n*
f huile *f* en usage
n afgewerkte olie
p olej zużyty

5279 utility dolly
d Schienentasche *f*
f tas *m* rail
n railtas
p klepadło uniwersalne

V

5280 vacuum
  d Unterdruck *m*
  f vide *m*
  n vacuüm
  p podciśnieniowy

  *   **vacuum actuator** → 5285

5281 vacuum advance
  d Unterdruckverstellung *f*
  f commande *f* d'avance à dépression
  n onderdrukverstelling
  p podciśnieniowe regulowanie zapłonu

5282 vacuum booster
  d Unterdruckverstelleinrichtung *f*
  f capsule *f* à dépression
  n onderdrukbekrachtiger
  p siłownik podciśnieniowy

5283 vacuum brake
  d Vakuumbremse *f*
  f frein *m* à dépression
  n vacuümrem; onderdrukrem
  p hamulec próżniowy

5284 vacuum chamber
  d Unterdruckkammer *f*
  f chambre *f* à dépression
  n onderdrukkamer
  p komora podciśnieniowa

5285 vacuum controller; vacuum actuator
  d Unterdruckversteller *m*
  f commande *f* à dépression
  n onderdrukregulateur
  p wybierak podciśnieniowy

5286 vacuum fuel feed
  d Unterdruckbrennstofffördenung *f*
  f alimentation *f* par le vide
  n onderdrukgestuurde brandstoftoevoer
  p podciśnieniowe zasilanie paliwem

5287 vacuum gauge
  d Unterdruckmesser *m*
  f indicateur *m* de dépression
  n onderdrukmeter; vacuümmeter
  p próżniomierz; wakuometr; manometr
      próżniowy

5288 vacuum hose
  d Unterdruckschlauch *m*
  f conduit *m* à dépression
  n vacuümslang
  p przewód podciśnieniowy

5289 vacuum modulator
  d Unterdruckmodulator *m*
  f modulateur *m* à dépression
  n klep die mate van onderdruk regelt
  p modulator podciśnieniowy

5290 vacuum pomp
  d Vakuumpumpe *f*
  f pompe *f* à vide
  n vacuümpomp
  p pompa podciśnieniowa

5291 vacuum relief valve
  d Unterdruckventil *n*
  f clapet *m* à dépression
  n onderdrukklep
  p zawór podciśnieniowy

5292 vacuum reservoir
  d Unterdruckspeicher *m*
  f réservoir *m* à dépression
  n onderdrukketel; vacuümketel
  p zasobnik podciśnienia

5293 vacuum sensor
  d Unterdruckfühler *m*
  f capteur *m* de charge
  n vacuümsensor
  p czujnik próżniowy

5294 vacuum servo brake
  d Saugluftbremse *f*
  f servofrein *m* à dépression
  n onderdrukrembekrachtiger
  p hamulec o napędzie próżniowym

5295 vacuum tank
  d Saugluftbehälter *m*
  f réservoir *m* à dépression
  n onderdruktank
  p zbiorniczek próżniowy

5296 valve
  d Ventil *n*
  f soupape *f*
  n ventiel; klep
  p zawór

5297 valve adjustment
  d Ventileinstellung *f*
  f rattrapage *m* de jeu des soupapes

   n kleppenstelling
   p regulowanie rozrządu

5298 valve arrangement
   d Ventilanordnung *f*
   f position *f* des soupapes
   n klepinrichting
   p usytuowanie zaworów rozrządu

5299 valve cage
   d Ventilkäfig *m*
   f bride *f* de soupape
   n kleppenkooi
   p gniazdo zaworu

5300 valve casing
   d Ventilgehäuse *n*
   f carter *m* de soupape
   n klepkast
   p skrzynia zaworowa

   * **valve clearance** → 5318

5301 valve core remover
   d Ventilschlüssel *m*
   f démonte-valves *m*
   n ventielsleutel
   p klucz do zaworów

5302 valve cover
   d Ventildeckel *m*
   f couvercle *m* des soupapes
   n kleppendeksel
   p pokrywa zaworu

5303 valve cover housing
   d Ventilgehäusedeckel *m*
   f cache-soupapes *m*
   n kleppendekselhuis
   p pokrywa skrzyni zaworowej

5304 valve face
   d Ventildichtfläche *f*; Ventilauflagefläche *f*
   f portée *f* de soupape
   n klepschoteldraagvlak; klepschotelrand
   p przylgnia uszczelniająca zaworu

5305 valve gear
   d Steuerungsantrieb *m*
   f commande *f* de la distribution
   n klepbediening
   p sterowanie zaworów

5306 valve grinder
   d Ventilschleifvorrichtung *f*
   f rode-soupapes *m*
   n kleppenslijper
   p urządzenie do szlifowania zaworów

5307 valve grinding
   d Ventilschleifen *n*
   f rodage *m* de soupapes
   n kleppenslijpen
   p szlifowanie zaworów

5308 valve guide
   d Ventilführung *f*
   f guide *m* de soupape
   n klepgeleider
   p prowadnica zaworu

5309 valve head
   d Ventilteller *m*
   f tête *f* de soupape
   n klepkop
   p grzybek zaworu

5310 valve leakage
   d Undichtheit *f* des Ventils
   f non-étanchéité *f* de la soupape
   n kleplekkage
   p nieszczelność zaworu

5311 valveless
   d ventillos
   f sans soupapes
   n kleploos
   p bezzaworowy

5312 valve lift
   d Ventilhub *m*
   f levée *f* de soupape
   n kleplichthoogte
   p wznios zaworu; skok zaworu

5313 valve lifter; valve tappet
   d Ventilstössel *m*
   f tige *f* poussoir
   n klepstoter; kleplifter
   p popychacz zaworu

5314 valve locking split cone
   d Ventilhalbkegel *m*
   f demi-clavette *f* conique
   n halvemaanvormige klepspie
   p półstożek zamka zaworu

5315 valve needle
   d Ventilnadel *f*
   f pointeau *m* de soupape
   n vlotterpen
   p iglica zaworu

5316 valve operating mechanism
   d Ventiltrieb *m*
   f commande *f* de soupape

n klepaandrijving
p napęd zaworu

5317 valve piston
  d Dosierkolben *m*
  f piston *m* de soupape; piston *m* de clapet
  n plunjerklep
  p tłoczek zaworu do dozowania

5318 valve play; valve clearance
  d Ventilspiel *n*
  f jeu *m* de soupape
  n klepspeling
  p luz zaworu

  * **valve reseating tool** → 5323

5319 valve retainer
  d Ventilhalter *m*
  f outil *m* de retenue de soupapes
  n klepschotelhouder
  p podpora talerza zaworowego

5320 valve rocker
  d Kipphebel *m*
  f calbuteur *m* de soupape
  n kleptuimelaar
  p dźwignia zaworu

5321 valve seat
  d Ventilsitz *m*
  f siège *m* de pointeau
  n vlotterpenzitting
  p gniazdo zaworu

5322 valve seat angle
  d Ventilsitzwinkel *m*
  f angle *m* de siège de soupapes
  n klepzittinghoek
  p kąt gniazda zaworu

5323 valve seat cutter; valve reseating tool
  d Ventilsitzfräser *m*
  f fraise *f* pour siège de soupape
  n klepzetelfrees
  p frez do gniazda zaworu

5324 valve spanner
  d Ventileinstellschlüssel *m*; Ventilschlüssel *m*
  f clé *f* pour soupapes
  n klepstelsleutel
  p klucz do zaworów

5325 valve spring
  d Ventilfeder *f*
  f ressort *m* de soupape
  n klepveer
  p sprężyna zaworu

5326 valve spring compressor
  d Ventilfederspanner *m*
  f lève-soupape *f*
  n klepveerspanner; klepveertang
  p przyrząd do sprawdzania sprężyn zaworowych

5327 valve spring removal tool
  d Ventilfederhebel *m*
  f démonte-soupapes *m*
  n klepveerhefboom
  p przyrząd do wciskania sprężyn zaworów

5328 valve spring retainer
  d Ventilfederteller *m*
  f cuvette *f*
  n klepveerschotel
  p miska sprężyny zaworu

5329 valve stem
  d Ventilschaft *f*
  f tige *f* de soupape
  n klepstang
  p trzonek zaworu

  * **valve tappet** → 5313

5330 valve timing
  d Ventilöffnungswinkel *m*
  f phases *fpl* de distribution
  n kleptiming
  p fazy rozrządu

5331 vane pump
  d Flügelzellpumpe *f*
  f pompe *f* à pailettes
  n vleugelpomp
  p pompa łopatkowa

5332 vane type fuel supply pump
  d Rotationsförderpumpe *f*
  f pompe *f* rotative d'alimentation
  n roterende opvoerpomp, vleugeltype
  p rotacyjna pompa zasilająca

5333 vane type limited rotary hydraulic motor
  d Drehflügelschwenkmotor *m*
  f moteur *m* hydraulique à palettes
  n vleugelzwenkmotor
  p silnik łopatkowy

5334 vapour lock
  d Dampfblasenbildung *f*
  f tampon *m* de vapeur
  n dampbelvorming
  p pęcherz parowy

  * **variable choke carburetter** → 1248

5335 **varnish**
d Lack *m*
f vernis *m*
n lak
p lakier

5336 **V belt**
d Keilriemen *m*
f courroie *f* trapézoïdale
n V-riem; V-snaar
p pas klinowy

5337 **V belt drive**
d Keilriementrieb *m*
f transmission *f* par courroie trapézoïdale
n V-snaaroverbrenging
p napęd pasowy klinowy

5338 **V belt tensioner**
d Riemenspanner *m*
f tendeur *m* de la courroie trapézoïdale
n riemspanner
p napinacz pasa klinowego

5339 **vehicle**
d Fahrzeug *n*
f véhicule *m*
n voertuig
p pojazd

5340 **vehicle autonomy**
d Autonomie *f* des Fahrzeuges
f autonomie *f* du véhicule
n autonomie van voertuig
p autonomia pojazdu

5341 **vehicle overall length**
d Fahrzeuglänge *f*
f longueur *f* hors tout de véhicule
n totale lengte van het voertuig
p długość pojazdu

5342 **vehicle overall width**
d Fahrzeugbreite *f*
f largeur *f* hors tout de véhicule
n totale breedte van het voertuig
p szerokość pojazdu

5343 **vehicle registry**
d Register *m* der Fahrzeuge
f immatriculation *f* de voiture
n voertuigregistratie
p ewidencja pojazdów

5344 **vehicle running weight**
d Gewicht *n* des vorbereiteten Fahrzeugs
f poids *m* à vide en ordre de marche

n rijklaar voertuiggewicht
p ciężar pojazdu gotowego do jazdy

5345 **vehicle spin**
d Schleudern *n*
f embardée *m*
n slippen
p zarzucenie

5346 **vehicle stability**
d Fahrstabilität *f*
f stabilité *f* de route
n voertuigstabiliteit
p stabilność pojazdu

5347 **velocity**
d Geschwindigkeit *f*
f vitesse *f*
n snelheid
p szybkość

5348 **velocity diagram**
d Geschwindigkeitsplan *m*
f plan *m* de vitesses
n snelheidsdiagram
p wykres prędkości

5349 **velocity fluctuation**
d Ungleichförmigkeitsgrad *m*
f irrégularité *f* du mouvement
n snelheidsschommeling
p stopień nierównomierności

5350 **ventilating pipe mounting flange**
d Flansch *m* der Entlüftungsleitung
f support *m* de ventilation de carter
n montageflens voor carterontluchting
p kołnierz rury odpowietrzającej

5351 **ventilation**
d Lüftung *f*
f ventilation *f*
n ventilatie
p wentylacja

5352 **ventilation duct**
d Leitung *f* für Lüftung
f conduit *m* d'aération
n ventilatiekanaal
p przewód wentylacyjny

5353 **ventilation grille**
d Lüftungsgrill *m*
f grille *f* d'aération
n ventilatierooster
p krata wentylacyjna

5354 **ventilator window**
d Ausstellfenster *n*

f aérateur *f*
n ventilatieraampje
p okienko wentylacyjne

5355 venturi tube
  d Venturirohr *n*
  f trompe *f* de venturi
  n venturibuis
  p zwężka Venturiego

5356 vent wire
  d Luftspiess *m*
  f aiguille *f* à air
  n prikker
  p szydło odpowietrzające; nakłuwak

5357 vertical bevel drive shaft
  d Königswelle *f*
  f arbre *m* vertical de commande d'arbre
  n distributietussenas
  p wałek królewski

5358 vertical engine
  d stehender Motor *m*
  f moteur *m* vertical
  n staande motor
  p silnik pionowy; silnik stojący

5359 vertical outline
  d Seitenansicht *f*
  f contour *m* en vue de côté
  n zijaanzicht
  p obrys pionowy

5360 vibration; oscillation
  d Vibration *f*; Schwingung *f*
  f vibration *f*; oscillation *f*
  n vibratie; trilling
  p wibracja; drganie

5361 vibration damper
  d Schwingungsdämpfer *m*
  f amortisseur *m* de vibration
  n trillingsdemper
  p tłumik drgań

5362 vice
  d Schraubstock *m*
  f étau *m*
  n bankschroef
  p imadło

5363 vice dolly
  d Schraubstockfaust *f*
  f tas *m* d'étau
  n bankschroef voor tegenhouder
  p kowadełko imadła

5364 vineyard pruning saw
  d Winzersage *f*
  f scie *f* de vigneron
  n snoeizaag voor wijnbouw
  p piła winogrodnika

5365 visco driven fan
  d Viskolüfter *m*
  f ventilateur *m* à coupleur viscothermostatique
  n ventilateur met viscokoppeling
  p wentylator ze sprzęglem lepkościowym

5366 viscosity
  d Viskosität *f*
  f viscosité *f*
  n viscositeit
  p lepkość

5367 viscosity index
  d Index *m* der Viskosität
  f indice *f* de viscosité
  n viscositeitindex
  p wskaźnik lepkości; indeks wiskozowy

5368 viscous coupling
  d Viskositätskupplung *f*
  f viscocoupleur *m*
  n viscokoppeling
  p sprzęgło lepkościowe

5369 visibility
  d Sichtbarkeit *f*
  f visibilité *f*
  n zichtbaarheid
  p widoczność

5370 volatility
  d Flüchtigkeit *f*
  f volatilité *f*
  n vluchtigheid
  p lotność

  * voltage → 4993

5371 voltage control
  d Spannungsregelung *f*
  f régulation *f* de voltage
  n spanningsregeling
  p regulowanie napięcia

5372 voltage drop
  d Spannungsabfall *m*
  f chute *f* de tension
  n spanningsvermindering; spanningsval
  p obniżenie napięcia

5373 voltage regulator
  d Spannungsregler *m*

   f régulateur *m* de tension
   n spanningsregelaar
   p regulator napięcia

5374 voltage regulator terminal
   d Spannungsreglerklemme *f*
   f borne *f* de régulateur de tension
   n aansluitklem van spanningsregelaar
   p zacisk regulatora napięcia

5375 voltmeter
   d Voltmeter *n*; Spannungsmesser *m*
   f voltmètre *m*
   n voltmeter
   p woltomierz

5376 volume
   d Rauminhalt *m*; Volumen *n*
   f volume *m*
   n inhoud; volume
   p objętość

5377 volumetric efficiency
   d volumetrischer Wirkungsgrad *m*
   f rendement *m* volumétrique
   n volumetrisch rendement
   p sprawność objętościowa; sprawność
      wolumetryczna

5378 vortex motion
   d Wirbelbewegung *f*
   f mouvement *m* de tourbillon
   n wervelbeweging
   p ruch wirowy

5379 vulcanization
   d Vulkanisation *f*
   f vulcanisation *f*
   n vulcanisatie
   p wulkanizacja

W

5380 wall panel set
d Werkzeugsatz *m* auf Tafel
f compositions *fpl* stock sur tableaux
n lambrizeringset
p komplet narzędziowy ścienny

5381 warding file
d Schlüsselfeile *f*
f lime *f* à clefs
n sleutelvijl
p pilnik kluczykowy; pilnik płaski zbieżny

5382 warm air duct
d Warmluftverteilerschacht *f*
f boîtier *m* d'air chaud
n verwarmingsleiding
p kanał ciepłego powietrza

5383 warm running compensator
d Warmlaufregler *m*
f régulateur *m* de pression de commande;
   correcteur *m* de réchauffage
n warmloopregelaar; warmdraairegelaar
p regulator ciśnienia sterującego

5384 warning flasher switch
d Warnblinkleuchtenschalter *m*
f interrupteur *m* de feux de détresse
n schakelaar voor noodknipperlichten
p włącznik postojowych świateł błyskowych

5385 warning lamp
d Warnleuchte *f*
f lampe-témoin *f*
n waarschuwingslicht
p lampka kontrolna

* **warning signal** → 174

5386 warranty period
d Garantiezeit *f*
f période *f* de garantie
n garantieperiode; garantietermijn
p okres gwarancji

5387 washer
d Unterlegscheibe *f*
f rondelle *f*
n onderlegplaat
p podkładka

5388 washer holder bit
d Mundstück *n* für Ringe
f embout *m* pâte-rondelle
n mondstuk voor ringen
p nasadka do pierścieni

5389 washer hook
d Zughaken *m* für Ringe
f crochet *m* tire-rondelle
n haak voor trekken aan ringen
p hak do wyciągania pierścieni

5390 washer jet; fluid jet
d Flüssigkeitszerstäuber *m*
f pulvérisateur *m* d'eau
n sproeier
p rozpylacz cieczy

5391 washing
d Waschen *n*
f lavage *m*
n wassen
p mycie

5392 wash leather
d Putzleder *n*
f chamoisine *f*
n poetsleder
p zamsz do czyszczenia

5393 waste oil collector
d Altölsammler *m*
f collecteur *m* pour huile usagée
n opvangtank voor afgewerkte olie
p kolektor zużytego oleju

5394 water cooled brake
d wassergekühlte Bremse *f*
f frein *m* à refroidissement par eau
n watergekoelde rem
p hamulec chłodzony wodą

5395 water cooled engine
d wassergekühlter Motor *m*
f moteur *m* à refroidissement d'eau
n watergekoelde motor
p silnik chłodzony wodą

5396 water cooled exhaust manifold
d wassergekühlte Auspuffleidung *f*
f tubulure *f* d'échappement à chemise d'eau
n watergekoeld uitlaatspruitstuk
p rura wylotowa chłodzona wodą

5397 water cooling
d Wasserkühlung *f*
f refroidissement *m* par eau

n waterkoeling
p chłodzenie wodą

5398 water heating
d Wasserheizung *f*
f chauffage *m* à eau
n waterverwarming
p ogrzewanie wodne

5399 water jacket
d Wassermantel *m*
f chemise *f* d'eau
n watermantel
p płaszcz wodny

5400 water level
d Wasserstand *m*
f niveau *m* d'eau
n waterniveau
p poziom wody

5401 waterproof
d wasserdicht
f imperméable
n waterdicht
p wodoszczelny

5402 water pump
d Wasserpumpe *f*
f pompe *f* à eau
n waterpomp
p pompa wody

5403 water pump gasket
d Wasserpumpendichtung *f*
f joint *m* de pompe à eau
n waterpomppakking
p uszczelka pompy wody

5404 water pump impeller
d Pumpenrad *n*
f roue *f* à ailettes
n waterpompschoepenrad
p wirnik pompy wody

5405 water pump pliers
d Wasserpumpenzangen *fpl*
f pinces *fpl* pélican à simple crémaillère
n waterpomptang
p obcęgi do pompy wodnej; szczypce do pompy wodnej

5406 water separator
d Wasserabscheider *m*
f séparateur *m* d'eau
n waterafscheider
p osuszacz

5407 water temperature gauge; coolant temperature indicator
d Kühlwasserthermometer *n*
f indicateur *m* de température d'eau de refroidissement
n thermometer voor koelvloeistof
p wskaźnik temperatury cieczy chłodzącej

5408 water tube
d Wasserrohr *n*
f tube *m* à eau
n waterpijpje
p rurka wody

5409 wattage of bulb
d Glühlampenwattstärke *f*
f wattage *m* de lampe à incandescence
n vermogen van gloeilamp
p moc żarówki

5410 weakened mixture
d abgemagertes Gemisch *n*
f mélange *m* pauvre
n verzwakt mengsel
p mieszanka uboga

5411 wear; consumption
d Abnutzung *f*; Verbrauch *m*
f usure *f*; consommation *f*
n slijtage
p zużycie

5412 wear resistance
d Abnutzungsbeständigkeit *f*
f résistance *f* à l'usure
n slijtvastigheid
p odporność na ścieranie; odporność na zużycie

5413 weatherstrip
d Fensterrand *m*
f encadrement *m* de fenêtre
n raamsponning
p obrzeże okna

5414 wedge dolly
d Komma *n*
f tas *m* virgule
n gebogen tas
p kowadełko klinowe

5415 weight bridge
d Brückenwage *f*
f balance *f* à bascule
n weegbrug
p mostek wagowy

5416 weight reduction
d Gewichtsverringerung *f*

f réduction *f* de poids
n gewichtsvermindering
p zmniejszenie ciężaru

5417 weld
d Schweissnaht *f*
f soudure *f*
n lasnaad
p spaw

5418 welded joint
d Schweissverbindung *f*
f soudure *f*
n lasverbinding
p szew spawalniczy

5419 welder
d Schweisser *m*
f soudeur *m*
n lasapparaat
p spawarka

5420 welder's hammer
d Elektroschweisserhammer *m*
f marteau *m* à piquer
n hamer voor laswerk
p młotek spawacza

5421 welding rod
d Schweissstab *m*
f baguette *f* de soudage
n lasstaaf
p pałeczka do spawania

* **well base rim → 1801**

5422 wet clutch; fluid clutch
d Nassekupplung *f*; Flüssigkeitskupplung *f*
f embrayage *m* fluide; embrayage *m* hydraulique
n natte koppeling; vloeistofkoppeling
p sprzęgło mokre; sprzęgło hydrauliczne

5423 wet cylinder liner
d nasse Zylinderbüchse *f*
f cylindre *m* à chemise humide
n cilinder met natte cilindervoering
p cylinder z mokrą tuleją

5424 wheel aligner
d Prüfvorrichtung *f* für die Radstellung; Ausrichtgerät *n*
f appareil *m* de contrôle des roues
n uitlijnapparaat
p przyrząd do ustawiania kół

5425 wheel alignment
d Lenkgeometrie *f*; Radstellung *f*

f géométrie *f* de direction; réglage *m* des roues
n stuurgeometrie
p geometria kół

5426 wheel arrangement
d Fahrwerk *n*
f disposition *f* des roues
n wielplan
p układ jezdny

5427 wheel base
d Achsstand *m*
f empattement *m*
n wielbasis
p rozstaw osi

5428 wheel bending
d Radverdrehung *f*
f voile *m* de roue
n slingering in wiel
p zwichrowanie koła

5429 wheel brake
d Radbremse *f*
f frein *m* de roue
n wielrem
p mechanizm hamulcowy koła

5430 wheel bump movement
d Rädersprung *m* zum Anschlag
f course *f* de roue à débattement
n wielinvering tot aanslag
p skok obciążania koła jezdnego

5431 wheel cover
d Raddeckel *m*
f chapeau *m* de moyeu de roue
n wieldop; wielsierdeksel
p pokrywa koła

5432 wheel cylinder
d Radbremszylinder *m*
f cylindre *m* de frein de roue
n wielremcilinder
p rozpieracz hamulca koła jezdnego

5433 wheel disc
d Radscheibe *f*
f disque *m* de roue
n wielschotel
p tarcza koła

* **wheel gap → 2729**

5434 wheel housing
d Radkasten *m*
f passage *m* de roue

n wielkast
p wnęka koła jezdnego

5435 wheel hub
d Radnabe *f*
f moyeu *m* de roue
n wielnaaf
p piasta koła

5436 wheel hub bearing
d Radnabenlager *n*
f roulement *m* de moyeu
n wielnaaflager
p łożysko piasty koła

5437 wheel hub bearing support
d Radlagerstütze *f*
f fusée *f* des roulements de moyeu de roue
n ondersteuning van het wiellager
p wspornik łożysk piasty koła

5438 wheel hub flange
d Radnabenantriebsflansch *m*
f flasque *m* motrice de moyeu de roue
n wielnaafflens
p kołnierz piasty koła

5439 wheel hub shaft
d Radnabenwelle *f*
f arbre *m* de moyeu de roue
n wielnaafas
p wał piasty koła

5440 wheel load
d Radlast *f*
f pression *f* de roue
n wielbelasting
p obciążenie koła

5441 wheel nut
d Radmutter *f*
f écrou *m* de roue
n wielmoer
p nakrętka mocująca koła

5442 wheel offset
d Radversetzung *f*
f déport *m* de roue
n naloop
p przesunięcie koła

5443 wheel puller
d Radabzieher *m*
f démonte-roue *m*
n wieltrekker
p ściągacz kół

5444 wheel reduction gear
d Radvorgelege *f*

f réducteur *m* de roue
n wielreductiebak
p zwolnica

5445 wheel rim; rim base
d Radfelge *f*
f jante *f* de roue
n wielvelg
p obręcz koła

5446 wheel stud; hub bolt
d Radbolzen *m*; Nabenbolzen *m*
f boulon *m* de moyeu; axe *m* de fixation de roue
n wielbout
p śruba mocująca koła

5447 wheel suspension
d Radaufhängung *f*
f suspension *f* des roues
n wielophanging
p zawieszenie kół

5448 wheel suspension lever
d Schwinggabel *f*
f bras *m* de suspension
n wielophanging type "swing"
p wahacz zawieszenia koła

5449 wheel track
d Radstand *m*
f voie *f*
n spoorbreedte
p rozstaw kół

5450 wheel tractor
d Radschlepper *m*
f tracteur *m* à roues
n wieltrekker
p ciągnik kołowy

5451 wheel wrench
d Radmutterschlüssel *m*
f clé *f* pour écrous de roues
n hoekdopsleutel
p klucz nasadowy do nakrętek kół

5452 white metal
d Weissmetall *n*
f métal *m* blanc
n witmetaal
p biały metal

5453 wide blade double spoon
d Hebel *m* und Ausbeuleisen *n*, breit
f palette *f* double à tables larges
n dubbele brede lepel
p szeroka łyżka blacharska

5454 wide spoon
d Löffeleisen *n* breit
f cuillère *f* large
n brede uitdeuklepel
p łyżka szeroka

5455 width indicator
d Peilstange *f*
f indicateur *m* d'encombrement
n breedteaanwijzer
p wskaźnik szerokości

* **wind ahead** → **2625**

* **wind box** → **105**

5456 winding
d Wicklung *f*
f enroulement *m*
n wikkeling
p uzwojenie

5457 winding insulation
d Wicklungsisolierung *f*
f isolement *m* des roulements
n wikkelingisolatie
p izolacja uzwojeń

5458 window
d Fenster *n*
f fenêtre *f*
n raam
p okno

5459 window frame
d Fensterrahmen *m*
f encadrement *m* de fenêtre
n raamsponning
p obramowanie okna

5460 window glass cleaning agent
d Scheibenreiniger *m*
f nettoyeur *m* vitres
n ruitenreiniger; ruitreinigingsmiddel
p środek czyszczący do szyb

5461 window handle removal tool
d Werkzeug *n* für Demontage von Fensterkurbel
f outil *m* pour déposer les manivelles de glaces
n vork voor demontage van de portierruitslinger
p przyrząd do demontażu korb okiennych

5462 window sun screen
d Hinterjalousie *f*
f rideaux *mpl* arrière
n zonnescherm achterruit
p zasłona okna tylnego

5463 window winder
d Fensterheber *m*; Scheibenheber *m*
f lève-vitre *m*
n raamopeningsmechanisme
p podnośnik szyby okna

5464 windscreen pillar
d Fenstersäule *f*
f pied *m* de la fenêtre
n voorruitstijl
p słupek okna

5465 windscreen washer
d Scheibenwascher *m*
f lave-glace *m*
n ruitensproeier
p spryskiwacz szyby przedniej

5466 windscreen washer pump
d Scheibenwascherpumpe *f*
f pompe *f* de lave-glace
n ruitensproeierpomp
p pompa spryskiwacza szyby przedniej

5467 windscreen wiper
d Scheibenwischer *m*
f essuie-glace *f*
n ruitenwisser
p wycieraczka szyby przedniej

5468 windscreen wiper motor
d Scheibenwischermotor *m*
f moteur *m* d'essuie-glace
n ruitenwissermotor
p silnik elektryczny wycieraczki

5469 windshield tools
d Werkzeug *n* zum Einsetzen von Frontscheiben mit Gummi
f outils *m* pour pare-brise à jonc
n gereedschap voor voorruit met ruitrubber
p narzędzia do przedniej szyby

5470 windshield wiper and washer control
d Wischwaschschalter *m*
f commande *f* essuie-glace et lave-glace
n schakelaar voor ruitenwisser en ruitensproeier
p wyłącznik wycieraczki i spryskiwacza szyby przedniej

5471 wing nut; butterfly nut; fly nut
d Flügelmutter *f*
f écrou *m* à ailettes
n vleugelmoer
p nakrętka motylkowa; nakrętka skrzydełkowa

5472 wing side inner panel
d Kotflügelinnenverkleidung *f*

f joue *f* d'aile
n binnenbekleding van het spatscherm
p osłona wnęki błotnika

5473 wing valance
d seitliches Schutzblech *n* am Kotflügel
f baroir *m* d'aile
n binnenspatscherm
p osłona blaszana błotnika

5474 wiper arm
d Wischerarm *m*
f porte-raclette *m*
n wisserarm
p ramię wycieraczki

5475 wiper arm pivot
d Scheibenwischerarmachse *f*
f axe *m* de bras de balai
n as van ruitenwisser
p ośka ramienia wycieraczki

5476 wiper blade
d Wischerblatt *n*
f balai *m* d'essuie-glace
n wisserbad
p wycierak

5477 wiper blade holder
d Wischerblattgriff *m*
f chape *f* de bras de balai
n vang van het wisserblad
p uchwyt wycieraka

5478 wiper blade rubber element
d Wischerblatteinlage *f*
f lampe *f* de balai
n wisserbladrubber
p wkładka wycieraka

5479 wiper blade stem
d Scheibenwischerfeder *f*
f raclette *f* de balai
n veer van ruitenwisser
p pióro wycieraczki

5480 wiper contact guide
d Kontaktführung *f*
f guide *m* de contact mobile
n geleider voor contactbus
p prowadnica zwory

5481 wiper motor mounting bracket
d Scheibenwischermotoraufhängung *f*
f suspension *f* de moteur d'essuie-glace
n montageplaat voor de ruitenwissermotor
p wieszak silnika wycieraczki

5482 wiper switch
d Scheibenwischerschalter *m*
f commutateur *m* d'essuie-glace
n ruitenwisserschakelaar
p wyłącznik wycieraczki

5483 wire brush
d Drahtbürste *f*; Stahldrahtbürste *f*
f brosse *f* métallique
n staalborstel
p szczotka druciana

5484 wire gauze filter
d Siebfilter *m*
f filtre *m* à tamis
n filterzeef
p filtr siatkowy

5485 wire tyre
d Drahtreifen *m*
f pneu *m* à tringles
n draadband
p opona drutowa

5486 wire wheel
d Drahtspeichenrad *n*
f roue *f* à rayons-fil
n draadspaakwiel
p koło szprychowe

5487 wiring diagram
d Schaltschema *n*
f schème *m* des connexions
n bedradingsschema
p schemat połączeń

5488 wiring harness
d Kabelnetz *n*
f faisceau *m* de fils
n kabelboom
p zespół przewodów

5489 wiring system
d elektrische Ausrüstung *f*
f équipement *m* électrique
n elektrische installatie
p instalacja elektryczna

* **wishbone → 5172**

5490 workbench
d Arbeitstisch *m*
f établi *m*
n werkbank
p stół warsztatowy

5491 workbench anvil
d Werkbankamboss *m*

f bigorne *m* d'établi
n aanbeeld voor werkbank; aambeeld voor
   werkbank
p kowadło ze stołem roboczym

5492 workbench press with hand pump
d Werkstattpresse *f* mit Handpumpe
f presse *f* d'établi à pompe manuelle
n werkbankpers met handpomp
p prasa warsztatowa z ręczną pompą

5493 workbench standard cutting snip
d Universalwerkbankblechschere *f*
f cisaille *f* universelle d'établi
n werkbankstandaardschaar
p uniwersalne nożyce blacharskie

5494 workbench vice
   (instrument which holds the strut firmly in
   position for shock absorber or spring removal)
d Spannzange *f* für Federbeine
f étau *m* spécial d'établi
n werkbankklem
p imadło stołowe warsztatowe

5495 working gauge
d Werkstattlehre *f*
f calibre *m* de fabrication
n werkplaatskaliber
p sprawdzian roboczy

* **working height** → 47

5496 working load
d Arbeitsbelastung *f*
f charge *f* d'utilisation
n arbeidsbelasting
p obciążenie robocze

5497 working pressure
d Arbeitsdruck *m*
f pression *f* motrice
n werkdruk
p ciśnienie robocze

* **workshop** → 4100

5498 workshop crane
   (crane for removal, fitting and transport of
   heavy objects around the workshop)
d Werkstattkran *m*
f grue *f* d'atelier
n werkplaatskraan
p dźwig warsztatowy stanowiskowy

5499 work station
d Arbeitsstelle *f*

f poste *f* de travail
n werkvloer
p stanowisko robocze

5500 worm
d Schnecke *f*
f vis *f* sans fin
n worm
p ślimak

5501 worm and nut steering gear
d Schraubenspindellenkung *f*
f direction *f* à vis et écrou
n schroef- en moerstuurinrichting
p śrubowa przekładnia kierownicza

5502 worm and peg steering
d Rollenfingerschraubenlenkung *f*
f direction *f* à vis et doigt
n worm-en-rolvingerstuurinrichting
p korbowa przekładnia kierownicza

5503 worm and sector steering gear
d Schneckensegmentlenkung *f*
f direction *f* à vis et secteur
n worm-en-wormwielsector
p ślimakowa przekładnia kierownicza

5504 worm wheel
d Schneckenrad *n*
f roue *f* hélicoïdale
n wormwiel
p ślimacznica

5505 woven fabric belt
d Geweberiemen *m*
f courroie *f* tissée
n geweven riem
p pas tkany

5506 wrench extension for strut nuts
d Schutzkappe *f* für Stossdämpfermuttern
f douille *f* pour écrous d'amortisseurs
n dop voor schokdempermoeren
p klucz do śrub amortyzatora

5507 wrought iron
d Schweisseisen *n*
f fer *m* soudé
n smeedijzer
p żelazo zgrzewne

Y

5508 yawing moment
 d Giermoment *n*
 f couple *m* d'embardée
 n giermoment
 p moment odchylający

5509 year of construction
 d Baujahr *n*
 f année *f* de construction
 n bouwjaar
 p rok budowy; rok konstrukcji; rok produkcji

5510 yield point
 d Streckgrenze *f*
 f limite *f* d'étirage
 n strekgrens
 p granica plastyczności

5511 yoke lever
 d Bügelarm *m*
 f chape *f* de bride
 n jukbeugel
 p ramię jarzma

Z

5512 zebra crossing
d Fussgängerstreifen *m*
f passage *m* pour piétons
n zebrapad
p przejście dla pieszych

5513 zero adjustment of the computer
d Nulleinstellung *f* von dem Computer
f réglage *m* du zéro de l'ordinateur
n nulstelling van de computer
p zerowanie komputera

5514 zinc
d Zink *n*
f zinc *m*
n zink
p cynk

5515 zip fastener
d Reissverschluss *m*
f fermeture *f* à glissière
n ritssluiting
p zamek błyskawiczny

Deutsch

319

Querwelle 1392

Radabzieher 5443
Radaufhängung 5447
Radaufhängungsbein 4850
Radbolzen 5446
Radbremse 5429
Radbremszylinder 5432
Raddeckel 2729, 5431
Rädersprung zum Anschlag 5430
Radfelge 5445
radiale Dichtleiste 3954
Radialkolbenmotor 3955
Radialkolbenpumpe 3956
Radioeinbausatz 3975
Radiostörung 3976
Radiozangen mit Seitenschneider
    und gebogenen Backen 496
Radkappe 2729
Radkasten 5434
Radlagerstütze 5437
Radlast 5440
Radmutter 5441
Radmutterschlüssel 5451
Radnabe 5435
Radnabenantriebsflansch 5438
Radnabenlager 5436
Radnabenwelle 5439
Radscheibe 5433
Radschlepper 5450
Radschraubenschlüssel 4156
Radstand 5449
Radstellung 5425
Radverdrehung 5428
Radversetzung 5442
Radvorgelege 5444
Rahmen 2314
Rahmenbrille 2316
Rahmengabel 2318
Rahmenhöhe 1013
Rahmenlängsträger 2321
Rahmenplatte 2320
Rahmenschere 2566
Rahmenverwindung 5227
Randstein 1402
Raspel 3983
Rauchbegrenzer 3922
Raum 990
Räumbahle 760
Rauminhalt 5376
Raupenschlepper 1377
Rautenstichel 1538
Reaktionsgrad 1503
rechnergesteuerte Maschine 1210
rechtes Spurstangenglied 5141
rechte Tür 4144
rechts angeordnete Lenkung 4145
rechtsschneidender Spiralbohrer
    1762

rechtwinkliger Schraubendreher
    3502
Reduktionsgetriebe 4057
Reflektor 4063
Regelhebel 2560
regelmassige Wartungsarbeiten
    3715
Regelstange 1285, 1287
Regelungssystem 1290
Regenrinne 1764
Register der Fahrzeuge 5343
Registervergaser 3405
registrierendes Instrument 4050
Registriernummer 4068
Regler 2505
Reglerfeder 2512
Reglergehäuse 2510
Reglergrundplatte 2506
Reglerhebel 2511
Reglerkegel 2508
Reglerkern 4073
Regler mit Drosselbetätigung
    5056
Reglermuffe 2507
Reglerschraubenfeder 359
Reglerwiderstand 4070
Regulierhülse 87
Regulierkante 3319
Regulierventil 4072
Regulierventil der Temperatur
    4984
Reibahle 760
Reibkörper 2344
Reiblager 2342
Reibradtrieb 2337
Reibscheibe 2335, 2338
Reibspindelpresse 2340
Reibstange 5
Reibung 2332
Reibungsfläche 2343
Reibungskraft 2336
Reibungskupplung 2334
Reibungsmoment 3371
Reibungsstossdämpfer 2341
Reibungsverlust 2333
Reibungszahl 1132
reiches Gemisch 4143
Reifen 5237
Reifendruckprüfer 5245
Reifengenerierung 5246
Reifengrösse 5248
Reifenheber 5243
Reifenluftdruck 2851
Reifenprofil 5242
Reifenschulter 5185
Reifenventil 5249
Reifenverschleiss 5251
Reifenwächter 3942
Reifenwechselmaschine 5240

Reifenwulst 438
Reifkloben 993
Reiheneinspritzpumpe 3397
Reihenmotor 2885
Reihenschaltung 4399
Reinheitsgrad 1502
reinigen 12
Reinigung 1079
Reinigungsmittel 1080
Reisegeschwindigkeit 1400
Reissnadel 4313
Reissverschluss 5515
Reisszwecke 1746
Relais 4078
Relaisnotbremsventil 4080
Relaissteuerung 4079
Relaisventil 4082
Relativbewegung 4077
Rentabilitätsgrenze 747
Reparatur 4097
Reparaturauftrag 4099
Reparatursatz 4098
Reparaturwerkstatt 4100
Reparaturwerkzeugkasten 1944
Reserve 4813
Reserveanlage 347
Reservetank 313
Restfederweg 4107
Restunwucht 4106
Retarder 4118
Richten 4832
Richtplatte 4901
Richtschraube 1933
Richtung 1583
Richtvorrichtung 4831
Richtwerkzeug 4834
Riechstoff 3496
Riegelfeder 2942
Riegelflansch 2273
Riegelkugel 2938
Riegelstopfen 2941
Riemen 472
Riemenabnützung 481
Riemenantrieb 473
Riemengabel 479
Riemenscheibe 475, 3923
Riemenscheibengetriebe 480
Riemenschlupf 478
Riemenspanner 5338
Riemenverschleiss 481
Riesenfahrzeug 3499
Rillenkugellager 228, 1483
Ring 4155
Ringbrennkammer 229
Ringfeder 4164
ringförmige Führung 1056
Ringkaliber 4157
Ringmutter 4161
Ringschlagschlüssel 4844

Français

petite tête de bielle 1242
phare à commande automatique
   de niveau 4380
phare de recul 340
phase 3727
phases de distribution 5330
pièce de distributeur 1638
pièce de fonte 920
pièce de rechange 4589
pièce originale 3594
pièce polaire d'inducteur 4219
pièce rapportée en caoutchouc
   4243
pied 3734
pied à coulisse 842, 4540
pied à profondeur 1518
pied avant 2367
pied de couvercle de filtre 2150
pied de la fenêtre 5464
pied de mouton 1085
pied milieu 954
pierre à affuter demi-ronde 2572
pierre abrasive 5
pierre à huile 3565
pigment 3733
pignon 1283, 1785
pignon à chaîne de distribution
   1372
pignon central 4879
pignon cylindrique 4706
pignon de commande de chaîne
   1796
pignon de démarreur 4740
pignon de direction 3953
pignon de l'arbre intermédiaire
   1340
pignon de marche arrière 4131
pignon de réglage 1830
pignon de renvoi d'arbre primaire
   1767
pignon du vilebrequin 1374
pignon extérieur 3603
pignon intérieur 2890
pignon intermédiaire de marche
   arrière 4129
pignon menant de première
   vitesse 2177
pignon mené de marche arrière
   4124
pignon rotor 3603
pignon satellite 3798
pignons de pompe à huile 3553
pile sèche 1811
pilier central 954
pilotage électronique de
   transmission automatique 302
pince 1393, 3451
pince à becs longs affilés 3139
pince à cintrer 3750

pince à circlip 1051
pince à creuset 1398
pince à dégauchir 4833
pince à dénuder à vis 73
pince à désassembler les tôles
   1482
pince à poser les œillets 2080
pince à segments 3780
pince à souder par points 4664
pince à soyer 2585
pince à soyer pneumatique 125
pince autoserreuse 4375
pince autoserreuse orientable
   4377
pince batterie 425
pince coupante de côté 4473
pince crocodile 178
pince de paveur 3701
pince gratte-laque 1126
pince marteau 2120
pincement des roues 5083
pince multiprise 3762
pince plate 4836
pince pour batterie 417
pince pour circlips 1051
pince pour déposer les pastilles
   des poussoirs 4442
pince pour fusibles 2439
pince pour le câble de frein à
   main 3675
pince pour l'électronique 1927
pince pour repouser et tourner les
   pistons d'étrier 685
pince pour soudure 3113
pince ronde 4230
pinces à cran demi-lune 3848
pinces coupe-câbles pour câbles
   électriques 817
pinces étaux 571
pinces pélican à simple
   crémaillère 5405
pinces pour ressorts de segments
   de freins 706
pinces radio coupantes coudées
   496
pioche 3262
piste de dérapage 4519
pistolet à air 547
pistolet de vernissage 4667
pistolet thermique 2635
piston 3763
piston à déflecteur 350
piston à jupe fendue 4555, 4656
piston amortisseur 2738
piston autothermique 307
piston bimétal 516
piston d'amortisseur 4448
piston de clapet 5317
piston de clapet de dosage 1683

piston de cylindre récepteur 4522
piston déflecteur 1491
piston de frein 682
piston de maître-cylindre 3252
piston de pompe d'accélération 25
piston de soupape 5317
piston de soupape de démarrage
   4755
piston différentiel 1561
piston façonné 162
piston hydraulique 2745
piston hydraulique annulaire 2732
piston libre 2325
piston plat 1488
piston primaire 3894
piston rotatif 4207
piston surdimensionné 3639
piston télescopant 4978
piston tout acier 181
piton d'enlèvement 1749
pivot 3740
pivot d'attelage 2131
pivot de charnière 2679
pivot de fusée 4851
pivot de fusée d'essieu 4928
pivot fixe 2187
pivot réglable 69
pivot rotule 377
place debout 4727
place pour les jambes 3064
plafonnier 934
plage à gants 2501
planche de tableau de bord 1461
plancher 3793
plan de base 453
plan de joint 4325
plan de la ville 1065
plan d'ensemble 2482
plan de vitesses 5348
planéité 2220
plane pour écorchage blanc 1748
planétaire de différentiel 1562
plan horizontal 2555
plan tangent 13
plaque à rouleaux 4188
plaque constructeur 2778
plaque d'ancrage 206
plaque d'appui de cric 2976
plaque d'écoulement 3605
plaque de couverture 4435
plaque de fabricant 3419
plaque de friction 2338
plaque de glissement 4546
plaque de glissement de ressort
   4701
plaque de guidage 2563
plaque d'électroaimant 3206
plaque de licence 3482
plaque de moteur 1995

plaque de nationalité 3424
plaque de rupteur 741
plaque de serrage 1620, 3126
plaque d'immatriculation 3482
plaque d'injecteur 3478
plaque indicatrice de commande 2844
plaque inférieure 598
plaque isolante de coupe-circuit 1053
plaque-modèle 3700
plaque négative 3437
plaque pare-chocs 787
plaque signalétique 1466
plaque support 2231
plaque tournante 5223
plaquette d'accoudoir 265
plaquette de fixation 4357
plaquette de frein 674
plaquette de protection 3913
plaquette de réglage 85
plaquette de ressort 4695
plaquette de support 622, 4894
plastifiant 4567
plateau cannelé 4403
plateau de fermeture 1098
plateau de frein 655
plateau de marchepied 4259
plateau de pression 3877
plateau de rangement 4828
plateau incliné 4914
plateau porte-balai 767
plateau souple 339
platine 3809
plâtroir à poignée ouverte 2166
plieuse 488
plomb 3040
plongeur 3817
plot intérieur de chapeau 1633
pneu à clous 4854
pneu à crampons 4854
pneu à crampons massifs 3250
pneu à tringles 5485
pneu ballon 380
pneu conventionnel 5203
pneu de génie civil 1861
pneu hiver 3390
pneu lisse 363
pneumatique 5237
pneumatique à basse pression 3160
pneumatique à carcasse diagonale 1390
pneumatique à carcasse diagonale ceinturée 512
pneumatique à haute pression 2668
pneumatique sans chambre à air 5201

poche d'huile 3541
poids à vide 4963
poids à vide en ordre de marche 5344
poids de carrosserie 570
poids de la voiture en ordre de marche 1403
poids du châssis 1018
poids maxi sur barres de toit 3265
poids maxi sur flèche 3268
poids non-suspendu 5271
poids remorque 5135
poids spécifique 4620
poids suspendu 4705
poids total en plein charge 2552
poignée 2588
poignée à cliquet 3986
poignée articulée 3488
poignée de frein 665
poignée de manivelle 1356
poignée de marteau 2578
poignée de porte 1662
poignée de réglage 3081
poignée de réglage de dossier de siège 4332
poignée de serrage préréglée 3862
poignée double 1704
poignée dynamométrique 5112
poignée extérieure 1678
poignée intérieure de porte 2900
point à graisser 3170
point d'éclair 2208
point de congélation 2329
point de cric 2974
point de freinage 686
point d'égouttement 1804
point de rotation 3790
point de transition 747
point d'injection 2862
point d'oscillation 2427
pointe à tracer 4313
pointeau 957
pointeau de soupape 5315
pointe de tour 3037
pointe du poinçon 957
point mort 2779, 5219
point mort haut 5106
point rond 4225
pointu 36
polarisable 3828
polarisation 3829
pôle magnétique 3198
pôle négatif 3438
poli fin 2429
polir 4561
pollution 2827
pommeau de levier de vitesses 4622
pompe 3930

pompe à air 157
pompe à basse pression 3159
pompe à combustible 2413
pompe à eau 5402
pompe à engrenage 2478
pompe à essence 2452
pompe à graisse 2530
pompe à haute pression 2667
pompe à huile 3550
pompe à huile à denture intérieure 2958
pompe à huile à haute pression 2665
pompe à huile à segments 4550
pompe à incendie automobile 2168
pompe à main 3242
pompe à membrane 1544
pompe antigel 243
pompe à pailettes 5331
pompe à piston axial 317
pompe à piston radial 3956
pompe aspirante refoulante 1612
pompe à tarer les injecteurs 2874
pompe auxiliaire 4886
pompe à vide 5290
pompe centrifuge 969
pompe d'alimentation 2395
pompe d'alimentation à piston 3819
pompe de balayage 4294
pompe de cric 2977
pompe de lave-glace 5466
pompe de reprise 22
pompe de suralimentation 1002
pompe de transfert 5162
pompe d'injection 2863
pompe d'injection à distributeur rotatif 1643
pompe d'injection en ligne 3397
pompe dynamique 4216
pompe électrique 1901
pompe-hélice 4311
pompe hydropneumatique 2767
pompe immersible 4862
pompe-injecteur 5261
pompe moyée 2822
pompe rotative 4209
pompe rotative d'alimentation 5332
pont 755
pont arrière 4003, 4005
pont arrière démultiplicateur 1714
pont banjo 393
pont de câble 813
pont élévateur 292
pont élévateur pour autos 903
pont-levis 1744
pont oscillant à deux pivots 1705

ressort de rappel de pédale 21
ressort de rappel de plongeur
    3818
ressort de refoulement 2112
ressort de régulateur 2512
ressort de rotule 383
ressort de rupteur 745
ressort de sélecteur 4365
ressort de soupape 5325
ressort de synchroniseur 4937
ressort de tarage 3479
ressort de torsion 5122
ressort de verrou 2942
ressort de verrouillage 3122
ressort de vitesse maximum 3273
ressort d'un embrayage 5061
ressort en anneau 4164
ressort en caoutchouc 542
ressort en spirale 2649, 4641
ressort équilibrant 359
ressort extérieur de soupape 3604
ressort façonné 4115
ressort hélicoïdal 1137
ressort hélicoïdal de compression
    1195
ressort hélicoïdal de poulie
    réceptrice 3612
ressort hélicoïdal de suspension
    4909
ressort hélicoïdal de traction 2070
ressort hydropneumatique 2768
ressort intérieur de soupape 2891
ressort libre 2326
ressort minibloc 3337
ressort pneumatique 3825
ressort taré 838
ressort tendeur 4996
ressort transversal à lames 5177
ressort transversal de
    compensation 5174
ressort travaillant en flexion 2240
retard 1476
retard de fermeture de la soupape
    d'admission 4874
retard d'ouverture de la soupape
    d'échappement 2054
retour d'allumage 336
retour de flamme 336
rétracteur 477
retrait 4462
rétroviseur extérieur 3602
rétroviseur intérieur 2936
revêtement 3094
revêtement de montant arrière
    4027
revêtement de pied 3736
revêtement de pied avant 2368
revêtement de pied milieu 956
revêtement de protection 3917

revêtement de siège 4338
revêtement du toit 2614
revêtement en mousse 2229
revêtement latéral de repose-pied
    4482
révision du moteur 1997
révision générale 3229
révision technique 2904
rhéomètre 2261
rideau de radiateur 3969
rideaux arrière 5462
ridelle 4471
rigide 4811
rigidité 4149
rinçage 4165
rinçage du moteur 1982
rivé à chaud 2721
rivet 4166
rivetage 4170
rivet à tête plate 1224
rivet creux à tête plate 2198
rivet de chaîne 985
rivet de garniture de frein 670
rivet explosé 2068
rivet faux 533
rivet noyé 2267
rivure à double rangée 1715
rivure à triple rangée 5195
rivure à une rangée 4507
rivure en quinconce 4167
robinet 1129
robinet à trois voies 5044
robinet d'arrêt 4819
robinet de batterie 424
robinet de carburant 2425
robinet de commande 5152
robinet de frein de stationnement
    3678
robinet de soutirage 4949
robinet de vidange 1739
robinet de vidange de carburateur
    889
robinet d'isolement 1411
robinet thermostatique 5022
rochet 3118
rodage de soupapes 5307
rodage du cylindre 1437
rodage du moteur 2540
roder 4257
rode-soupapes 5306
rodoir de soupapes pneumatique
    149
rondelle 3791, 5387
rondelle à aileron 4941
rondelle à ergot 4564
rondelle à ressort 4690
rondelle Belleville 1603
rondelle contact 1265
rondelle crantée 2095

rondelle d'appui 348, 3890
rondelle de butée 801, 5062
rondelle découpée 5100
rondelle de feutre 2117
rondelle de l'échappement 2030
rondelle en amiante 273
rondelle en caoutchouc 4249
rondelle en cuir 3060
rondelle isolante 2913
rondelle plate 2223
rotation 4212
rotor 4218
rotor à griffes 4219
rotor de soufflante 545
rotor du compresseur 4883
rotule 2466
rotule de fourchette 3580
rotule de positionnement 378
rotule de poussée 5109
rotule de timonerie 3096
rotule sphérique 4629
roue à ailettes 5404
roue à denture extérieure 4708
roue à disque 1604
roue à rachet 3989
roue à rayons-fil 5486
roue arrière 4035
roue avant 2378
roue à voile ajouré 3713
roue conique 508
roue conique hypoïde 2775
roue cul-de-sac 1472
roue de distribution 864, 5071
roue de friction 2344
roue démontable 1527
roue dentée cylindrique à denture
    hélicoïdale 2647
roue dentée de commande 1793
roue dentée de piston 3768
roue dentée intérieure 232
roue de secours 4591
roue en alliage léger 192
roue hélicoïdale 5504
roue libre 2327
roue pour chaîne 988
roue sans disque 1601
roues jumelles 5224
roue type artillerie 272
rouille 4262
rouleau 4183
rouleau aiguille 3431
rouleau compresseur 4765
rouleau cylindrique 1446
rouleau de chaîne 986
rouleau de coincement 2980
rouleau de papier 3668
rouleau en hélice 2648
roulement 461
roulement à aiguilles 3432

Nederlands

borstelhouder 766
borstelhouderplaat 767
borstelveer 769
borstel voor accupool en -klem 3173
botsing 1156
botsproef 2823
botssignaleringssensor 1376
botte beitel 552
bougie 4601
bougieborstel 4602
bougiegat 4606
bougiehuis 4608, 4612
bougiesleutel 4598, 4609
bougiespanningzoeker 4611
Bourdon-buis 601
bout 573
boutenschaar 575
bout met conische kop 4312
bout met ronde kop 1022
bout met vierkante kop 4714
bout met vierkante kraagkop 4710
bout met vierkante tapkop 4712
bout met zeskantige kop 2659
boutveer 577
boutverbinding 576, 3294
bout zonder moer 4305
bouwjaar 5509
bouwtekening 2907
bovenaanzicht 2555, 2704, 3802
bovenliggende nokkenas 3628
bovensmeermiddel 5276
bovenste dode punt 5106
bovenste ligger van chassis 2322
bovenste scharnier 5277
bovenste zuigerveer 5107
bowdenkabel 604, 605
brandbaar mengsel 1169
brandblusser 2169
brandstof 2384
brandstofaccumulator 2385
brandstofbesparing 2392
brandstofdichtheid 2387
brandstofdoorvoerleiding 2388
brandstofdoseerinrichting 2390
brandstofdrukkanaal 2404
brandstofdrukpijp 2411
brandstofduurzaamheid 2391
brandstoffilter 2396
brandstofinspuiting 2400
brandstofinspuitpompregelaar 2864
brandstofinspuitsysteem 2394
brandstofkolf 2495
brandstofkraantje 2425
brandstofleiding 2410
brandstof-luchtmengsel 137
brandstofmengsel 2408
brandstofniveau 2406

brandstofniveaumeter 2397
brandstofopvoerpomp 2395
brandstofopvoerpompnok 2416
brandstofoverstroomkanaal 2405
brandstofpersklep 2412
brandstofpomp 2413
brandstofpompcontact 2417
brandstofpomphuis 2414
brandstofpomplichaam 2414
brandstofreservevoorraadlicht 2420
brandstofretourleiding 2421
brandstoftank 2423
brandstoftankvuldop 2424
brandstoftoevoerbuis 2403
brandstoftoevoerstroming 2399
brandstoftoevoersysteem 2422
brandstofverbruiknorm 2386
brandstofverstuiver 2402, 2409
brandstofvoorraadmeter 2397
brandstofvoorverhitter 2398
brandweerauto 2168
brede gebolde lepel 1654
brede uitdeuklepel 5454
breedstraler 761
breedteaanwijzer 5455
breedte van zitkussen 4343
breekbaarheid 759
breekpen 4428
breeksterkte 748
breuksterkte 748
Brinell-hardheid 758
brons 762
bronslagerschaal 764
bronzen huls 763
brug 755
brugschakeling 756
buffer 772
buffertank 4904
buigbelasting 491
buigijzer 4833
buiging 487
buigkracht 490
buigmachine 488
buigmoment 489
buigspanning 491
buigtang 3750
buigtest 494
buigzame manchetinzet 2235
buigzame veer 2240
buikzaag 1409
buis 3749, 5200
buisboor 1320
buisdwarsdrager 5205
buiskoeler 5209
buislamp 5208
buis van het achterashuis 331
buisvormige cardanas 1778

buitendiameter van de draaicirkel 3615
buitenkogelschaal 3601
buitenloopvlak van de remtrommel 1809
buitenomtrek van tandwiel 1394
buitenpasser 841
buitenrotor 3603
buitenschroefdraad 2076
buitenspiegel 3602
buitenste conische schijf 3600
buitenste klepveer 3604
buitentemperatuurmeter 3599
buitentrekker 4376
buitenverlichting 2074
buizenbundel 3439
buizenchassis 5206
buizenframe 5206
bumperhoorn 785
bumperklem 784
bumperplaat 787
bumpersteun 783
bus 1220
busje 3338
busketting 795

calorie 845
calorische waarde 843, 844
cantileverveer 870
canvas kap 871
caravan 877
carborundumvijl 6
carburateur 884
carburateurafstelling 886
carburateurdeksel 888
carburateurhuis 887
carburateurregelaar 5056
carburateurstartklep 896
carburateurvoet 891
cardanaandrijving 2992
cardanastunnel 5170
cardankoppeling 1788, 5265
cardanpijp 899
cardantunnel 2465, 3910
carrosserie 553
carrosseriedecoratie 559
carrosseriedelen 561
carrosseriegereedschap 569
carrosseriegewicht 570
carrosserie-isolatie 564
carrosserienummer 566
carrosserieonderhoudsmiddel 557
carrosserieplaat 878
carrosseriescharnier 562
carrosseriesteun 555
carrosserievorm 560
carrosserie zonder kooiconstructie 3665
carter 1353

smering 3171
smering met de hand 3241
smeringstabel 3172
smoorklep 5052
smoorklepas 5053
smoorklephefboom 5049
smoorklepschakelaar 5054
smoorring 5051
smoortap 3748
smoren 5055
snede 1419
snedediepte 1519
sneeuwblazer 4562
sneeuwketting 251
snelheid 4621, 5347
snelheid beperken 4056
snelheidsdiagram 5348
snelheidsmeter 4625, 4628, 4942
snelheidsmeterkabel 4626
snelheidsregelaar 4623
snelheidsschommeling 5349
snelheid verminderen 4056
snelsluitklep 2102
snelwerkende terugschakelklep
    2670
snijbrander 1422
snijdiepte 1519
snijgereedschap 1876
snijijzer 1418
snijplaat 5035
snijplaathouder 1553
snijring 1420
snoeizaag voor wijnbouw 5364
snoer 1318
sok 3753
soldeerbout 4571
soldeerlamp 548
solderen 4570
soortelijke warmte 4619
soortelijk gewicht 4620
spanas 1070
spanband 2103
spanhefboom 5063
spanhoek 3980
spaninrichting van de remband
    631
spanklauw 2544
spankracht 1069
spanning 4993
spanningsregelaar 5373
spanningsregeling 5371
spanningsval 5372
spanningsvermindering 5372
spanrol 4995
spanrol van ventilatorriem 2090
spanstift 4694
spanveer 4996
spanveerzitting 4679
spanzaag 4586

spatbord 2121
spatlap 3392
spatscherm 2121
spatsmering 409
speciaal gereedschap 4615
speciale uitrusting 4614
specifiek brandstofverbruik 4618
speelruimte 1083
speling 1083
speling tussen tanden 337
speling van de inlaatklep 2883
speling van de uitlaatklep 2055
sperdifferentieel met gedeeltelijke
    blokkering 3090
sperdiode 539
spernok voor achteruitrijden 346
sperrelais 540
sperstroom 2964
spie 2993
spieas 4648
spiebaanstaal 3003
spiebanen 4647
spiebout 4645
spiegel 3346
spieverbinding 3000
spieverbindingen en vertandingen
    2994
spijkerbuigtang 3418
spijkerzak 3417
spilboor 606
spindelregelaar 4071
spiraalboor 5225
spiraalveer 1137, 2649, 4641
spiraalvormige rol 2648
spiraalvormig huis 4640
spits 36
spitsuren 4261
spleetfilter 1873
spleetzuiger 4656
splijtwig 2108, 2383
splijtzaag 4657
splitpen 4654
spoed van schroefdraad 3043
spoel 1134
spoeldruk 4297
spoellucht 4295
spoelolie 2266
spoelpomp 4294
spoiler 4658
spontane ontsteking 4660
spoorbreedte 5449
spoorbreedte achter 5137
spoorbreedte voor 5136
spoormal 1618
spoorstang 4784
spoorstangarm 1803, 4802
spoorstanggeleidingshefboom
    5139
spreider 4673

sproeier 4668, 5390
sproeierkamer 2984
sproeiernaald 3320
sproeier van carburateur 894
sproeiwagen 4842
spuitbus met drijfgas 95
spuitcabine 3655
spuitdruk 2260
spuiten 4665, 4669
spuiter 3654
spuiterij 3658
spuitgieten 3870
spuitgietmachine 1550
spuitinrichting 4670
spuitpistool 4667
spuittap 3745
staaf 397
staaffilter 1084
staaffrees 4228
staal 4766
staalborstel 5483
staaldraad 4773
staalplaat 4432
staande motor 5358
staanplaats 4727
stabilisator 4719
stabilisatorstangspanner voor het
    verwijderen van de
    stabilisatorstang 4994
stabiliteit 4718
stalen bus 4770
stalen diafragma 4767
stalen doos 4826
stalen draagbak 908
stalen lineaal 4769
stalen ring 4768
stand 4757
standaard 5236
standaardkaliber 4061
standaardmaat 4725
standaardmotor 406
standaarduitrusting 3593
standaardvelg 4048
standby 4726
stang 397
stangenstelsel 3077
stangenstelsel voor rembediening
    671
stang met kogeleind 375
stappenmotor 3929
starre leiding 4150
startaccu 4734
starterkrans 4741
starterrondsel 4740
starterstang 4751
startgenerator 4737
starthulp 2798
startkabel 422
startknop 4735

Polski

cewka obrotowa 4206
cewka uzwojenia wzbudzającego
    2027
cewka wtórna 4347
cewka zapłonowa 2801
cęgi do cięcia rur 5204
charakterystyka 999
charakterystyka biegu jałowego
    2784
charakterystyka dynamiczna 1848
chlorokauczuk 1032
chłodnica międzystopniowa 2932
chłodnica oleju 3515
chłodnica oleju przekładniowego
    2464
chłodnica pochyła 4824
chłodnica powierzchniowa 4903
chłodnica sekcyjna 541
chłodnica ulowa 2693
chłodnica wodnorurkowa 5209
chłodnica z rurkami
    użebrowanymi 2537
chłodzenie powietrzem 123
chłodzenie tłoku 3767
chłodzenie wodą 5397
chłodzenie wyparne 1866
chłodziarka 4065
chromowany 1044
chwytacz iskier 4595
chwyt ręczny uniwersalny 1296
ciągarka 1743
ciągłe zasilanie wtryskowe
    benzyny 1268
ciagnik gąsienicowy 1377
ciągnik kołowy 5450
ciągnik rolniczy 2101
ciecz chłodząca 1299
ciecz niskokrzepnąca 241
cienkie kowadełko z rynną 5024
ciepło właściwe 4619
cięcie poosiowe 944
cięgło 1737
cięgło elastyczne opancerzone
    605
cięgło rozruchowe 1034
cięgło włączające 50
cięgło zsania 1034
cięgno blokady zamka 1675
cięgno dźwigni hamulca ręcznego
    3677
cięgno klamki drzwi 1663
cięgno pędne 1790
cięgno rozruchu 4751
cięgno ssania 1036
cięgno sztywne 4440
cięgno wyprzedzające 1115
cięgno zębatki sterowniczej 1287
ciężarek wirujący 2268
ciężar holowany 5135
ciężar jednostkowy 4620

ciężar nadwozia 570
ciężar nieuresorowany 5271
ciężar podwozia 1018
ciężar pojazdu gotowego do jazdy
    5344
ciężar własny 4963
ciężar własny samochodu
    gotowego do jazdy 1403
ciśnienie 3864
ciśnienie akustyczne 44
ciśnienie atmosferyczne 283
ciśnienie dynamiczne 1850
ciśnienie dźwięku 44
ciśnienie hamowania 689
ciśnienie indykowane 2839
ciśnienie kinetyczne 1850
ciśnienie na ząb 5103
ciśnienie obliczone 837
ciśnienie ogumienia 2851
ciśnienie oleju 3542
ciśnienie płukania 4297
ciśnienie płynu chłodzącego 1301
ciśnienie początkowe 2856
ciśnienie przewodu 3238
ciśnienie robocze 5497
ciśnienie spalania 1171
ciśnienie spalin 2037
ciśnienie sprężania 1199
ciśnienie ssania 4870
ciśnienie średnie 3280
ciśnienie tłoczenia 2260
ciśnienie w ogumieniu 5244
ciśnienie wsteczne 341
ciśnieniomierz 155
ciśnieniomierz do ogumienia 146
ciśnieniomierz do pomiaru
    sprężania 1196
cokół prądnicy 2487
cokół żarówki 780
cykl spalania 2804
cylinder aluminiowo-chromowy
    189
cylinder bezpiecznika 2437
cylinder chłodzony powietrzem
    121
cylinder do pchania 3944
cylinder do wyciągania 3927
cylinderek obciążenia do
    odpowietrzania obwodu
    chłodzenia 1001
cylinderek pompy 3935
cylinder hamulca koła
    mechanizmu różnicowego
    1563
cylinder hamulcowy 648
cylinder hydrauliczny 2737
cylinder hydrauliczny
    jednostronnego działania 4497
cylinder hydrauliczny
    obustronnego działania 1686

cylinder owalny 3616
cylinder poziomy 2702
cylinder profilowany 3904
cylinder zamka 3112
cylinder zamka drzwiowego 1671
cylinder zamknięty 531
cylinder z komorą spalania o
    kształcie półokrągłym 1655
cylinder z mokrą tuleją 5423
cynk 5514
cyrkiel drążkowy 443
cyrkiel pomiarowy 3286
cyrkiel warsztatowy prosty 1646
cyrkulacja 1062
cysterna lotniskowa 124
czas hamowania 719
czas przestoju 1734
czas rozruchu 4753
czas suszenia 1818
czas trwania reakcji 3999
czasza 3601
czasza wyciskowa 4083
czerpak oleju 3517
częstotliwość drgań własnych
    3425
częstotliwość obrotów silnika
    1999
częstotliwość własna 3425
części maszyn 3184
części plastykowe 3806
części podwozia 1012
część oryginalna 3594
część ruchoma 3389
część składowa 1186
część toczna 4195
część zamienna 4589
część zapasowa 4589
człon 3095
człon bierny 3611
człon czynny 2893
człon napędzający 2893
człon napędzany 3611
czołowa przekładnia główna 4707
czop 3740
czop ciężarka 2269
czop główny 1956, 3218
czop główny grzebieniowy 1148
czop główny stożkowy 4953
czopik iglicy 3745
czop kontrolny 1019
czop końcowy wału 4417
czop korbowy 1357
czop krótki 4852
czop kulisty 377
czop łączący 614
czop łożyskowy zwrotnicy 4851
czop nastawny 69
czop nośny 4637
czop osi 323
czop poprzeczny środkowy 3426

instalacja ogrzewcza 2639
instalacja paliwowa 2422
instalacja taśmowa 1298
instalować 2906
instrukcja montażowa 281
instrukcja obsługi 4405
instrukcja regulacji 79
instruktor jazdy 1787
integrator 2927
intensywność 2929
intensywność światła oślepiania
    1467
iskiernik zabezpieczający 4276
iskra 4593
iskra zetknięcia 768
iskrownik 3204
iskrownik w kole zamachowym
    2274
iskrzenie 4596
izolacja cieplna 2640
izolacja drutówki 435
izolacja nadwozia 564
izolacja rdzenia 3741
izolacja termiczna 2640
izolacja uzwojeń 5457
izolacja włóknista 2125
izolator 2917
izolator świecy zapłonowej 4607
izolujący 2910
izomer 2970

jarzmo hamulca 704
jarzmo napędzane 1770
jarzmo nastawne 76
jarzmo osadnika 608
jarzmo przełączenia przekładni
    4439
jarzmo resoru 3046
jarzmo rury wydechowej 2033
jarzmo wtryskiwacza 2873
jazda próbna 5188
jednorurowy amortyzator
    hydrauliczny 3571
jednostkowe zużycie paliwa 4618
jednostopniowy 4511
jednozaciskowy hamulec
    tarczowy 4500

kabel 811
kabel jednożyłowy 4501
kabel niskiego napięcia 3164
kabel połączeniowy do masy
    1863
kabel przystawki 54
kabel rozruchowy 422
kabel sterowniczy 1274
kabel ułatwiający rozruch 422
kabel wielożyłowy 3395
kabel wysokociśnieniowy 2672
kabel zasilający 3844

kadłub gaźnika 887
kadłub pompy oleju 3554
kadłub pompy paliwa 2414
kadłub rozdzielacza 1634
kadłub silnika 1965
kadłub świecy 4608
kalibrowanie 840
kalkomania 3304
kaloria 845
kamień szlifierski 2542
kanalik dopływowy 2404
kanalik nadmiarowy paliwa 2405
kanalik rozdzielczy 1639
kanał 996
kanał ciepłego powietrza 5382
kanał dolotowy 2920
kanał korka 3812
kanał odpowietrzający 172
kanał oleju 3521
kanał podciśnieniowy 4866
kanał powietrzny 132
kanał przelotowy 5160
kanał rewizyjny 2905
kanał rozdzielczy 1628
kanał wylotowy 2043
kanister 869
kapa 872
kapa przeciwzakłóceniowa 4758
kaptur zaworu powietrza 5250
karburacja 883
karburator 884
karetka pogotowia 194
karkas z zamkiem 771
karoseria 553
karoseria samonośna 4516
karta pracy 2987
karta smarowania 3172
katalizator 925
katalizator ceramiczny 971
katalizator metaliczny 3306
katalizator monolityczny 1821
katoda 930
kauczuk chlorowany 1032
kąpiel olejowa 3508
kąt gniazda zaworu 5322
kąt nachylenia 216
kąt natarcia 215, 3980
kąt obrotu 218
kątomierz 212
kątomierz uniwersalny 1167
kątowa oś nośna 227
kątownik podporowy 3908
kątownik stalowy 211
kątownik stały 3351
kątownik z przekątną 946
kąt pochylenia koła 847
kąt pochylenia sworznia
    zwrotnicy 3009
kąt przesunięcia fazowego 217
kąt przodovania 213

kąt przyporu 3865
kąt rozwarcia styków
    przerywacza 738
kąt ukosu 503
kąt wyprzedzenia 919
kąt zejścia 214
kąt znoszenia opony 5241
kąt zwarcia styków przerywacza
    735
kief rdzenia twornika 4219
kielnia tynkarska 2166
kieł tokarski 3037
kierowalność 4748
kierownica dmuchawy 546
kierunek 1583
kierunek jazdy 1782
kierunek napędu 1766
kierunek obrotu 1590
kierunek przepływu prądu 1588
kierunek ruchu 1589
kierunek wirowania 1590
kierunkowskaz 1585
kierunkowskaz migowy 535
kieszeń bagażnika 5199
kieszeń na dokumenty pokładowe
    4830
kieszeń na mapy 3244
kilof 3262
kit 3949
kit do znakowania 3248
kit stale elastyczny 3717
kit uszczelniający 4319, 4320
klakson 2705
klamka drzwi 1662
klamka drzwiowa 1681
klamka drzwi wewnętrznych
    2900
klamka zewnętrzna drzwi 1678
klamra głowicy rozdzielczej 1627
klamra pasów bezpieczeństwa
    3030
klapa odpowietrzająca 753
klapa tylna 4020
klej do tworzywa piankowego
    2279
klej kauczukowy 4251
klepać 467
klepadło do krawędzi 2645
klepadło kuliste 1652
klepadło płaskie 2212
klepadło płaskie w kształcie klina
    2224
klepadło uniwersalne 5279
klepadło wklęsłe 1606
klepak 5074
kleszcze bezpiecznikowe 2439
kleszcze nastawne do ściągania
    izolacji 73
kleszcze płaskie 4836
klimatyzator 116

koło zębate daszkowe 2654
koło zębate hypoidalne 2775
koło zębate rozrządu 5071
koło zębate strzałkowe 2654
koło zębate tłoka 3768
koło zębate walcowe skośne 2647
koło zębate wału pośredniego
     1340
koło zębate wału rozrządu 864
koło z metalu lekkiego 192
kołpak 907
kołpak karbowany 4141
kołpak koła 2729
kołpak piasty 2729
kołpak prądnicy 2486
kombinacja barw 1158
kombinezon roboczy 3618
komfort jazdy 5183
komora 990
komora ciśnienia 3867
komora do piaskowania 4283
komora dyszowa 2984
komora hydropneumatyczna 2766
komora mieszalna 3353
komora natryskowa 3655
komora osadowa 4358
komora pływakowa 2242
komora podciśnieniowa 5284
komora podgrzewcza 2723
komora pompowania 3937
komora powietrza 130
komora powietrza katalizatora
     152
komora spalania 1194
komora suszarnicza 1816, 1817
komora wirowa 5215
komora wstępna 3856
komora wyrównawcza 2111
komórka fotoelektryczna 3730
komparator 1176
komparator elektroniczny 1921
kompensacja hamulca 632
kompensator 3911
komplet do wyciągania 2545
komplet narzędzi do usuwania
     zawleczek 4655
komplet narzędziowy ścienny
     5380
komputer pokładowy 5194
komutator 1152
kondensator 1214
kondensator klimatyzacji 117
kondensować 1213
koniec dolotu 2876
koniec skrętu sprężyny
     przyłożonej 4711
koniec wylotu 3606
koniec zakazu 1958
konserwacja 3226
konserwacja przestrzeni

zamkniętych 4583
konserwacja zapobiegawcza 3861
konsola 620
konsola radiowa 3974
konsola środkowa 959
konstrukcja skorupowa podwozia
     3373
konstrukcja zespolona 2925
kontener 1266
kontrola jakości 3951
kontrola końcowa 2158
kontrola poślizgu 17
kontrola żarówki 776
kontrolka rozrusznika 4739
kontur 1270
konwekcja 1293
konwertor 1294
końcówka cięgna 4437
końcówka do założenia klucza
     nasadowego 4565
końcówka przewodu 832
końcówka wału 4420
końcówka wału napędowego
     1777
końcówka wtryskiwacza 4671
korba do wierteł 1753
korba ręczna 1356
korba rozruchowa 4750
korbowa przekładnia kierownicza
     5502
korbowód 1236
korbowód widlasty 2299
korek 3813
korek bezpieczeństwa 4275
korek blokujący 2941
korek chłodnicy 3959
korekcja mieszanki 3356
korek gumowy 4253
korek spustowy 1739
korek spustowy miski olejowej
     3538
korektor zawieszenia 4907
korek uliczny 5145
korek wlewu pokrywy ogniwa
     akumulatora 937
korek załadowania 1006
korek zamykający 1938
korozja 1330
korozja końcówek 4999
korpus cylindra hamulcowego
     649
korpus osi tylnej 4004
korpus rozpylacza 2983
korpus siodła 4541
korpus spiralny 4640
korpus wtryskiwacza 3476
koszta garażu 2444
koszty produkcji 3577
koszyczek igieł 3428
koszyczek łożyska 371

koszyk satelitów 3799
kowadełko blacharskie 1094
kowadełko do wgnieceń 4064
kowadełko imadła 5363
kowadełko klinowe 5414
kowadełko promieniowe 3977
kowadełko ręczne 4661
kowadełko uniwersalne 4069
kowadełko zagięte 5082
kowadło piły 4292
kowadło ręczne do gięcia 1651
kowadło sześcienne 4709
kowadło w postaci szyi łabędzia
     4912
kowadło ze stołem roboczym
     5491
kozioł łożyskowy 446
kozioł resoru tylnego 4031
krajak 2497
krakowanie katalityczne 927
krata 2536
krata bagażowa na dachu 3177
krata wentylacyjna 5353
kratownica płaska 3796
krawędź pilnika 1875
krawędź radełkowana 3331
krawędź skrawająca 1419
krawędź sterująca 3319
krawędź uszczelniająca 4321
krawędź wlotowa 2919
krawężnik kamienny 1402
krążek izolacyjny 2914
krążek prowadzący 2565
krążek sprężysty 2648
krokodylek 178
króciec ciepłego powietrza 2715
króciec odpływu cieczy
     chłodzącej 1300
króciec odpowietrzania skrzyni
     korbowej 1354
króciec podłączeniowy gaźnika
     891
króciec przewodu nadmiarowego
     3055
króciec redukcyjny 4060
króciec rurowy 5260
króciec rury 2713
króciec ssawny 2403, 2923
króciec tłoczny pompy 2388
króciec wlewu oleju 3525
krótki satelit 3797
kryza trójramienna 5038
krzywa charakterystyczna 1000
krzywa rozładunku 1597
krzywka 846
krzywka blokująca 3114
krzywka dekompresyjna 1479
krzywka ograniczająca 3089
krzywka pompy paliwowej 2416
krzywka przerywacza 734

ściągacz do zacisków
    akumulatorów 3174
ściągacz kół 5443
ściągacz koła pasowego wału
    korbowego 3925
ściągacz łożyska kulkowego 369
ściągacz łożyska tylnego 4011
ściągacz piasty koła 2730
ściągacz sworzni 4856
ściągacz śrub 4310
ścieranie bieżnika opony 5186
ścieranie okładzin 2
ścieranie opon 5251
ścieranie się stopki opony 432
ściernica do zdzierania 1474
ściernica szmerglowa 1948
ściernica talerzowa 1607
ściernica tarczowa 1607
ścięcie 992
ściskacz pierścieni tłokowych
    3776
ściskana sprężyna śrubowa 1195
ściski szybkodziałające 4288
ścisk stolarski 1877
ślad hamowania 4518
ślepe złącze na czop 534
ślimacznica 5504
ślimak 4778, 5500
ślimak cylindryczny 1449
ślimak globoidalny 2724
ślimakowa przekładnia
    kierownicza 5503
ślimak walcowy 1449
śliska jezdnia 4551
ślizgacz kulisy 3097
ślizgaczowa pompa oleju 4550
ślizgacz prowadzący 2119
ślizgacz resoru 4701
ślizgacz sworznia tłokowego
    2558
ślizgacz środkujący 4636
ślusarz 3128
śmieciarka 4066
średnia prędkość pizepływu 3281
średnia prędkość tłoka 3300
średnica cylindra 1427
średnica czopu głównego 3209
średnica czopu korbowego 1358
średnica grzybka zaworu
    dolotowego 2881
średnica grzybka zaworu
    wylotowego 2053
średnica tłoka 3766
średnica toczna 3788
średnica wewnętrzna gwintu 2899
średnica wewnętrzna otworu 592
średnie ciśnienie efektywne 3279
średnie ciśnienie użyteczne 3279
środek antyadhezyjny 4846
środek ciężkości 952

środek czyszczący 1080
środek czyszczący chłodnicę
    3960
środek czyszczący do materiałów
    włókienniczych 5008
środek czyszczący do plastyku
    3804
środek czyszczący do szyb 5460
środek do czyszczenia styków
    1257
środek do konserwacji karoserii
    557
środek do polerowania nadwozia
    samochodu 905
środek do usuwania oleju 3506
środek impregnujący 2825
środek myjący 1530
środek parcia 953
środek przeciwkorozyjny 4263
środek przeciwpieniący 2281
środek przeciwstukowy 239
środek rdzochronny 4263
środek usuwający owady 2895
środek usuwający plamy 4722
środek uszczelniający 3056
środek utwardzający 2601
środek wiążący 522
środek zabezpieczający 3116
środek zapobiegający oblodzeniu
    233
środek zapobiegający przyleganiu
    4846
środki zapobiegawcze 3854
środkowa prowadnica siedzenia
    4337
środkownik 946
środkowy człon drążka 5138
środkowy hamulec taśmowy 2944
śruba 573, 4304
śruba cylindra 1426
śruba dociskowa 3083
śruba dociskowa o łbie
    sześciokątnym 2659
śruba do kołnierza 2200
śruba głowicy cylindra 1432
śruba hamulca 634
śruba koła zębatego tarczowego
    1396
śruba kołnierzowa 3735
śruba korbowodu 1239
śruba korekcyjna 1933
śruba kotwowa 203
śruba lewoskrętna 3063
śruba mimośrodowa 1871
śruba mocująca 287
śruba mocująca koła 5446
śruba napędowa 3850
śruba o czopikowym łbie
    czworokątnym 4712
śruba odpowietrzająca 530

śruba oporowa koła talerzowego
    4160
śruba pasowana 4000
śruba pociągowa toczna 373
śruba przekładni kierowniczej
    4796
śruba regulacyjna 84
śruba regulacyjna drążka 4088
śruba regulacyjna dźwigni
    wyłączającej 4085
śruba regulacyjna popychacza
    4959
śruba regulująca 4412
śruba resoru 4676
śruba samozabezpieczająca 4383
śruba stożkowa płaska 1342
śruba typu Bendix 493
śruba w kształcie U 5253
śruba wydrążona 2690
śruba zabezpieczająca 3120
śruba zabezpieczająca koło przed
    kradzieżą 253
śruba zaciskająca 522, 523
śruba zaciskowa 523
śruba zawieszenia silnika 1966
śruba z dwustronnym gwintem
    4305
śruba ze sworzniem rozwidlonym
    2297
śruba zębnika 3486
śruba z kołnierzem 1150
śruba złączna o łbie
    czworokątnym 4714
śruba złączna o łbie
    młoteczkowym 2579
śruba złączna o łbie oczkowym
    2079
śruba złączna o łbie
    sześciokątnym 2658
śruba złączna o łbie walcowym i
    gnieździe sześciokątnym 2660
śruba złączna o łbie wieńcowym
    4710
śruba złączna płyt 3808
śruba z łbem 876
śruba z łbem kulistym 1022
śruba z łbem stożkowym 4312
śrubokręt odbezpieczony 5268
śrubowa przekładnia kierownicza
    5501
śrubowa sprężyna resorowa 1137
śruby podniesieniowe 3084
śruby podnośne 3084
śrutowanie 381
świadectwo próby 5004
światła cofania 4137
światła gabarytowe 4479
światła mijania 237
światła postojowe 3679
światło gabarytowe 3246